Lecture Notes in Computer Science 14641

The series Lecture Notes in Computer Science (LNCS), including its subseries Lecture Notes in Artificial Intelligence (LNAI) and Lecture Notes in Bioinformatics (LNBI), has established itself as a medium for the publication of new developments in computer science and information technology research, teaching, and education.

LNCS enjoys close cooperation with the computer science R & D community, the series counts many renowned academics among its volume editors and paper authors, and collaborates with prestigious societies. Its mission is to serve this international community by providing an invaluable service, mainly focused on the publication of conference and workshop proceedings and postproceedings. LNCS commenced publication in 1973.

Ioanna Miliou · Nico Piatkowski ·
Panagiotis Papapetrou
Editors

Advances in Intelligent Data Analysis XXII

22nd International Symposium on Intelligent Data Analysis, IDA 2024
Stockholm, Sweden, April 24–26, 2024
Proceedings, Part I

 Springer

Editors
Ioanna Miliou 🆔
Stockholm University
Kista, Sweden

Nico Piatkowski 🆔
Fraunhofer IAIS
Sankt Augustin, Germany

Panagiotis Papapetrou 🆔
Stockholm University
Kista, Sweden

ISSN 0302-9743   ISSN 1611-3349 (electronic)
Lecture Notes in Computer Science
ISBN 978-3-031-58546-3   ISBN 978-3-031-58547-0 (eBook)
https://doi.org/10.1007/978-3-031-58547-0

This Springer imprint is published by the registered company Springer Nature Switzerland AG
The registered company address is: Gewerbestrasse 11, 6330 Cham, Switzerland

Paper in this product is recyclable.

Preface

We are delighted to introduce the proceedings of the 22nd International Symposium on Intelligent Data Analysis, held from April 24–26, 2024, in Stockholm, Sweden. Originating in 1995, the symposium was organized biennially until 2009. Starting from 2010, the symposium shifted its focus towards encouraging submissions that present groundbreaking and innovative ideas, even if they might not be as fully developed as those presented at other conferences. The 2024 edition of IDA maintained this tradition, inviting submissions that, while possibly considered preliminary in other contexts, promise significant research advancements. This edition also marked the return of the Industrial Challenge track, encouraging both academic and industrial researchers to tackle a machine learning challenge focused on predicting the imminent failure of specific vehicle engine components. Participants worked with data from Scania trucks operating under demanding conditions. The IDA Symposium welcomes a broad range of modeling and analysis methodologies from all disciplines, aiming to be an interdisciplinary forum that fosters discussions on intelligent data analysis that span various fields. We invite contributions that address intelligent support for modeling and analyzing data from complex, dynamic systems.

Within this context, IDA 2024 extended its support for data analysis beyond the conventional algorithmic solutions typically discussed in academic literature. Only those submissions that integrated established technologies within intelligent data analysis frameworks or applied such technologies in innovative ways to the analysis and/or modeling of complex systems were considered. The traditional review process, which tends to favor small, incremental improvements over existing research, might have deterred the submission of the innovative research that IDA 2024 aimed to attract. To counter this, reviewers and senior PC members were encouraged to favor innovative ideas to complex solutions and extensive experimental evaluations. The outcome was a highly compelling program. We received 94 submissions, from which 40 were accepted as regular papers and 3 as short papers, including those from the industrial challenge. Each submission underwent a rigorous single-blind review by three members of the program committee and one senior member.

We were pleased to have the following distinguished invited speakers at IDA 2024:

- Dimitrios Gunopulos, National and Kapodistrian University of Athens, Greece on the topic "Computing counterfactuals with feasibility and compactness guarantees"
- Dino Pedreschi, University of Pisa, Italy on the topic "Social Artificial Intelligence: Challenges of the Human-AI Ecosystem"
- Danica Kragic, KTH Royal Institute of Technology, Sweden on the topic "Representation learning and foundation models in robotics".

The conference was held at the Department of Computer and Systems Sciences of Stockholm University in Stockholm.

We wish to express our gratitude to all authors of submitted papers for their intellectual contributions; to the Program Committee members and advisors and additional reviewers for their effort in reviewing, discussing, and commenting on the submitted papers and to the members of the IDA Steering Committee for their ongoing guidance and support. We thank Zed Lee for running the conference website. Special thanks go to the industrial challenge chair Tony Lindgren for handling the submission and reviewing process of the industrial challenge papers, as well as to Luis Galárraga for handling the PhD Poster Track. We gratefully acknowledge those who were involved in the local organization of the symposium: Sindri Magnusson, Ali Beikmohammadi, Franco Rugolon, and Maria Bampa. We thank our Frontier Prize chair, Jaakko Hollmen, our advisory chairs Matthijs van Leeuwen and Siegfried Nijssen for their precious guidance during the preparation of IDA, and our Social Media chair, Zhendong Wang who together with Ioanna Miliou took care of all strategic social media communications related to IDA. Finally, we are grateful to our sponsors: Stockholm University, Scania AB, Digital Futures, Springer, and The Artificial Intelligence Journal. We are especially indebted to KNIME, who funded the IDA Frontier Prize for the most visionary contribution presenting a novel and surprising approach to data analysis in the understanding of complex systems. Last but not least we thank the City of Stockholm for hosting IDA's reception evening at the Stockholm City Hall.

April 2024

Ioanna Miliou
Nico Piatkowski
Panagiotis Papapetrou

Organization

General Chair

Panagiotis Papapetrou   Stockholm University, Sweden

Program Committee Chairs

Ioanna Miliou   Stockholm University, Sweden
Nico Piatkowski   Fraunhofer IAIS, Germany

Senior Program Committee

Hendrik Blockeel	KU Leuven, Belgium
Henrik Bostrom	KTH Royal Institute of Technology, Sweden
Peggy Cellier	INSA Rennes and IRISA, France
Bruno Cremilleux	Université de Caen Normandie, France
Wouter Duivesteijn	TU Eindhoven, Netherlands
Benoît Frénay	University of Namur, Belgium
Elisa Fromont	Université Rennes 1 and IRISA/Inria, France
Esther Galbrun	University of Eastern Finland, Finland
Joao Gama	INESC TEC - LIAAD, Portugal
Sibylle Hess	TU Eindhoven, Netherlands
Frank Höppner	Ostfalia University of Applied Science, Germany
Arno Knobbe	Leiden University, Netherlands
Georg Krempl	Utrecht University, Netherlands
Charlotte Laclau	Polytechnique Institute and Télécom Paris, France
Siegfried Nijssen	Université Catholique de Louvain, Belgium
Celine Robardet	INSA Lyon, France
Arno Siebes	Universiteit Utrecht, Netherlands
Stephen Swift	Brunel University London, UK
Maryam Tavakol	TU Eindhoven, Netherlands
Allan Tucker	Brunel University London, UK
Matthijs van Leeuwen	Leiden University, Netherlands
Veronica Vinciotti	University of Trento, Italy
David Weston	Birkbeck University of London, UK

Program Committee

Pedro Henriques Abreu	CISUC, Portugal
Thiago Andrade	University of Porto and INESC TEC, Portugal
Ali Ayadi	University of Strasbourg, France
Paulo Azevedo	Universidade do Minho, Portugal
Maria Bampa	Stockholm University, Sweden
Mitra Baratchi	LIACS - University of Leiden, Netherlands
José Borges	INESC TEC - FEUP, Portugal
Tassadit Bouadi	Universite de Rennes 1, France
Paula Brito	University of Porto and INESC TEC - LIAAD, Portugal
Dariusz Brzezinski	Poznan University of Technology, Poland
Mirko Bunse	TU Dortmund, Germany
Sebastian Buschjäger	TU Dortmund, Germany
Rui Camacho	University of Porto, Portugal
Paulo Cortez	University of Minho, Portugal
Thi-Bich-Hanh Dao	University of Orléans, France
Maria Demidik	DESY, Germany
Hadi Fanaee-T	Halmstad University, Sweden
Brígida Mónica Faria	Polytechnic Institute of Porto and LIACC, Portugal
Ad Feelders	Universiteit Utrecht, Netherlands
Sébastien Ferré	Univ Rennes and CNRS Inria IRISA, France
Carlos Ferreira	INESC TEC, Portugal
Françoise Fessant	Orange, France
Mikel Galar	Universidad Pública de Navarra, Spain
Luis Galárraga	Inria, France
Benoit Gauzere	INSA Rouen, France
Rui Gomes	Universidade de Coimbra, Portugal
Zhijin Guo	University of Bristol, UK
Thomas Guyet	Inria Centre de Lyon, France
Barbara Hammer	CITEC and Bielefeld University, Germany
Alberto Fernandez Hilario	University of Granada, Spain
Tomáš Horváth	ELTE, Hungary
Dino Ienco	Irstea, France
Szymon Jaroszewicz	Polish Academy of Sciences, Poland
Baptiste Jeudy	Laboratoire Hubert Curien, France
Bo Kang	Ghent University, Belgium
Frank Klawonn	Ostfalia University, Germany
Jiri Klema	Czech Technical University, Czech Republic
Maksim Koptelov	University of Caen Normandy, France

Alejandro Kuratomi   Stockholm University, Sweden
Christine Largeron   Université Jean Monnet Saint-Etienne, France
Nada Lavrač   Jožef Stefan Institute, Slovenia
Zed Lee   Stockholm University, Sweden
Vincent Lemaire   Orange Innovation, France
Sindri Magnusson   Stockholm University, Sweden
Joao Mendes-Moreira   INESC TEC, Portugal
Vera Miguéis   University of Porto, Portugal
Rita Nogueira   INESC TEC, Portugal
Slawomir Nowaczyk   Halmstad University, Sweden
Andreas Nürnberger   Magdeburg University, Germany
Kaustubh Patil   Forschungszentrum Jülich, Germany
Ruggero Pensa   University of Turin, Italy
Pedro Pereira Rodrigues   University of Porto, Portugal
João Pimentel   Dalhousie University, Canada
Marc Plantevit   EPITA, France
Luboš Popelínský   Masaryk University, Czech Republic
Filipe Portela   UMINHO, Portugal
Luis Reis   University of Porto, Portugal
Justine Reynaud   University of Caen Normandy, France
Rita Ribeiro   University of Porto, Portugal
Duncan Ruiz   Pontifícia Universidade Católica do Rio Grande
                              do Sul, Brazil
Amal Saadallah   TU Dortmund, Germany
Akrati Saxena   Leiden University, Netherlands
Jørg Schløtterer   University of Marburg and University of
                              Mannheim, Germany
Roberta Siciliano   University of Naples Federico II, Italy
Paula Silva   University of Porto and INESC TEC, Portugal
Amina Souag   Canterbury Christ Church University, UK
Arnaud Soulet   University of Tours, France
Myra Spiliopoulou   Otto-von-Guericke-University Magdeburg,
                              Germany
Jerzy Stefanowski   Poznan University of Technology, Poland
Shazia Tabassum   INESC TEC, Portugal
Sónia Teixeira   INESC TEC, Portugal
Alicia Troncoso   Pablo de Olavide University, Spain
Peter van der Putten   Leiden University, Netherlands
Bruno Veloso   University of Porto and INESC TEC - FEP,
                              Portugal
Tom Viering   Delft University of Technology, Netherlands

João Vinagre	Joint Research Centre - European Commission, Spain
Sheng Wang	University of Bristol, UK
Hilde Weerts	Eindhoven University of Technology, Netherlands
Pascal Welke	TU Wien, Austria
Zhaozhen Xu	University of Bristol, UK
Paul Youssef	University of Marburg, Germany
Leishi Zhang	Canterbury Christ Church University, UK
Albrecht Zimmermann	Université de Caen Normandie, France

Sponsors

We thank our sponsors for their support:

- Stockholm University
- The City of Stockholm
- Digital Futures
- Scania
- KNIME
- Springer

Contents – Part I

Applications

Natural Language Processing

Contents – Part II

Optimization

XAI

Industrial Challenge

Foundations of AI and ML

Tackling the Abstraction and Reasoning Corpus (ARC) with Object-Centric Models and the MDL Principle

Sébastien Ferré[(✉)]

Univ Rennes, CNRS, Inria, IRISA, Campus de Beaulieu, 35042 Rennes, France
ferre@irisa.fr

Abstract. The Abstraction and Reasoning Corpus (ARC) is a challenging benchmark, introduced to foster AI research towards human-like intelligence. It is a collection of unique tasks about generating colored grids, specified by a few examples only. In contrast to the transformation-based programs of existing work, we introduce object-centric models that are in line with the natural programs produced by humans. Our models can not only perform predictions, but also provide joint descriptions for input/output pairs. The Minimum Description Length (MDL) principle is used to efficiently search the large model space. A diverse range of tasks are solved, and the learned models are similar to natural programs.

1 Introduction

Artificial Intelligence (AI) has made impressive progress in the past decade at specific tasks, sometimes achieving super-human performance: e.g., image recognition [11], board games [15]. However, AI still misses the generality and flexibility of human intelligence to adapt to novel tasks with little training. To foster AI research beyond narrow generalization [8], F. Chollet [4] introduced a measure of intelligence that values *skill-acquisition efficiency* over *skill performance*, i.e. what matters is the amount of prior knowledge and experience that an agent needs to reach a reasonably good level at a range of tasks (e.g., board games), not its absolute performance at any specific task (e.g., chess). Chollet also introduced the Abstraction and Reasoning Corpus (ARC)[1] benchmark in the form of a psychometric test to measure and compare the intelligence of humans and machines. ARC is a collection of tasks that consist in learning how to transform an input colored grid into an output colored grid, given very few examples (3.3 on average). Figure 1 shows two ARC tasks (with the expected test output grid missing). ARC is a very challenging benchmark. While humans can solve more than 80% of the tasks [10], the winner of a Kaggle contest[2] could only solve 20%

---

[1] Data and testing interface at https://github.com/fchollet/ARC.
[2]https://www.kaggle.com/c/abstraction-and-reasoning-challenge.

S. Ferré: This research is supported by Labex Cominlabs (ANR-10-LABX-07-01).

I. Miliou et al. (Eds.): IDA 2024, LNCS 14641, pp. 3–15, 2024.
https://doi.org/10.1007/978-3-031-58547-0_1

of the tasks (with a lot of hard-coded primitives and brute-force search). At the ARCathon'22 contest[3] the winner solved 6% of the tasks, and we ranked 4th by solving 2% of them. The published approaches [2,3,7,17], and also the Kaggle winner, tackle the ARC challenge as a *program synthesis* problem [12], where a program is a composition of primitive transformations, and learning is done by searching the large program space. In contrast, psychological studies [1,10] have shown that, when asked to verbalize instructions on how to solve a task, participants produce object-centric instructions, called *natural programs*. They typically first describe what to expect in the input grid, and then how to generate the output grid based on the elements found in the input grid.

We make two contributions to the ARC problem: (1) *object-centric models* that enable to both parse and generate data (here grids) in terms of object patterns and computations on those objects; and (2) an efficient search of object-centric models based on the *Minimum Description Length (MDL) principle* [14]. A model for an ARC task combines two *grid models*, one for the input grid, and another for the output grid. This closely matches the structure of natural programs. Compared to the transformation-based programs that can only predict an output grid from an input grid, our models can also provide a joint description for a pair of grids. The MDL principle comes from information theory, and says that *"the model that best describes the data is the model that compress them the more"* [9,14]. It has for instance been applied to pattern mining [16]. The MDL principle is used at two levels: (a) to choose the best parses of a grid according to a grid model, and (b) to efficiently search the large model space by incrementally building more and more accurate models. The two contributions support each other because existing search strategies could not handle the large number of elementary components of our grid models, and because the transformation-based programs are not suitable to the incremental evaluation required by MDL-based search. We report promising results based on grid models that are still far from covering all knowledge priors assumed by ARC. Correct models are found for 96/400 varied training tasks with a 60 s time budget. Many of those share similarity with the natural programs produced by humans.

Section 2 discusses related work. Section 3 defines our object-centric models, and Sect. 4 explains how to learn them with the MDL principle. Section 5 reports on experimental results, comparing to existing approaches. Further details and illustrations can be found in the companion paper on arXiv [6].

2   Related Work

The ARC benchmark is recent and not many approaches have been published so far. All those we know define a DSL (Domain-Specific Language) of programs – based on function composition – that transform an input grid into an output grid, and search for a program that is correct on the training examples [2,3,7, 17]. The differences mostly lie in the primitive functions (prior knowledge) and in the search strategy. To guide the search in the huge program space, those

[3] https://lab42.global/past-challenges/arcathon-2022/.

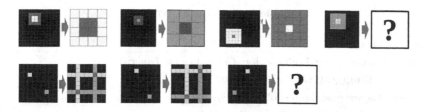

Fig. 1. Training tasks b94a9452 (top) and 23581191 (bottom).

approaches use either grammatical evolution [7], neural networks [3], search tree pruning with hashing and Tabu list [17], or stochastic search trained on solved tasks [2]. A difficulty is that the output grids are generally only used to score a candidate program so that the search is kind of blind. Alford [3] improves this with a neural-guided bi-directional search that grows the program in both directions, from input and output. Xu [17] compares the in-progress generated grid to the expected grid but this assumes that output grids are comparable to input grids, which is not true for all tasks. Function-based DSL approaches have a scaling issue because the search space increases exponentially with the number of primitive functions. For this reason, search depth is often bounded by 3 or 4. Ainooson [2] alleviates this difficulty by defining high-level functions that embody specialized search strategies.

Johnson *et al.* [10] report on a psychological study of ARC. It reveals that humans use object-centric mental representations to solve ARC tasks. This is in contrast with existing solutions that are based on grid transformations. Interestingly, the tasks that are found the most difficult by humans are those based on logics (e.g., an exclusive-or between grids) and symmetries (e.g., rotation), precisely those most easily solved by transformation-based approaches. The study exhibits two challenges: (1) the need for a large set of primitives, especially about geometry; (2) the difficulty to identify objects, which can be only visible in part due to overlap or occlusion. A valuable resource is LARC, for Language-annotated ARC [1], collected by crowd-sourcing. It provides for most training tasks one or several *natural programs*.

Beyond the ARC benchmark, a number of work has been done in the domain of *program synthesis*, which is also known as program induction or programming by examples (PbE) [12]. PbE is used in the FlashFill feature of Microsoft Excel 2013 to learn complex string processing formulas from a few examples [13]. Dreamcoder [5] alternates a *wake* phase that uses a neurally guided search to solve tasks, and a *sleep* phase that extends a library of abstractions to compress programs found during wake.

3 Object-Centric Models

We introduce *object-centric grid models* as a DSL mixing *patterns* and functions, in contrast to functions only. Unlike function-based programs that are evaluated

Table 1. Patterns by type

type	patterns
Grid	**Layers**(size: *Vector*, color: *Color*, layers: *Layer*[]), **Tiling**(grid: *Grid*, size: *Vector*)
Layer	**Layer**(pos: *Vector*, object: *Object*)
Object	**Colored**(shape: *Shape*, color: *Color*)
Shape	**Point**, **Rectangle**(size: *Vector*, mask: *Mask*)
Mask	**Bitmap**(bitmap: *Bitmap*), **Full**, **Border**, **EvenCheckboard**, ...
Vector	**Vec**(i: *Int*, j: *Int*)

like expressions, our grid models are used to *parse* a grid, i.e. to understand its contents according to the model, and also to *generate* a grid, using the model as a template. Moreover, parsing and generation can be non-deterministic. A *task model* comprises two grid models that enable to predict an output grid or to describe a pair of grids.

3.1 Mixing Patterns and Functions

The purpose of a grid model is to distinguish between invariant and variant elements across the grids of a task. The syntax of grid models is defined by the following EBNF grammar, where M stands for *model* and E for *expression*.

$$M ::= ? \mid pattern(arg_1 : M_1, \ldots, arg_k : M_k) \mid E$$
$$E ::= value \mid !path \mid func(E_1, \ldots, E_k)$$

This definition is actually generic and independent of ARC tasks. What is domain-specific is the choice of patterns, values, and functions. A model M can be one of: an *unknown* ? for a totally unconstrained model; a *pattern* for specifying some constraint; or an *expression* E for a completely constrained model, resulting from a computation. For example, the model **Rectangle**($size$: ?, $mask$: **Full**) specifies a full rectangular object of any size. The model uses two patterns: **Rectangle** with two arguments, and **Full** with no argument. This example shows that models can be nested, and that models are not only about grids but also about any kind of grid components exhibited by patterns, e.g. 2D vectors for sizes. Expressions E are defined in the usual way, as a combination of values, functions, and variables. Variables have the form !*path*, and are used in the output grid model in order to retrieve information from the input grid. A path $path = arg_1.arg_2...$ is a chain of pattern arguments that identifies a component of a grid model by navigating from the root of the model, through the nested patterns. Function arguments are not reachable because functions are not invertible, hence they do not need a name. We call *description* a fully specified and computed grid model, it is a nesting of patterns and values only: e.g., **Rectangle**($size$: **Vec**(2, 2), $mask$: **Full**).

$M^i = $ **Layers**(?, black, [**Layer**(?, **Colored**(**Rectangle**(?, **Full**), ?)),
                               Layer(?, **Colored**(**Rectangle**(?, **Full**), ?))])
$M^o = $ **Layers**(!lay[1].object.shape.size, !lay[0].object.color, [
        Layer(!lay[0].pos - !lay[1].pos, *coloring*(!lay[0].object, !lay[1].object.color))])

Fig. 2. A correct model for task b94a9452 (omitting argument names).

We now define the concrete grid models that we have used in our experiments by defining the values, patterns, and functions. Four types of values are used: integers, colors, bitmaps (i.e., Boolean matrices), and grids (i.e., color matrices). Table 1 lists the patterns. Each pattern has a result type, and typed arguments. The argument types constrain which values/patterns/functions can be used in arguments. The names of arguments are used to reference the components of a grid model or grid description. Our grid models describe a grid as either a stack of layers on top of a background having some size and color, or as the tiling of a grid up to covering a grid of given size. A layer is an object at some position. An object is so far limited to a one-color shape, where a shape is either a point or some mask-specified shape fitting into a rectangle of some size. A mask is either specified by a bitmap or by one of a few common shapes such as a full rectangle or a rectangular border. Positions and sizes are 2D integer vectors. The 30 available functions (not detailed here for lack of space) essentially cover arithmetic operations on integers and on vectors, where vectors represent positions, sizes, and moves; and geometric notions such as measures (e.g., area), translations, symmetries, scaling, and periodic patterns (e.g., tiling).

A *task model* $M = (M^i, M^o)$ is made of an input grid model M^i and an output grid model M^o. Figure 2 shows a correct model for task b94a9452.

3.2 Parsing and Generating Grids with a Grid Model

We introduce two operations that must be defined for any grid model M: the *parsing* of a grid g into a description π and the *generation* of a grid description π, and thus of a grid g. These operations are analogous to the parsing and generation of sentences from a grammar, where syntactic trees correspond to our descriptions π. In both operations, the expressions present in the model M are first evaluated and reduced to their value, using a description as the evaluation context, called *environment* and written ε. Concretely, each variable is a path in ε and is replaced by the sub-description at the end of this path. The functions are then evaluated. The result is a reduced model M' consisting of unknowns, patterns, and values only.

Parsing. The parsing of a grid g consists in replacing the unknowns of the reduced model M' by descriptions corresponding to the content of the grid. Part of this content may be left undescribed, which can be seen as *noise* from the point of view of the model. This noise is taken into account in Sect. 4.1 when defining description lengths. For example, the parsing of the first input grid in Fig. 1 (top) by the input grid model M^i in Fig. 2 results in the following description:

π^i = **Layers(Vec(12,13)**, black, [**Layer(Vec(2,4)**, **Colored(Rectangle(Vec(2, 2)**, **Full)**, yellow), **Layer(Vec(1,3)**, **Colored(Rectangle(Vec(4,4)**, **Full)**, red))]).

Generation. The generation of a grid consists in replacing the remaining unknowns in the reduced model M' by random descriptions of the right type, in order to obtain a grid description, which can then be converted into a concrete grid. For example, the output model M^o of Fig. 2, applied with environment $\varepsilon = \pi^i$ (the above input grid description), generates the following description π^o = **Layers(Vec(4,4)**, yellow, [**Layer(Vec(1,1)**, **Colored(Rectangle(Vec(2,2)**, **Full)**, red)]). This description conforms to the expected output grid.

An important point is that these two operations are *multi-valued* and *ranked* to reflect their non-determinism, i.e. they return an ordered list of descriptions. Indeed, there are often several ways of parsing or generating a grid according to a model, and some are preferable to others. For example, parsing a grid that contains several objects when the model specifies a single object.

3.3   Predict and Describe Grids with Task Models

We demonstrate the versatility of task models by showing that they can be used in two different modes: to *predict* the output grid from the input grid, and to *describe* a pair of grids jointly. We use below the notation $\rho, \pi \in parse(M, \varepsilon, g)$ to say that π is the ρ-th parsing of the grid g according to the model M with environment ε; and the notation $\rho, \pi, g \in generate(M, \varepsilon)$ to say that π is the ρ-th description generated by the model M with environment ε, and that g is the concrete grid described by π. The rank ρ is motivated by the fact that parsing and generation are multi-valued and ranked.

The *predict* mode is used after a model has been learned, in the evaluation phase with test cases. It consists in first parsing the input grid with the input model and the *nil* environment in order to get an input description π^i, and then to generate the output grid by using the ouput model and the input grid description as the environment.

$$predict(M, g^i) = \{(\rho^i, \rho^o, g^o) \mid \rho^i, \pi^i \in parse(M^i, \varepsilon^i = nil, g^i),$$
$$\rho^o, \pi^o, g^o \in generate(M^o, \varepsilon^o = \pi^i)\}$$

The *describe* mode is used in the learning phase of the model (see Sect. 4). It allows to obtain a description of a pair of grids. It consists in the parsing of the input grid and the output grid. Let us note that the parsing of the output grid depends on the result of the parsing of the input grid, hence the term "joint description".

$$describe(M, g^i, g^o) = \{(\rho^i, \rho^o, \pi^i, \pi^o) \mid \rho^i, \pi^i \in parse(M^i, \varepsilon^i = nil, g^i),$$
$$\rho^o, \pi^o \in parse(M^o, \varepsilon^o = \pi^i, g^o)\}$$

In the two modes, the *nil* environment is used with the input model because the input grid comes first, without any prior information. Note also that both modes inherit the multi-valued property of parsing and generation.

These two modes highlight an essential difference between our object-centric models and the function-based programs of existing approaches. The latter are designed for prediction (computation of the output as a function of the input), they do not provide a description of the grids.

4   MDL-Based Model Learning

MDL-based learning works by searching for the model that compresses the data the more. The data to be compressed is here the set of training examples. We have to define two things: (1) the description lengths of models and examples, and (2) the search space of models and the learning strategy.

4.1   Description Lengths

A common approach in MDL is to define the overall description length (DL) as the sum of two parts (*two-parts MDL*): the model M, and the data D encoded according to the model [9].

$$L(M, D) = L(M) + L(D \mid M)$$

The model is here a task model $M = (M^i, M^o)$ composed of two grid models, and the data is the set of training examples, i.e. pairs of grids (g^i, g^o). To compensate for the small number of examples, and to allow for sufficiently complex models, we use a *rehearsal factor* $\alpha \geq 1$, like if each example were seen α times.

$$L(M) = L(M^i) + L(M^o) \quad L(D \mid M) = \alpha \sum_{(g^i, g^o)} L(g^i, g^o \mid M)$$

The DL of an example is based on the most compressive joint description.

$$L(g^i, g^o \mid M) = min_{\rho^i, \rho^o, \pi^i, \pi^o \in describe(M, g^i, g^o)}$$
$$[\ L(\rho^i, \pi^i, g^i \mid M^i, \varepsilon^i = nil) + L(\rho^o, \pi^o, g^o \mid M^o, \varepsilon^o = \pi^i) \ ]$$

Terms of the form $L(\rho, \pi, g \mid M, \varepsilon)$ denote the DL of a grid g encoded according to a grid model M and an environment ε, via the ρ-th description π resulting from the parsing. We can decompose these terms by using π as an intermediate representation of the grid.

$$L(\rho, \pi, g \mid M, \varepsilon) = L(\rho) + L(\pi \mid M, \varepsilon) + L(g \mid \pi)$$

The term $L(\rho)$ encodes the choice of the parsed description beyond rank 1, penalizing higher ranks. The term $L(\pi \mid M, \varepsilon)$ measures the amount of information that must be added to the model and the environment to encode the description, typically the values of the unknowns. The term $L(g \mid \pi)$ measures the differences, if any, between the original grid and the grid produced by the description. A correct model is obtained when, for $\rho^i = 1$ and $\rho^o = 1$, $L(\rho^o, \pi^o, g^o \mid M^o, \varepsilon^o = \pi^i) = 0$ for all examples, i.e. when using the first description for each grid, there is nothing left to code for the output grids, and therefore the output grids can be perfectly predicted from the input grids.

Three elementary domain-specific DLs therefore have to be defined: $L(M)$, $L(\pi \mid M, \varepsilon)$, $L(g \mid \pi)$. We sketch those definitions for the grid models defined in Sect. 3. We recall that description lengths are generally derived from probability distributions with the equation $L(x) = -\log P(x)$, corresponding to an optimal coding [9]. Defining $L(M)$ amounts to encode a syntax tree with unknowns, patterns, values, paths, and functions as nodes. Thanks to types, only a subset of those are actually possible at each node: e.g. type *Layer* has only one pattern. We use uniform distributions across possible nodes, and universal encoding for non-bounded ints. Defining $L(\pi \mid M, \varepsilon)$ amounts to encode the description components that are unknowns in the model. As descriptions form a subset of models, the above definitions for $L(M)$ can be reused, only adjusting the probability distributions to exclude unknowns, paths and functions. Defining $L(g \mid \pi)$ amounts to encode which cells in grid g are wrongly specified by description π. We also have to encode the number of differing cells.

4.2   Search Space and Strategy

The search space for models is characterized by: (1) an initial model, and (2) a *refinement* operator that returns a list of refined models $M_1 \ldots M_n$ given a model M. The refinement operator has access to the joint descriptions, so it can be guided by them. Similarly to previous MDL-based approaches [16], we adopt a greedy search strategy based on the description length of models. At each step, starting with the initial model, the refinement that reduces the more $L(M, D)$ is selected. The search stops when no model refinement reduces it. The search necessarily terminates because the DL must decrease at each step but the found model may not be a solution to the task. To compensate for the fact that the input and output grids may have very different sizes, we actually use a *normalized description length* $\hat{L}$ that gives the same weight to the input and output components of the global DL, relative to the initial model.

Our initial model uses the unknown grid ? for both input and output, i.e. $M_{init} = (?, ?)$. The available refinements are the following:

- the insertion of a new layer in the list of layers – one of **Layer**(?,**Col.**(**Point**,?)), **Layer**(?,**Col.**(**Rectangle**(?,?),?)), **Layer**(?,!*object*), and !*layer* – where !*object* (resp. !*layer*) is a reference to an input object (resp. an input layer);
- the replacement of an unknown at path p by a pattern $P = pattern(?, \ldots, ?)$ when for each example, there is a parsed description π s.t. $\pi.p$ matches P;
- the replacement of a model component at path p by an expression e when for each example, there is a description π s.t. $\pi.p = e$.

4.3 Pruning Phase

The learned model sometimes lacks generality, and fails on test examples. This is because the goal of MDL-based learning as defined above is to find the most compressive task model on pairs of grids. This is relevant for the description mode but, in the prediction mode, the input grid model is used as a pattern to match the input grid, and it should be as general as possible provided that it captures the correct information for generating the output grid. For example, if all input grids in training examples have height 10, then the model will fail on a test example where the input grid has height 12, even if that height does not matter at all for generating the output.

We therefore add a *pruning phase* as a post-processing of the learned model. The principle is to start from this learned model, and to repeatedly apply *inverse refinements* while this does not break correct predictions. Inverse refinements can remove a layer or replace a pattern/value by an unknown.

5 Evaluation

We evaluated our approach on the 800 public ARC tasks, and we also took part in the ARCathon 2022 challenge (as team MADIL). In ARC, a prediction is successful only if the predicted output grid is *strictly equal* to the expected grid for *all* test examples, there is no partial success. However, three trials are allowed for each test example to compensate for potential ambiguities in the training examples. To ensure a good balance of the computational time between parsing and learning, we set some limits that remained stable across our experiments. The number of descriptions produced by the parsing of a grid is limited to 64 and only the 3 most compressive are retained for the computation of refinements. At each step, at most 100,000 expressions are considered and only the 20 most promising refinements, according to a DL estimate, are evaluated. The rehearsal rate α is set to 10. The tasks are processed independently one of each other. The results are given for a learning time per task limited to 60 s plus 10 s for the pruning phase. We used one run per task set as there is no randomness involved. Our experiments were run with a single-thread implementation[4] on Fedora 32, Intel Core i7 × 12 with 16GB memory. The learning and prediction logs and the screenshots of the solved training tasks are available at https://www.irisa.fr/LIS/ferre/pub/ida2024/ (Table 2).

Task Sets and Baselines. We consider four task sets for which results have been reported: the 400 training and 400 evaluation public tasks, the 100 secret tasks of Kaggle'20, and the 100 secret tasks of ARCathon'22. We presume that those secret tasks are taken from the 200 secret ARC tasks. As baselines, we consider published methods that report results on the considered task sets [2, 3, 7, 17]. We also include the winners of the two challenges for reference. Unfortunately, the reported results are scarce, and the papers do not provide their code.

---

[4] Open source available at https://github.com/sebferre/ARC-MDL.

Table 2. Number and percentage of solved tasks and average learning time for solved tasks, for different methods on different task sets

task set	method	solved tasks		runtime
ARC training (400 tasks)	Fischer *et al.*, 2020	31	7.68%	
	Alford *et al.*, 2021	22	5.50%	
	Xu *et al.*, 2022	57	14.25%	
	Ainooson *et al.*, 2023	104	26.00%	178.7 s
	OURS	96	24.00%	4.6 s
ARC evaluation (400 tasks)	Ainooson *et al.*, 2023	26	6.50%	
	OURS	23	5.75%	11.4 s
Kaggle'20 (100 tasks)	Icecuber (winner)	20	20.6%	
	Fischer *et al.*	3	3.0%	
ARCathon'22 (100 tasks)	pablo (winner)	6	6%	
	Ainooson *et al.*	2	2%	
	OURS (4th ex-aequo)	2	2%	

Success Rates. On the training tasks, for which we have the more results to compare with, our method solves 24% tasks, almost on par with the best method, by Ainooson *et al.* (26%). Both methods also solved a similar number of evaluation tasks (5.75% vs 6.50%), and both solved 2% tasks at ARCathon'22, and ranked 4th ex-aequo. Comparing the different task sets, the evaluation tasks appear to be significantly more difficult than the training tasks, and the secret tasks of ARCathon seem even more difficult as the winner could only solve 6% tasks. Icecuber managed to correctly solve an amazing 20.6% tasks at Kaggle'20, but at the cost of the hand-coding of 142 primitives, 10k lines of code, and brute-force search (millions of computed grids per task).

The ARC evaluation protocol allows for three predictions per test example. However, the first prediction of our method is actually correct in 90 of the 96 solved training tasks. This shows that our learned models are accurate in their understanding of the tasks. To better evaluate the generalization capability of learned models, we also measured the generalization rate as the proportion of models that are correct on training examples that are also correct on test examples: 92% (94/102) on training tasks, and 72% (23/32) on evaluation tasks. This again suggests that the evaluation tasks feature a higher generalization difficulty. Without the pruning phase, this rate decreases to 89% (91/102) on training tasks. This shows that the pruning phase is useful, although description-oriented model learning is already good at generalization. Reasons for failures to generalize are: e.g., the test example has several objects while all training examples have a single object; the training examples have a misleading invariant.

Efficiency and Model Complexity. Intelligence is the efficiency at acquiring new skills, according to Chollet. Although ARC enforces data efficiency by having only a few training examples per task, and unique tasks, it does not enforce efficiency in the amount of priors, nor in the computation resources. It is therefore useful to assess the latter. We already mentioned Icecuber's method that relies on a large number of primitives, and intense computations. The method of Ainooson *et al.*, which has comparable performance to ours, uses 52 primitives and about 700 s on average per solved task. In comparison, our method uses 30 primitives and 4.6 s per solved training task (21.7 s over all training tasks). Those short runtimes are made possible by our greedy strategy. Doubling the learning timeout at 120 s does not lead to solving more tasks, so 60 s just seems to be enough to find a solution if there is one.

Another way to evaluate efficiency is to look at the complexity of learned models, typically the number of primitives composing the model in program synthesis approaches. A good proxy for this complexity is the depth of search that was reached in the allocated time. In our case, it is equal to the number of refinements applied to the initial empty model. Few methods provide this information: Icecuber limits depth to 4, and Ainooson's best results are achieved with a brute-force search with maximum depth 3. Methods based on DreamCoder [3] have similar limits but can learn more complex programs by discovering and defining new operations as common compositions of primitives, and reusing them from one task to another. Our method can dive much deeper in less computation time, thanks to its greedy strategy. The number of refinement steps achieved in a timeout of 60 s on the training tasks ranges from 4 to 57, with an average of 19 steps. This demonstrates the effectiveness of the MDL criteria to guide the search towards correct models. This claim is reinforced by the fact that a beam search (width=3) did not lead to solving more tasks.

Learned Models. The learned models for solved tasks are very diverse despite the simplicity of our models. They express various transformations: e.g., moving an object, extending lines, putting one object behind another, order objects from largest to smallest, remove noise, etc. Note that none of these transformations is a primitive in our models, they are learned in terms of objects, basic arithmetics, simple geometry, and the MDL principle.

We compared our learned models to the natural programs of LARC [1]. Remarkably, many of our models involve the same objects and similar operations than the natural programs. For example, the natural program for task b94a9452 is: *"[The input has] a square shape with a small square centered inside the large square on a black background. The two squares are of different colors. Make an output grid that is the same size as the large square. The size and position of the small inner square should be the same as in the input grid. The colors of the two squares are exchanged."* For other tasks, our models miss some notions used by natural programs but manage to compensate them: e.g., topological relations such as "next to" or "on top" are compensated by the three attempts; the majority color is compensated by the MDL principle selecting the largest object. However, in most cases, the same objects are identified.

These observations demonstrate that our object-centric models align well with the natural programs produced by humans, unlike approaches based on the composition of grid transformations. An example of a program learned by [7] on the task 23b5c85d is strip_black; split_colors; sort_Area; top; crop, which is a sequence of grid-to-grid transformations, without explicit mention of objects.

6 Conclusion

We have presented a novel and general approach to efficiently learn skills at tasks that consist in generating structured outputs as a function of structured inputs. Our approach is based on descriptive task models that combine object-centric patterns and computations, and on the MDL principle for guiding the search for models. We have detailed an application to ARC tasks on colored grids. We have shown promising results, especially in terms of efficiency, model complexity, and model naturalness. Going further on ARC will require a substantial design effort as our current models cover so far a small subset of the knowledge priors that are required by ARC tasks (e.g., missing goal-directedness).

References

1. Acquaviva, S., et al.: Communicating natural programs to humans and machines. Adv. Neural. Inf. Process. Syst. **35**, 3731–3743 (2022)
2. Ainooson, J., Sanyal, D., Michelson, J.P., Yang, Y., Kunda, M.: An approach for solving tasks on the abstract reasoning corpus. arXiv preprint arXiv:2302.09425 (2023)
3. Alford, S., et al.: Neural-guided, bidirectional program search for abstraction and reasoning. CoRR abs/2110.11536 (2021)
4. Chollet, F.: A definition of intelligence for the real world. J. Artif. Gen. Intell. **11**(2), 27–30 (2020)
5. Ellis, K., et al.: DreamCoder: bootstrapping inductive program synthesis with wake-sleep library learning. In: PLDI 2021: Proceedings of the 42nd ACM SIGPLAN International Conference on Programming Language Design and Implementation, pp. 835–850 (2021)
6. Ferré, S.: Tackling the abstraction and reasoning corpus (ARC) with object-centric models and the MDL principle. arXiv preprint arXiv:2311.00545 (2023)
7. Fischer, R., Jakobs, M., Mücke, S., Morik, K.: Solving abstract reasoning tasks with grammatical evolution. In: LWDA, pp. 6–10. CEUR-WS 2738 (2020)
8. Goertzel, B.: Artificial general intelligence: concept, state of the art, and future prospects. J. Artif. Gen. Intell. **5**(1), 1 (2014)
9. Grünwald, P., Roos, T.: Minimum description length revisited. Int. J. Math. Ind. **11**(01), 1930001 (2019)
10. Johnson, A., Vong, W.K., Lake, B., Gureckis, T.: Fast and flexible: human program induction in abstract reasoning tasks. arXiv preprint arXiv:2103.05823 (2021)

11. Krizhevsky, A., Sutskever, I., Hinton, G.E.: ImageNet classification with deep convolutional neural networks. Adv. Neural. Inf. Process. Syst. **25**, 1097–1105 (2012)
12. Lieberman, H.: Your Wish is My Command. The Morgan Kaufmann series in interactive technologies, Morgan Kaufmann / Elsevier (2001)
13. Menon, A., Tamuz, O., Gulwani, S., Lampson, B., Kalai, A.: A machine learning framework for programming by example. In: International Conference on Machine Learning, pp. 187–195. PMLR (2013)
14. Rissanen, J.: Modeling by shortest data description. Automatica **14**(5), 465–471 (1978)
15. Silver, D., Huang, A., Maddison, C.J., et al.: Mastering the game of go with deep neural networks and tree search. Nature **529**(7587), 484–489 (2016)
16. Vreeken, J., Van Leeuwen, M., Siebes, A.: Krimp: mining itemsets that compress. Data Min. Knowl. Disc. **23**(1), 169–214 (2011). https://doi.org/10.1007/s10618-010-0202-x
17. Xu, Y., Khalil, E.B., Sanner, S.: Graphs, constraints, and search for the abstraction and reasoning corpus. arXiv preprint arXiv:2210.09880 (2022)

RMI-RRG: A Soft Protocol to Postulate Monotonicity Constraints for Tabular Datasets

Iko Vloothuis and Wouter Duivesteijn$^{(\boxtimes)}$

Technische Universiteit Eindhoven, Eindhoven, The Netherlands
i.vloothuis@student.tue.nl, w.duivesteijn@tue.nl

Abstract. Ensuring that a predictive model respects monotonicity constraints can enhance societal acceptance of such models. Literature on monotone classification shows that it can even improve classifier performance. However, a set of applicable monotonicity constraints is often assumed as input for the model. We propose RMI-RRG: a soft protocol that can be employed to postulate monotonicity constraints for any tabular dataset. The protocol encompasses consensus from scientific literature, aggregating the strength of (anti-)monotonicity relations in an RMI Table, aggregating the effect of imposing more constraints on the number of relabelings required to fully monotonize the dataset in a Required Relabelings Graph (RRG), and inspecting the effect on the comparability rate. We illustrate the deployment of the protocol on six datasets, arriving at some conclusions that deviate from conclusions from (mutually disagreeing) existing literature, and showing how individual steps in the protocol each have their role to play in arriving at a final postulate.

Keywords: Monotone Classification · Monotonicity Constraints ·
RMI-RRG Protocol · Rank Mutual Information · Required Relabelings
Graph

1 Introduction

If Person A and Person B both apply for a mortgage at the same mortgage lender, if the data we have on both persons is identical except that we know that Person A has a higher income, then it would be strange if Person B is approved for the mortgage while Person A is not. In data mining we call such an event a *violation* of a *monotonicity constraint*: all else equal, if the value of certain input variables increase, we cannot have a decrease in the value of the target. Literature (e.g. [6]) on monotone classification and regression has shown that making it mandatory for a prediction model to respect given monotonicity constraints can keep predictive performance at the same level while guaranteeing no monotonicity violations, or even increase predictive performance.

Humans react quite viscerally to monotonicity constraints and their violations: we automatically connect this to a feeling of (un-)fairness, sometimes in

I. Miliou et al. (Eds.): IDA 2024, LNCS 14641, pp. 16–27, 2024.
https://doi.org/10.1007/978-3-031-58547-0_2

an almost axiomatic way (such as in the mortgage example above). However, given any tabular dataset, it is not necessarily obvious which monotonicity constraints are reasonable to postulate. Two paths exist in literature. The *subjective* approach inspects the domain of the dataset, reasons about the intrinsic meaning of the target variable, connects this to reasoning about the intrinsic meaning of input variables, and draws a conclusion about which monotonicity constraints to impose. In the mortgage example this makes sense, but it comes with two drawbacks. On the one hand, many datasets exist where monotonicity constraints might be helpful but it is not that easy to reason about the domain of the dataset, so this is hard to put in practice. On the other hand, the subjective nature of this process may lead different researchers to draw different conclusions. The *objective* approach measures something objective about the relation between input variables and the output variable, and puts a threshold on that measurement: any input variables whose relation to the output variable surpasses the threshold, are deemed to have a monotonicity relation with the output variable. Such a procedure is clear, objective, and ensures that anybody who runs this procedure will always find the same answer; said answer can also be completely arbitrary (due to the need to fix the threshold level somewhere) and lead to suboptimal results. No existing solution is satisfactory.

In this paper, we introduce the *RMI-RRG protocol*: a soft protocol to answer the question: "Given any tabular dataset, which monotonicity constraints should we postulate?" It incorporates aspects of both approaches: the subjective and the objective. We do not claim that all researchers that follow this protocol will derive the exact same conclusions; this is why we call it a soft protocol. But we show how a scientific consensus can give us a good first insight. We show that a summary of the direction and strength of monotonicity relations as provided in an *RMI Table* (cf. Sect. 5) can sharpen the image. We show how we can correct misleading conclusions from RMI Tables, by inspecting the *Required Relabelings Graph* (RRG, cf. Sect. 6): a visualization of the effect that imposing additional constraints has on the number of relabelings required to fully monotonize the dataset. We show how inspection of comparability rates and mRMR values can finetune the conclusions. In the end, we deploy the RMI-RRG protocol on six datasets, resulting in postulating monotonicity constraints on each dataset.

2   Related Work

Monotonicity is a fundamental concept spanning mathematics and various disciplines, denoting the preservation of order. In simple terms, a function or relation is deemed to satisfy (anti-)monotonicity if it maintains the (reversed) order of its inputs in the outputs. Specifically, for any pair of values, if one value is greater than or equal to the other, the corresponding output must retain the same (or, in case of antimonotonicity, reversed) order.

Monotonicity finds applications across diverse fields such as economics, statistics, optimization, mathematics, and computer science. In economics, it primarily models preferences and utility functions [5], while statistics employs it for non-parametric regression and analyzing ordered categorical data [21]. In optimization, monotonicity often serves as a foundation for convergence and efficiency [26]. In machine learning and pattern recognition, monotonicity assumptions can enhance model interpretability and generalization. By introducing monotonicity constraints, meaningful trends and relationships in data can be captured. This proves invaluable in domains like credit scoring, risk assessment, or fraud detection, where monotonic patterns are expected [17].

Classification is one domain that notably benefits from monotonicity. Leveraging monotonic relationships empowers classifiers to align with domain-specific expectations, refine model performance, and furnish more dependable and interpretable predictions [15,19].

For a survey on classification with monotonicity constraints, see [4]. The paper encompasses extensively employed datasets, monotonic preprocessing, relabeling techniques, and a diverse set of classifiers respecting monotonicity constraints. One of the earlier applications of monotone constraints to kNN classification [6] showed that adapting kNN to respect monotonicity constraints can be achieved without loss of predictive accuracy; in fact, predictive accuracy increases on some datasets. Monotonicity constraints have also been imposed when employing kNN as a kernel method [16], yielding markedly superior results compared to the baseline kernel estimators. Other notable monotone classifiers are MonoBoost [2] and MonGel [8].

3   Preliminaries and Notation

Let X denote the $n \times p$ matrix of input variables, where each row represents an observation with values x. Let $\pi_S(X)$ denote the projection of X on any subset $S \subseteq \{1, \ldots, p\}$ of the input variables. Let Y denote an $n \times 1$ vector representing the output variable, taking on values from a one-dimensional space Y, where $Y \subseteq \mathbb{R}$. Let $D = \{(x_i, y_i)\}_{i=1}^{n}$ denote the bag of observed data points.

A function f is said to be monotone if it preserves order, meaning that the ordering of the input values is preserved in the output values. Two types of monotone functions exist: increasing or decreasing (also known as non-decreasing or non-increasing, respectively). Specifically, a function is monotone increasing if, for any two values x and x' in the domain of f, it holds that $x \leq x' \Rightarrow f(x) \leq f(x')$, or, for a decreasing monotone function: $x \leq x' \Rightarrow f(x) \geq f(x')$.

A partial order is assumed for X, denoted by $\leq_X$. The partial order establishes criteria for determining the order between input values in X. A pair of points (x_i, x_j) in a dataset are said to be *comparable* if $x_i \geq x_j \vee x_j \geq x_i$, and *incomparable* if not. A pair of comparable data points (x_i, y_i) and (x_j, y_j) is considered monotone if the ordering of the input variables is preserved in the output labels. In other words, if $x_i \geq_X x_j$, then it must hold that $y_i \geq y_j$. Dataset D is considered monotone if all combinations of pairs of data points are either monotone or incomparable. We overload the partial order notation with any subset S of input variables, such that $x_i \leq_S x_j$ evaluates the comparability of x_i and x_j only on $\pi_S(X)$.

Definitions stated above assume a direct monotone restriction on the data, but an inverse relation is also possible. In that case the definition for an anti-monotone pair of data points would be $x_i \geq x_j \Rightarrow y_i \leq y_j$.

For any data point x_i and subset S of input variables, let the *upset* $\uparrow_{(i,S)}$ and *downset* $\downarrow_{(i,S)}$ be:

$$\uparrow_{(i,S)} = \{x_j \in D \mid x_i \leq_S x_j\} \qquad\qquad \downarrow_{(i,S)} = \{x_j \in D \mid x_j \leq_S x_i\}$$

The upset of x_i consists of all points whose input values are considered 'higher', when limited to variables in S, than x_i. The downset consists of all data points 'lower' than x_i. Again, for variables with an antimonotone relation these definitions would also be inverted.

3.1 Rank Mutual Information

Rank Mutual Information (RMI) [12] evaluates the monotonic consistency between variables. The RMI between variables A and B is:

$$\mathrm{RMI}(\{A\}, \{B\}) = -\frac{1}{n} \sum_{i=1}^{n} \log \frac{\mid \uparrow_{(i,\{A\})} \mid \cdot \mid \uparrow_{(i,\{B\})} \mid}{n \cdot \mid \uparrow_{(i,\{A\})} \cap \uparrow_{(i,\{B\})} \mid}$$

Applying this formula to an input and an output variable will deliver a value that can range from $[-\infty, \infty]$. In practice this value rarely ranges beyond $(-1.5, 1.5)$. The RMI gives the properties of strength (sign of the RMI) and direction (magnitude of the RMI) between one or more variables and a target. RMI measures the type of information needed while also being fairly robust against noise [11].

3.2 Relabeling

Some data mining algorithms (e.g., [6]) that guarantee respecting monotonicity constraints in their predictions, require as input a dataset that fully respects those same constraints. If the training data respects monotonicity constraints, any prediction such a method makes on test/validation data will necessarily also respect the constraints. A problem is that many datasets contain some monotonicity violations. A typical approach then is to relabel values in the affected target columns, so as to not violate monotonicity. Several optimal relabeling methods have been proposed, including Feelders relabeling [7,10], Single-pass Optimal Ordinal relabeling [20], and Optimal Flow Network relabeling [6,7]. These methods minimize the number of instances to relabel, and the relabeled dataset can be viewed as a monotone classifier that minimizes the error rate on the training data [6].

4 Main Direct Competitors

We will briefly go over our main competitors in this section. Keep in mind that the final goal of each method in this section is to come up with a set of *postulates*:

these are the input variables for which we have decided that a monotonicity constraint can be imposed between the input and the output variable. Along the way, we will often measure some *monotonicity relations* between input variables and the output variable that can come with a given *strength*. The final outcome, however, will have to turn these fuzzy monotonicity relations into a set of hard postulated monotonicity constraints.

4.1   Subjective Approaches

The traditional approach is the subjective one, where scientists reason about the domain of the datasets under consideration. One typically goes by the common sense approach as exemplified by the first paragraph of Sect. 1 of this paper: if two persons are identical except that Person A has a higher income than Person B, it would be ridiculous if Person B gets a mortgage while Person A does not. This "it stands to reason" test has been the gold standard in classification under monotonicity constraints. An example is provided by [6]. The most extensive example of this procedure, involving a careful evaluation of each dataset and the identification of variables based on domain knowledge, is given in [24].

4.2   Objective Approaches

Most of the objective strategies assume the existence of monotonicity relations between input variables and the output variable, and will then use a combination of the strength and direction of this relation to decide whether to include it as a postulate. The most common and simple strategy here is to assume that all variables have a relation with the target and assign a direction to them. This can be seen in [22] and [27], where tables containing information on the datasets clearly show the inclusion and direction of all chosen datasets; how the direction of the relation was established remains opaque.

Rank Mutual Information (RMI) can represent the strength of all monotonicity relations in the dataset; subsequently, a hard constraint on its values can be imposed to arrive at a set of monotonicity constraints. The authors of [4] have used $|RMI| > 0.1$ as the cutoff value: a monotonicity constraint will be postulated on an input variable and the output variable if and only if their absolute RMI value surpasses 0.1. Personal communication with the authors revealed that there seems to be no substantial reason for choosing this particular cutoff value.

The authors of [13] propose to use the RMI in combination with the *min-Redundancy Max-Relevance* (mRMR) algorithm to calculate optimal subsets of variables. The goal here is to iteratively add the variable with the highest ranked RMI value to the chosen set, while taking into account whether doing so would merely introduce redundant information. When to stop adding variables is not specified: the authors recommend to choose a fixed number of variables and apply that to each dataset, which strikes us as a blunt, arbitrary instrument.

Table 1 lists methods used in selecting which monotonicity constraints to postulate during experimentation, with an example of a paper deploying that strategy. Clearly, there is a wide range of strategies without any clear consensus.

5 RMI Tables and Required Relabelings Graphs

Our protocol employs two forms of data aggregation. On the one hand, for each dataset, we will order the input variables by decreasing absolute RMI values (cf. Table 3 for examples of such *RMI Tables* on real datasets); denote this ordering of input variables by $x_{(1)}, x_{(2)}, \ldots, x_{(p)}$. It stands to reason that in terms of monotonicity relations with the target variable, $x_{(1)}$ will be the input variable with the strongest relation, and $x_{(p)}$ will be the input variable with the weakest relation. On the other hand, subsequently, for each dataset, we will generate the *Required Relabelings Graphs* (RRGs). This is a line graph consisting of p observations. For each observation $1 \leq i \leq p$, we compute the number of relabelings required to make the dataset fully monotone, postulating a monotonicity relation between the output variable and *all* of the top-i strongest-related variables $\{x_{(1)}, \ldots, x_{(i)}\}$. Informally, as our index i increases, we keep expanding the set of input variables for which we postulate a monotonicity constraint (starting with the strongest such relations). As this set expands, one could reasonably expect the constraints to become more pressing, but the number of comparable pairs of observations will reduce. See Fig. 1 for examples of RRGs on real datasets. From RRGs one would hope to observe behavior such as in Fig. 1a: the number of required relabelings decreases, but the curve flattens beyond some point, indicating that postulating more constraints likely has limited benefits.

The point of these data aggregations is: we think RMI has something interesting to say about monotonicity relations, but boiling that message down to a threshold on the RMI value is likely too blunt an instrument. Inspecting the RMI Tables and the RRGs should provide more information.

6 The RMI-RRG Protocol

The RMI-RRG Protocol is a soft protocol for postulating monotonicity constraints on any tabular dataset, encompassing the following four steps:

1. Literature consistency check: identify input variables that are consistently recognized in the existing literature.
2. RMI alignment: cross-reference the input variable with their respective RMI values (cf. Table 3).

Table 1. Monotonicity constraint postulation strategies.

Strategy	Example reference
Selecting constraints based on a priori information	[2]
Selecting constraints based on other literature	[24]
Postulating all constraints	[9]
Selecting constraints based on RMI	[4, 18]
No explanation given	[1]

Table 2. Dataset metadata.

D	n	p	Y	Y type	Relabels	comp. %	Source
Breast Cancer	683	9	Class	bin	12	0.821	[14]
Car	1728	6	Decision	cat (4)	466	0.528	[14]
CMC	1473	9	contraceptive	cat (3)	865	0.657	[14]
Pasture	36	22	pasture-prod-class	cat (3)	2	0.694	[23]
PIMA	768	8	Outcome	bin	198	1.0	[14]
Windsor	546	11	price	num	241	0.414	[14]

3. RRG analysis: address any discrepancies between literature consensus and RMI values by inspecting the Required Relabelings Graph (cf. Fig. 1).
4. Comparability assessment: evaluate whether the inclusion of a variable impacts the comparability of the dataset or necessitates a substantial number of relabels; exclude those variables with an outsized negative impact.

Whether input variables are included as a monotone or an antimonotone relation is decided by the sign of the RMI value.

7 Experimental Results

We collected a set of datasets, including the ten most commonly used ordinal datasets in monotone classification literature [4], the datasets used in the original kNN classification under monotonicity constraints paper [6], and four additional ordinal datasets [3]. From this set of 16 datasets, we report results on a selection of six datasets in this paper: these are the six datasets from which the most interesting conclusions could be drawn. Results on the other ten datasets are available in the MSc thesis [25, Chapter 5] from which this paper was derived.

Table 2 provides metadata on the datasets in this paper. The column 'relabels' contains the number of relabels necessary to make the dataset fully monotone (cf. Sect. 3.2), and 'comp. %' shows the fraction of all possible pairs in the dataset that are comparable. The last column indicates the source from which the dataset was obtained.

7.1 Breast Cancer

The RMI values of the first few attributes (cf. Table 3a) are fairly close together and seem promising. Figure 1a reveals that imposing constraints on variables after *Bland_ Chromatin* flattens the curve, which could be an indication of these subsequent variables being redundant. Checking the redundancy between the variables with the highest RMI scores and these latter ones confirms this. We postulate monotonicity constraints on *Uniformity_ Size, Uniformity_ Shape, Bare_ Nuclei*, and *Bland_ Chromatin*.

Table 3. RMI Tables for every dataset in Table 2; only a limited number of attributes with highest absolute RMI are shown, due to space limitations.

Variable	Uniform_Size	Uniform_Shape	Bare_Nuclei	Bland_Chromatin	Single_Cell_Size
RMI	0.501	0.499	0.473	0.459	0.442

(a) Breast Cancer.

Variable	safety	no. persons	buying price	maintenance cost	lug_boot	no. doors
RMI	0.246	0.175	-0.107	-0.095	0.078	0.032

(b) Car.

Variable	wife_age	wife_edu	husband_occup	num_child	SOL_index	wife_working	husband_edu
RMI	-0.194	0.052	0.035	0.028	0.025	0.023	0.021

(c) CMC.

Variable	HFRG-pct-mean	OlsenP	Leaf-P	Eworms-main-3	MinN	NFIX-mean
RMI	0.659	0.598	0.592	0.561	0.54	0.526

(d) Pasture.

Variable	Glucose	BMI	Pregnancies	Age	DPF	Insulin	BloodPressure	SkinThickness
RMI	0.316	0.187	0.139	0.138	0.121	0.096	0.093	0.087

(e) PIMA.

Variable	lotsize	stories	bathrooms	garage	bedrooms	aircon	prefer	recreation
RMI	0.631	0.349	0.347	0.316	0.299	0.291	0.192	0.134
mRMR	0.63	0.31	0.30	0.29	0.29	0.28	0.14	0.13

(f) Windsor.

7.2   Car

This dataset proved to be rather controversial. Most literature [4, 22] claims that all input variables have a direct relation with the target variable. The RMI values (cf. Table 3b, Fig. 1b) disagree: only two or three variables seem even remotely fit to consider. Assuming constraints on all input variables, the fraction of observation pairs that is comparable is 14%. Limiting the constraints to only *safety, number of persons*, and *buying price*, increases the comparability rate to 52%. Hence, we postulate only those monotonicity constrains.

7.3   CMC

The CMC dataset illustrates the added value of the RRG on top of the RMI Table. From Table 3c, the *wife_age* variable seems to be the only sensible candidate for a monotonicity constraint: existing objective approaches [4] would draw that conclusion. But Fig. 1c shows that something more interesting is happening. While assuming constraints on *wife_edu* and *husband_occup* does not influence the number of required relabels much, subsequent variables do have an outsized

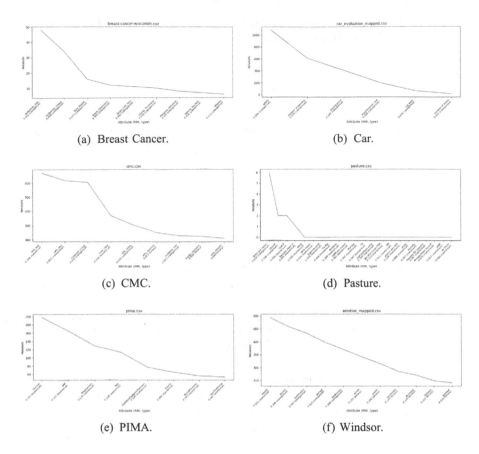

(a) Breast Cancer.   (b) Car.

(c) CMC.   (d) Pasture.

(e) PIMA.   (f) Windsor.

Fig. 1. Required Relabelings Graphs (RRGs) for every dataset in Table 2.

influence. Hence, we postulate monotonicity constraints on *wife_ age*, *wife_ edu*, *husband_ occup*, and *num_ child*. As part of a reason why this RRG moves in this strange shape, we hypothesize this is caused by *wife_ edu* and *husband_ occup* being categorical variables while *wife_ age* and *num_ child* are numerical. Postulating monotonicity constraints on numerical variables will most likely lead to smoother behavior than postulating such constraints on categorical variables, since the changes in the values for numerical variables can be much more subtle.

7.4   Pasture

This dataset only contains 37 observations, which means that the comparability and the number of relabels becomes more important than the raw values in Table 3d. Figure 1d shows that imposing constraints on variables after *FRG-pct-mean*, *OlsenP*, and *Leaf-P*, would lower the number of relabels to 0. This suggests that the number of comparable pairs of observations decreases drastically, which is indeed confirmed by measurement: comparability rate drops from 69% to

43% upon inclusion of a fourth constraint. Hence, we postulate monotonicity constraints on *FRG-pct-mean*, *OlsenP*, and *Leaf-P*.

7.5 PIMA

Literature disagrees on treatment of this dataset: [24] assumes monotonicity constraints on four variables, where [4] uses six. Only the *Glucose* variable has a substantially outlying RMI value (cf. Table 3e). Figure 1e shows that constraints on additional variables do not impact the number of required relabels in any unusual way. By lack of reason to decide otherwise, we exclude from consideration all variables whose absolute RMI value is lower than 0.1, hence postulating monotonicity constraints on *Glucose, BMI, Pregnancies, Age*, and *DPF*.

7.6 Windsor

There is no literature that specifically states which variables and directions are to be constrained in this dataset. The RMI values (cf. Table 3f) indicate that any variable lower than 0.2 is probably best excluded. Figure 1f shows that there are no unusual hitches for any specific variable. The mRMR values do show that redundant information is introduced for any variable after *aircon*, which can be seen in the drop in value. Hence, we postulate monotonicity constraints on all variables except *prefer* and *recreation*.

8 Conclusions

Monotonicity constraints make sense to human intuition. Ensuring that predictive models respect monotonicity constraints can increase societal acceptability of such models, and even increase their predictive performance. However, there is no consensus in scientific literature on how to determine which monotonicity constraints to assume for a given dataset. This paper provides RMI-RRG, a soft protocol to postulate monotonicity constraints for any tabular dataset.

Section 5 introduces two forms of data aggregation employed in the protocol. On the one hand, the RMI Table orders the input variables of a dataset by their perceived strength of (anti-)monotone relation to the target variable; it also provides the perceived direction of this relation. On the other hand, the Required Relabelings Graph (RRG) shows more detailed effects, of imposing constraints on additional variables, on the number of relabelings required to fully monotonize the dataset. The RMI-RRG protocol (cf. Sect. 6) incorporates these two aggregation methods, along with consensus from literature and assessment of comparability rates, into a final judgment.

In Sect. 7 we illustrate application of the RMI-RRG protocol on six datasets. Results encompass fairly typical behavior (*Breast cancer* dataset), a clear conclusion that disagrees with literature (*Car*), the added value of the RRG over existing methods (*CMC*), the necessity to look at comparability rates (*Pasture*) and mRMR values (*Windsor*), and a decisive conclusion where existing literature draws multiple distinct conclusions (*PIMA*).

RMI-RRG is a soft protocol: while we think it will often lead in a fairly clear direction, it is still possible that multiple scientists can arrive at multiple distinct conclusions. We think that this is unavoidable: monotonicity is, to a degree, in the eye of the beholder. We claim that the ambiguity of RMI-RRG is better than the objective arbitrariness of setting a hard threshold of 0.1 on the absolute RMI values [4], while RMI-RRG provides more information leading to more informed decisions than existing subjective approaches [6,24]. Future improvements might include separate treatment for categorical and numerical input variables (cf. Sect. 7.3), and a tradeoff between the number of required relabelings and the comparability rate.

References

1. Alcala-Fdez, J., Alcala, R., Gonzalez, S., Nojima, Y., Garcia, S.: Evolutionary fuzzy rule-based methods for monotonic classification. IEEE Trans. Fuzzy Syst. **25**(6), 1376–1390 (2017)
2. Bartley, C., Liu, W., Reynolds, M.: A novel framework for constructing partially monotone rule ensembles. In: Proceedings ICDE, pp. 1320–1323 (2018)
3. Bellmann, P., Lausser, L., Kestler, H.A., Schwenker, F.: A theoretical approach to ordinal classification: feature space-based definition and classifier-independent detection of ordinal class structures. Appl. Sci. **12**(4), 1815 (2022)
4. Cano, J.R., Gutiérrez, P.A., Krawczyk, B., Woźniak, M., García, S.: Monotonic classification: an overview on algorithms, performance measures and data sets. Neurocomputing **341**, 168–182 (2019)
5. Carlier, G., Dana, R.A.: Law invariant concave utility functions and optimization problems with monotonicity and comonotonicity constraints. Stat. Risk Model. **24**(1), 127–152 (2006)
6. Duivesteijn, W., Feelders, A.: Nearest neighbour classification with monotonicity constraints. In: Daelemans, W., Goethals, B., Morik, K. (eds.) Machine Learning and Knowledge Discovery in Databases. Lecture Notes in Computer Science(), vol. 5211, pp. 301–316. Springer, Berlin (2008). https://doi.org/10.1007/978-3-540-87479-9_38
7. Feelders, A.: Monotone relabeling in ordinal classification. In: Proceedings ICDM, pp. 803–808 (2010)
8. García, J., Fardoun, H.M., Alghazzawi, D.M., Cano, J.R., García, S.: MoNGEL: monotonic nested generalized exemplar learning. Pattern Anal. Appl. **20**(2), 441–452 (2015)
9. González, S., García, S., Li, S.T., John, R., Herrera, F.: Fuzzy k-nearest neighbors with monotonicity constraints: moving towards the robustness of monotonic noise. Neurocomputing **439**, 106–121 (2021)
10. González, S., Herrera, F., García, S.: Monotonic random forest with an ensemble pruning mechanism based on the degree of monotonicity. New Gener. Comput. **33**(4), 367–388 (2015)
11. Hao, H., Wang, M., Tang, Y.: Feature selection based on improved maximal relevance and minimal redundancy. In: Proceedings IMCEC, pp. 1426–1429 (2016)

12. Hu, Q., Che, X., Zhang, L., Zhang, D., Guo, M., Yu, D.: Rank entropy-based decision trees for monotonic classification. IEEE Trans. Knowl. Data Eng. **24**(11), 2052–2064 (2012)
13. Hu, Q., Pan, W., Song, Y., Yu, D.: Large-margin feature selection for monotonic classification. Knowl.-Based Syst. **31**, 8–18 (2012)
14. Kelly, M., Longjohn, R., Nottingham, K.: The UCI machine learning repository. http://archive.ics.uci.edu/ml
15. Kotlowski, W., Slowinski, R.: Rule learning with monotonicity constraints. In: Proceedings ICML, pp. 537–544 (2009)
16. Li, Z., Liu, G., Li, Q.: Nonparametric kNN estimation with monotone constraints. Economet. Rev. **36**(6–9), 988–1006 (2017)
17. Pan, W.: Fraudulent firm classification using monotonic classification techniques. In: Proceedings ITAIC, pp. 1773–1776 (2020)
18. Pei, S., Hu, Q.: Partially monotonic decision trees. Inf. Sci. **424**, 104–117 (2018)
19. Potharst, R., Feelders, A.J.: Classification trees for problems with monotonicity constraints. ACM SIGKDD Explor. Newslett. **4**(1), 1–10 (2002)
20. Rademaker, M., De Baets, B., De Meyer, H.: Optimal monotone relabelling of partially non-monotone ordinal data. Optim. Methods Softw. **27**(1), 17–31 (2012)
21. Saarela, O., Rohrbeck, C., Arjas, E.: Bayesian non-parametric ordinal regression under a monotonicity constraint. Bayesian Anal. **18**(1), 193–221 (2023)
22. Suárez, J.L., González-Almagro, G., García, S., Herrera, F.: A preliminary approach for using metric learning in monotonic classification. In: Fujita, H., Fournier-Viger, P., Ali, M., Wang, Y. (eds.) Advances and Trends in Artificial Intelligence. Theory and Practices in Artificial Intelligence. Lecture Notes in Computer Science(), vol. 13343, pp. 773—784. Springer, Cham (2022). https://doi.org/10.1007/978-3-031-08530-7_65
23. Vanschoren, J., van Rijn, J.N., Bischl, B., Torgo, L.: OpenML: networked science in machine learning. SIGKDD Explor. **15**(2), 49–60 (2013)
24. Verbeke, W., Martens, D., Baesens, B.: RULEM: a novel heuristic rule learning approach for ordinal classification with monotonicity constraints. Appl. Soft Comput. **60**, 858–873 (2017)
25. Vloothuis, I.: Imputing missing data using k-Nearest Neighbors under monotonicity constraints. Master's thesis, Technische Universiteit Eindhoven (2023)
26. Zhou, J., Mayne, R.W.: Monotonicity analysis and the reduced gradient method in constrained optimization. J. Mech., Transmiss., Autom. Des. **106**(1), 90–94 (1984)
27. Zhu, H., Liu, H., Fu, A.: Class-weighted neural network for monotonic imbalanced classification. Int. J. Mach. Learn. Cybern. **12**(4), 1191–1201 (2021)

A Structural-Clustering Based Active Learning for Graph Neural Networks

Ricky Maulana Fajri[1(✉)], Yulong Pei[1], Lu Yin[1,2], and Mykola Pechenizkiy[1]

[1] Eindhoven University of Technology, Eindhoven, The Netherlands
{r.m.fajri,y.pei.1,mykola.pechenizkiy}@tue.nl
[2] The University of Aberdeen, Aberdeen, UK
lu.yin@abdn.ac.uk

Abstract. In active learning for graph-structured data, Graph Neural Networks (GNNs) have shown effectiveness. However, a common challenge in these applications is the underutilization of important structural information. To address this problem, we propose the **S**tructural-Clustering **P**ageRank method for improved **A**ctive learning (SPA) specifically designed for graph-structured data. SPA integrates community detection using the SCAN algorithm with the PageRank scoring method for efficient and informative sample selection. SPA prioritizes nodes that are not only informative but also central in structure. Through extensive experiments, SPA demonstrates a higher accuracy and macro-F1 score over existing methods across different annotation budgets and achieves prominent reductions in query time. In addition, the proposed method only adds two hyperparameters, ϵ and μ in the algorithm to finely tune the balance between structural learning and node selection. This simplicity is a key advantage in active learning scenarios, where extensive hyperparameter tuning is often impractical.

Keywords: Active Learning · Structural-Clustering · PageRank · Graph Neural Network

1 Introduction

Graph Neural Networks (GNNs) [8,15] have emerged as a powerful tool for learning from graph-structured data, effectively capturing complex relationships and interdependencies between nodes. This progress has largely impacted areas where data inherently take the form of graphs, including social networks, biological networks, and communication systems. Concurrently, active learning [14], a subset of machine learning, has gained traction for its ability to efficiently utilize limited labeled data. In scenarios where labeling data is expensive or time-consuming, active learning strategically selects the most informative samples for labeling. This approach aims to maximize model prediction with a minimal amount of labeled data. The integration of active learning with GNNs presents a promising opportunity to enhance learning efficiency in graph-based learning tasks. Recent approaches in active learning, particularly in

I. Miliou et al. (Eds.): IDA 2024, LNCS 14641, pp. 28–40, 2024.
https://doi.org/10.1007/978-3-031-58547-0_3

the context of graph-structured data [6,16], have focused on various strategies to identify the most informative nodes. These methods often revolve around uncertainty sampling, diversity sampling, or a combination of both, aiming to select nodes that are either uncertain under the current model or are representative of the underlying data distribution. This integration has shown a great potential in improving the efficiency of GNNs, especially in semi-supervised learning settings where labeled data are scarce. However, these methods primarily leverage node features or embeddings, often overlooking the rich structural information inherent in graphs. Thus, despite the advancements, there remains a notable gap in research concerning the optimal exploitation of graph topology in active learning for GNNs. This gap highlights the need for novel active learning strategies that can harness both the feature and structural information in graph-structured data. To address this research gap, we propose a novel method that integrates community detection with active learning in GNNs. The proposed method employs the SCAN algorithm [17], recognized for effective community detection, alongside the PageRank algorithm [11] for node selection. By focusing on the community structures identified by SCAN and the node relevance ascertained by PageRank, we aim to select nodes that are informative in terms of features and pivotal in the graph's structure. In essence, the synergy of the SCAN algorithm and PageRank enables the selection of samples that are meaningful. Specifically, SCAN identifies local structures, whereas PageRank sheds light on the broader, global structures of graph data. Through extensive experiments, we demonstrate that the proposed method outperforms existing active learning methods for GNNs for a wide range of annotation budgets. Furthermore, the experiment on computational complexity indicates that the proposed approach leads to a reduction in query time. Thus, we summarize the contributions of the study as follows:

- We propose a novel active learning method for Graph Neural Networks (GNNs) that integrates the SCAN method [17] for community detection with the PageRank algorithm [11].
- Through extensive experiments, we demonstrate that the proposed method substantially outperforms existing active learning methods in GNNs across various annotation budgets.
- Additionally, the proposed method shows a notable reduction in computational complexity compared to recent active learning approaches for GNNs [9], which is crucial in real active learning implementation where the waiting time during the annotation process is one of the important factors.

2 Related Work

Active learning is a field in machine learning that focuses on reducing the cost of annotation while keeping the model performance stable and it has been comprehensively studied by [14]. In this section, we focus on active learning for graph-structured data. Early works in active learning for graph-structured data primarily focused on leveraging graph topology for selecting informative samples

[1,6]. These methods typically relied on measures like node centrality, degree, and cluster-based sampling, under the assumption that nodes with higher centrality or those bridging clusters are more informative. For example, AGE [3] evaluates the informativeness of nodes by linearly combining centrality, density, and uncertainty. Furthermore, ARNMAB extends this approach by dynamically learning the weights with a multi-armed bandit mechanism and maximizing the surrogate reward [5]. The other type of approach in active learning for graph neural networks is implementing partition or clustering as part of community detection. For example, FeatProp [15] combines node feature propagation with K-Medoids clustering for sample selection. The study was supported by a theoretical bound analysis showing an improvement in performance over other methods. Recently, Ma et al. [9] introduced the partition-based methods GraphPart and GraphPart-Far, which align with active learning algorithms in GNNs by focusing on selecting nodes for optimal coverage of feature or representation spaces, typically through clustering algorithms. On the other hand, The proposed method improves the conventional community detection approach by specifically employing clustering as a community detection algorithm. The proposed method works by capturing the local structures of nodes through community detection, while the PageRank scoring assesses their global significance.

3   Problem Formulation

3.1   Node Classification on Attributed Graphs

Graph theory offers a robust framework for modeling complex systems through structures known as attributed graphs. Specifically, an attributed graph is denoted as $G = (V, E, X)$, where V represents the set of nodes, $E \subseteq V \times V$ denotes the set of edges, and X is the set of node attributes. Each node $v \in V$ is associated with an attribute vector $\mathbf{x}_v \in \mathbb{R}^d$. The adjacency matrix $A \in \{0,1\}^{n \times n}$, where $n = |V|$, encodes the connectivity between nodes, with $A_{ij} = 1$ if there is an edge between nodes i and j, and $A_{ij} = 0$ otherwise.

Node classification in attributed graphs aims to assign labels to nodes based on their attributes and structural positions in the graph. This involves learning a function $f : V \rightarrow L$, where L is the set of possible labels. The challenge is to effectively leverage the information encoded in the graph structure and node attributes for accurate classification.

3.2   Graph Neural Networks (GNNs)

Graph Neural Networks (GNNs) are a class of neural networks designed for processing graph-structured data. They operate by aggregating information from a node's neighbors to update its representation. Formally, a GNN learns a function $f(G, X) \rightarrow Y$, where Y is the output matrix representing node-level predictions.

The learning process in a GNN involves updating node representations through successive layers. Let $H^{(k)}$ be the matrix of node representations at the k-th layer, with $H^{(0)} = X$. The update rule at each layer is given by:

$$H^{(k+1)} = \sigma(\tilde{A}H^{(k)}W^{(k)}) \tag{1}$$

where $\tilde{A}$ is the normalized adjacency matrix, $W^{(k)}$ is the weight matrix for layer k, and σ is a non-linear activation function.

The objective in training a GNN for node classification is often to minimize a loss function, typically the cross-entropy loss for the classification problem, defined as:

$$\mathcal{L} = -\sum_{v \in V_L} \sum_{l \in L} y_{vl} \log \hat{y}_{vl} \tag{2}$$

where $V_L \subseteq V$ is the set of labeled nodes, y_{vl} is the true label of node v for label l, and $\hat{y}_{vl}$ is the predicted probability of node v being in class l.

3.3 Active Learning Task for Graph Neural Networks

In the active learning scenario for GNNs, the objective is to select a subset of nodes $V_S \subseteq V$ to be labeled that maximizes the performance of the GNN. The selection process is guided by an acquisition function $\mathcal{A} : V \rightarrow \mathbb{R}$, which scores each unlabeled node based on its expected utility for improving the model. The challenge is to design $\mathcal{A}$ to account for both the graph structure and node features. The active learning process iteratively selects nodes, updates the model, and re-evaluates the remaining unlabeled nodes. In this study, we incorporate community detection into the active learning framework. We define a community structure C within the graph, and the acquisition function $\mathcal{A}$ is designed to preferentially select nodes that are central or informative within their communities, based on the hypothesis that such nodes provide more valuable information for the GNN model.

4 Proposed Method

4.1 Community Detection Using the SCAN Algorithm

The initial phase of community detection in graphs involves partitioning the network into distinct communities. This task is accomplished using the SCAN algorithm, a method recognized for its capability to identify densely connected subgraphs or communities in a network. Unlike modularity-based approaches, the SCAN algorithm relies on structural similarity and a shared neighbor approach for community detection.

Structural Similarity Measure. The core of the SCAN algorithm is the structural similarity measure between nodes, defined as follows:

$$S(i,j) = \frac{|N(i) \cap N(j)|}{\sqrt{|N(i)| \cdot |N(j)|}} \tag{3}$$

In this equation, $N(i)$ and $N(j)$ represent the neighbor sets of nodes i and j, respectively. The measure $S(i,j)$ quantifies the similarity based on the shared neighbors of the two nodes, normalized by the geometric mean of their degrees.

Community Detection Criteria. The SCAN algorithm employs two parameters, ϵ and μ, to determine community membership. A node i is in the same community as node j if the following conditions are met:

$$S(i,j) \geq \epsilon \quad \text{and} \quad |N(i) \cap N(j)| \geq \mu \tag{4}$$

where ϵ is a similarity threshold and μ is the minimum number of shared neighbors required for community formation. These parameters allow the SCAN algorithm to classify nodes into clusters, hubs, or outliers, based on their structural roles within the network.

4.2   Node Selection Based on PageRank

Upon successfully partitioning the graph into communities, the next step is to select representative nodes from each community. This selection is based on the PageRank algorithm, which assigns a numerical weighting to each node in the network. The weight of a node is indicative of the probability of arriving at that node by randomly walking through the network. The PageRank $PR(u)$ of a node u is defined as:

$$PR(u) = \frac{1-d}{N} + d \sum_{v \in B(u)} \frac{PR(v)}{L(v)} \tag{5}$$

where d is the damping factor, N is the total number of nodes, $B(u)$ is the set of nodes that link to u, and $L(v)$ is the number of links from node v. In the implementation, we use a damping factor of 0.95. For each community detected by the SCAN algorithm, the node with the highest PageRank score is selected. If the number of communities is less than the labeling budget b, additional nodes with the next highest PageRank scores are selected until b nodes are chosen. The final output is a set of b nodes, each representing its respective community, selected based on their significance within the network as determined by the PageRank algorithm.

4.3   SPA Algorithm

The proposed SPA algorithm combines the strengths of SCAN and PageRank algorithms. Algorithm 1 illustrates the detailed process of SPA algorithm.

5   Experiments

5.1   Experiment Settings

Experiments were conducted using various GNN models, including a 2-layer Graph Convolutional Network (GCN) [8] and a 2-layer GraphSAGE [7], both equipped with 16 hidden neurons. These models were trained using the Adam optimizer, starting with a learning rate of 1×10^{-2} and a weight decay of 5×10^{-4}. In this study, we adopted a straightforward batch-active learning framework, labeling each sample within a batch, which corresponds to the defined budget in

---

Algorithm 1: SPA Algorithm

Require: Adjacency matrix of the graph A, damping factor d for PageRank,
labeling budget b, clustering threshold ϵ, and minimum neighbors μ

Ensure : Sample of nodes S to be labeled, not exceeding budget b

1: Initialize $S = \emptyset$
2: Convert adjacency matrix A to graph data structure G
3: Calculate PageRank scores for all nodes in G with damping factor d
4: Apply the SCAN algorithm to G to detect communities $\{C_1, C_2, \ldots, C_k\}$
5: **for** each community C_i in $\{C_1, C_2, \ldots, C_k\}$ **do**
6:     Find node $n_{\max}$ in C_i with the highest PageRank score
7:     **if** $|S| < b$ **then**
8:         Add $n_{\max}$ to S
9:     **end**
10: **end**
11: **return** S to be labeled

---

each experiment. The budget varies from 10 to 160 for smaller datasets and from 80 to 1280 for larger ones. Experiment results are presented as the mean of 10 independent runs, each with different random seeds. For transparency and reproducibility, the code is available on GitHub[1]. Furthermore, in the experimental analysis of the SPA Algorithm, we carefully adjusted the parameters μ (minimum number of neighbors) and ϵ (clustering threshold) to optimize performance, finding optimal values at $\mu = 3$ and $\epsilon = 0.5$ for balancing community detection. Additionally, we selected a damping factor d of 0.95, based on empirical testing, to ensure an effective exploration of the graph's structure while maintaining a fast convergence rate.

5.2   Dataset

We conduct experiments on two standard node classification datasets: Citeseer and Pubmed [13]. Additionally, we use Corafull [2] and WikiCS [10] to add diversity of the experiments. Furthermore, to assess the proposed method's performance on more heterophilous graphs, we experiment with the Heterophilous-GraphDataset, which includes Minesweeper and Tolokers [12]. Table 1 presents a summary and the statistics of the datasets used in this research. We use the #Partitions parameter to determine the optimal number of communities in a graph, following the guidelines set by Ma et al. [9].

5.3   Evaluation Metrics

We use standard metrics such as accuracy and Macro-F1 score for evaluating the proposed method. Accuracy is measured as the ratio of correctly predicted

---

[1] https://github.com/rickymaulanafajri/SPA.

Table 1. Summary of datasets

Dataset	#Nodes	#Edges	#Features	#Classes	#Partitions
Citeseer	3,327	4,552	3,703	6	14
Pubmed	19,717	44,324	500	3	8
Corafull	19,793	126,842	8,710	70	7
WikiCS	19,793	126,842	8,710	70	7
Minesweeper	10,000	39,402	10	2	8
Tolokers	11,758	519,000	10	2	7

class and it is mathematically represented as:

$$\text{Accuracy} = \frac{TP + TN}{TP + TN + FP + FN} \tag{6}$$

where TP, TN, FP, and FN correspond to true positives, true negatives, false positives, and false negatives, respectively. On the other hand, the Macro-F1 score is computed by taking the arithmetic mean of the F1 scores for each class independently. It is defined as:

$$\text{Macro-F1} = \frac{1}{N} \sum_{i=1}^{N} \frac{2 \cdot \text{Precision}_i \cdot \text{Recall}_i}{\text{Precision}_i + \text{Recall}_i} \tag{7}$$

where N is the number of classes, and Precision_i and Recall_i denote the precision and recall for each class i, respectively. The precision and recall for each class are defined as follows:

$$\text{Precision}_i = \frac{TP_i}{TP_i + FP_i} \tag{8}$$

$$\text{Recall}_i = \frac{TP_i}{TP_i + FN_i} \tag{9}$$

5.4 Baselines Methods

We compare the proposed approach with various active learning methods, divided into two types. 1. A general active learning method which works regardless of the graph neural network architecture such as Random, Uncertainty, and PageRank. 2. An active learning method that is specifically designed for graph-structured data such as Featprop, Graphpart, and GraphPartFar. The query strategy of each method is defined as follows:

- Random: An active learning query strategy that selects a sample at random.
- Uncertainty [14]: Select the nodes with the highest uncertainty based on the prediction model.
- PageRank [3]: Query the sample from a subset of points with the highest PageRank centrality score.

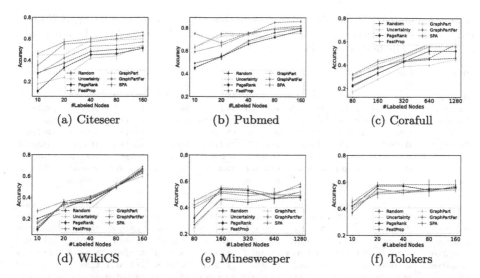

Fig. 1. Comparative Evaluation of SPA and Baseline Methods Across Multiple Datasets: Accuracy Versus Number of Labeled Nodes in GCN Architecture.

- FeatProp [15]: First use KMeans to cluster the aggregated node features and then query the node closest to the center of the clustering.
- GraphPart and GraphPartFar [9]: A recent state-of-the-art method on active learning for graph structured data. GraphPart divides the graph into several partitions using Clauset-Newman-Moore greedy modularity maximization [4] and selects the most representative sample in each partition to query. Graph-PartFar increases the diversity of samples by selecting nodes that are not too close and similar to each other.

6   Results

6.1   Experiment Results of SPA on GCN

In this study, we present the active learning result of the Graph Convolutional Network (GCN) across various datasets. First, we analyze the accuracy score of the proposed method compared to the baseline. Figure 1 illustrates the comparative evaluation of the proposed and baseline methods. It is evident that while all methods exhibit a general trend of improved accuracy with an increased budget, the proposed approach consistently maintains a higher accuracy rate. For example, in Citeseer, Pubmed, and Corafull, the proposed method excels in accuracy from the smallest labeling budget to the largest one. Although, in the other dataset the accuracy of the proposed method only shows a marginal improvement, it is still higher compared to all the baselines.

Next, we examine the Macro-F1 score of the proposed method. The results of these experiments are summarized in Table 2. Notably, the SPA method demonstrates consistently high results across various datasets, including Citeseer, Pubmed, and Corafull. Its superiority is particularly evident in situations

Table 2. Summary of the Macro-F1 score of the proposed approach using GCN architecture. The numerical values indicate the mean Macro-F1 score derived from 10 separate trials. The best score is in bold marker.

Baselines	Citeseer			Pubmed			Corafull		
	20	40	80	10	20	40	10	20	40
Random	28.4 ±12.6	37.6±6.7	48.9±5.8	49.1±11.4	55.7±10.6	69.5±6.2	23.8±2.0	33.2±1.8	43.2±2.1
Uncertainty	18.8±7.1	24.2±5.7	42.8±12.5	46.0±11.5	54.7±10.4	64.8±9.2	17.3±1.7	28.1±2.1	39.1±1.2
PageRank	27.2±6.6	36.3±7.8	49.6±7.5	45.3±8.5	55.7±13.0	66.3±8.6	22.8±1.4	33.4±1.1	43.7±0.5
FeatProp	27.9±5.5	42.9±4.5	53.7±4.5	59.1±5.5	65.4±5.2	75.1±2.8	29.6±1.0	37.6±0.8	46.7±0.8
GraphPart	45.0±0.7	45.4±4.1	59.0±2.0	63.0±0.7	73.2±1.0	74.0±1.3	31.0±1.3	**41.2±1.4**	**48.6±0.5**
GraphPartFar	35.1±0.6	55.2±2.4	57.5±2.0	75.7±0.3	67.5±0.5	76.2±0.9	28.0±1.2	38.4±0.6	44.3±0.7
SPA	**46.3±0.5**	**57.1±0.1**	**60.8±2.4**	**63.1±0.1**	**75.1±0.4**	**77.5±3.2**	**31.3±0.2**	40.4±0.2	47.7±0.2
Baselines	WikiCS			Minesweeper			Tolokers		
	20	40	80	160	320	640	20	40	80
Random	30.2 ± 2.1	51.3 ± 4.5	50.2±1.2	46.7±6.7	44.4±6.5	47.1±4.5	55.2±8.9	52.4±0.66	55.4±0.30
Uncertainty	26.4±1.2	29.3±0.7	51.8± 2.4	47.1±0.55	46.0±1.41	47.3±0.02	52.4±0.77	52.8±0.93	54.3±0.81
PageRank	35.4±4.3	35.7±0.8	50.4±5.4	54.6±0.41	53.9±0.49	47.3±0.07	57.9±0.56	57.1±0.41	54.4±0.99
FeatProp	34.2±2.1	40.1±0.41	51.4±0.06	51.3± 0.24	49.0±0.05	51.3±0.02	52.3±0.45	54.3±0.72	53.3±0.45
GraphPart	34.2±0.45	39.0± 0.21	**53.2±0.44**	52.2±0.03	51.0±0.93	49.8±0.84	53.2±0.48	53.3±0.32	53.4±0.21
GraphPartFar	33.4±0.05	38.0±0.1	50.1±0.21	52.3±6.0	51.7±0.16	50.3±0.29	51.7± 0.24	52.2±0.18	53.4±0.07
SPA	**36.5±0.78**	**41.5±0.02**	51.2±0.45	**55.1±0.07**	54.3±0.45	**52.2±0.45**	**58.0±0.06**	**58.0± 0.04**	**58.0 0.02**

with diverse sample sizes. This is most clearly observed in the Citeseer dataset, where SPA excels within a 40-label budget, highlighting its effectiveness in moderately sized sample environments. While GraphPart shows a competitive edge, particularly in the 20-label budget scenario of the Corafull dataset, SPA still maintains a slight but notable advantage in the 10-label budget.

6.2   Experiment Result of SPA on GraphSAGE

In addition, we conducted further experimental analysis using another Graph Neural Network (GNN) architecture, specifically GraphSAGE. We use all the datasets from the previous experiment. The initial focus was on evaluating the accuracy score of the proposed methods within the GraphSAGE architecture. Figure 2 illustrates the effectiveness of the proposed method in achieving higher accuracy. For instance, in the Citeseer dataset, the proposed method begins with an accuracy score of 0.32 at a 10-label budget, and it reaches its peak accuracy score with a score of 0.68 at a 1280-label budget. Secondly, we compare the Macro-F1 score of each method. Table 3 shows the Macro-F1 score of the baseline methods compared with the proposed approach. Table 3 illustrates that the proposed method consistently outperformed existing models in terms of Macro-F1 score in GraphSAGE architecture. Notably, it demonstrated a substantial improvement in Macro-F1 scores in WikiCS, Minesweeper, and Tolokers.

6.3   Complexity Analysis

The computational complexity of PageRank is typically $\mathcal{O}(n + m)$ per iteration in a graph G with n nodes and m edges. Additionally, the SCAN algorithm has a complexity of $\mathcal{O}(m\sqrt{m})$. This complexity is primarily dictated by the clustering process involving each edge and its neighboring nodes. While selecting

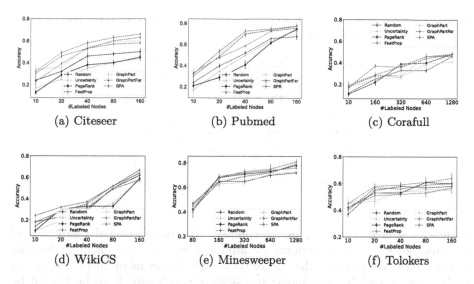

Fig. 2. Comparative Evaluation of SPA and Baseline Methods Across Multiple Datasets: Accuracy vs. Number of Labeled Nodes in GraphSAGE Architecture.

Table 3. Summary of the Macro-F1 score of the proposed approach using GraphSAGE architecture.

Baselines	Citeseer			Pubmed			Corafull		
	10	20	40	20	40	80	160	320	640
Random	24.1 ± 11.9	32.4 ± 6.6	46.1 ± 5.6	40.6 ± 13.0	52.3 ± 12.2	66.3 ± 7.9	15.4 ± 1.1	23.4 ± 1.4	33.5 ± 1.3
Uncertainty	17.6 ± 6.3	25.1 ± 6.1	35.7 ± 4.4	35.2 ± 6.6	50.5 ± 10.1	64.1 ± 10.5	15.3 ± 0.8	27.0 ± 1.9	39.7 ± 1.1
PageRank	13.0 ± 1.0	29.8 ± 1.3	38.3 ± 2.2	29.7 ± 0.3	41.7 ± 0.6	62.9 ± 0.3	12.4 ± 0.3	19.4 ± 0.8	30.3 ± 0.4
Featprop	23.4 ± 4.3	39.9 ± 6.2	53.5 ± 3.3	48.0 ± 5.9	59.1 ± 6.0	73.6 ± 1.7	18.7 ± 0.8	25.8 ± 0.7	35.0 ± 1.6
GraphPart	**34.1 ± 6.4**	36.1 ± 6.4	54.0 ± 4.6	52.0 ± 0.8	71.5 ± 0.5	74.6 ± 1.1	**19.7 ± 0.9**	**28.3 ± 0.7**	36.1 ± 1.0
GraphPartFar	30.7 ± 2.3	46.9 ± 5.0	53.1 ± 4.0	49.7 ± 3.1	70.7 ± 1.6	74.2 ± 0.4	17.5 ± 1.0	26.2 ± 1.4	34.1 ± 0.9
SPA	32.6 ± 0.2	**49.5 ± 1.3**	**58.2 ± 2.1**	**54.1 ± 2.1**	**73.20 ± 2.4**	**75.0 ± 0.3**	19.2 ± 0.2	27.9 ± 2.3	**36.7 ± 0.2**
Baselines	WikiCS			Minesweeper			Tolokers		
	20	40	80	160	320	640	20	40	80
Random	23.2 ± 2.3	29.03 ± 0.3	50.3 ± 0.4	65.2 ± 1.2	65.4 ± 0.3	70.2 ± 0.2	55.3 ± 0.6	58.2 ± 0.8	57.4 ± 0.5
Uncertainty	19.2 ± 0.9	24.4 ± 0.2	36.2 ± 0.8	63.2 ± 0.7	64.6 ± 1.3	71.4 ± 2.4	47.1 ± 8.3	54.2 ± 5.2	56.5 ± 6.5
PageRank	30.2 ± 3.2	33.6 ± 2.4	33.8 ± 7.4	68.1 ± 3.3	71.2 ± 2.8	73.0 ± 0.2	53.2 ± 6.4	53.6 ± 5.6	61.0 ± 5.3
Featprop	29.2 ± 0.7	31.2 ± 4.5	49.5 ± 3.2	**69.6 ± 3.2**	72.8 ± 2.1	74.5 ± 3.4	57.2 ± 2.4	56.4 ± 3.2	59.7 ± 6.5
GraphPart	29.7 ± 4.3	32.4 ± 2.3	50.1 ± 4.8	64.2 ± 3.6	70.4 ± 5.8	72.5 ± 3.5	53.6 ± 7.8	54.2 ± 0.6	54.2 ± 3.4
GraphPartfar	28.2 ± 0.7	35.4 ± 2.9	52.2 ± 9.8	64.4 ± 1.5	69.4 ± 3.9	72.2 ± 1.5	52.1 ± 3.6	54.0 ± 6.8	52.2 ± 2.6
SPA	**32.8 ± 2.9**	**37.2 ± 2.5**	**52.8 ± 1.6**	68.8 ± 1.8	**73.2 ± 0.1**	**75.2 ± 2.1**	**58.2 ± 1.5**	**59.6 ± 3.4**	**61.2 ± 1.5**

the highest PageRank node in each community incurs additional computational overhead, this is generally less important compared to the overall complexities of PageRank and SCAN. The overhead primarily depends on the number of communities and the size of the graph. Therefore, the overall performance of this combined approach is influenced by the size and connectivity of the input graph, with larger and more connected graphs incurring higher computational costs. When compared with recent state-of-the-art models like GraphPart and GraphPartFar, the proposed method demonstrates lower computational costs,

Table 4. Query Time Comparison: Proposed Method vs. State-of-the-Art (Average over 10 Runs, Measured in Seconds)

Dataset	Graph Architecture = GCN		
	GraphPart	GraphPartFar	SPA
Citeseer	10	20	0.21
Pubmed	13.3	17	4
Corafull	316	397	25
Wikics	45	52	23
Minesweeper	25	30	3.2
Tolokers	516	621	115

leading to reduced query times. To compare the computational complexity of the proposed method, we perform a computational cost experiment against the recent model. We measure the cost of each model in terms of query time which is the time needed by each model to calculate which sample to be labeled by the active learning process. Table 4 presents the query times for each method. The proposed method notably reduces query time, with the most substantial reduction seen in the Corafull dataset. In the Corafull dataset, SPA reduced the execution time down to 25 s, compared to 319 s for GraphPart and 397 s for GraphPartFar.

7   Discussion and Conclusion

This paper introduced a Structural-Clustering based Active Learning (SPA) approach for Graph Neural Networks (GNNs), which combines community detection with the PageRank scoring method. The SPA method strategically prioritizes nodes based on their information content and centrality within the graph's structure, leading to a more representative sample selection and enhancing the robustness of active learning outcomes. This is particularly effective in real-world applications like social network analysis and financial networks, which typically struggle with large amounts of labeled data requirements. SPA's efficiency across varying annotation budgets is an important advantage in scenarios with limited resources for labeling.

Furthermore, SPA integrates the structural clustering abilities of the SCAN algorithm with the PageRank scoring system. SCAN uses both feature and structural information in graphs to identify community-based local structures, while PageRank focuses on the global importance of nodes. The proposed method has demonstrated improved execution times and superior Macro-F1 scores across various datasets, yet it may face potential challenges in extremely large or complex graph structures.

In conclusion, the SPA method represents a substantial advancement in the field of active learning for GNNs. It not only improves performance but also enhances execution efficiency, marking an important step in applying active learning to graph-structured data. While showing promising results in various scenarios, we acknowledge the need for further research in optimizing the method for large-scale or complex graphs. The insights from this research lay the groundwork for future developments in comprehensive and efficient active learning models, catering to a wide range of active learning applications in graph-structured data.

References

1. Bilgic, M., Mihalkova, L., Getoor, L.: Active learning for networked data. In: Proceedings of the 27th International Conference on International Conference on Machine Learning, pp. 79-86 (2010)
2. Bojchevski, A., Günnemann, S.: Deep gaussian embedding of graphs: Unsupervised inductive learning via ranking. In: 6th International Conference on Learning Representations, ICLR (2018)
3. Cai, H., Zheng, V.W., Chang, K.C.C.: Active learning for graph embedding. arXiv preprint: arXiv:2007.02901 (2017)
4. Clauset, A., Newman, M.E.J., Moore, C.: Finding community structure in very large networks. Phys. Rev. E **70**, 066111 (2004)
5. Gao, L., Yang, H., Zhou, C., Wu, J., Pan, S., Hu, Y.: Active discriminative network representation learning. In: International Joint Conference on Artificial Intelligence (2018)
6. Gu, Q., Han, J.: Towards active learning on graphs: an error bound minimization approach. In: 2012 IEEE 12th International Conference on Data Mining, pp. 882–887 (2012)
7. Hamilton, W.L., Ying, Z., Leskovec, J.: Inductive representation learning on large graphs. In: Neural Information Processing Systems (2017)
8. Kipf, T., Welling, M.: Semi-supervised classification with graph convolutional networks. In: International Conference on Learning Representations (ICLR) (2017)
9. Ma, J., Ma, Z., Chai, J., Mei, Q.: Partition-based active learning for graph neural networks. Trans. Mach. Learn. Res. (2023)
10. Mernyei, P., Cangea, C.: WIKI-CS: a Wikipedia-based benchmark for graph neural networks. arXiv Preprint: arXiv:2007.02901 (2020)
11. Page, L., Brin, S., Motwani, R., Winograd, T.: The PageRank citation ranking: bringing order to the web. In: The Web Conference (1999)
12. Platonov, O., Kuznedelev, D., Diskin, M., Babenko, A., Prokhorenkova, L.: A critical look at the evaluation of GNNs under heterophily: are we really making progress? arxiv Preprint: arXiv:2302.11640 (2023)
13. Sen, P., Namata, G., Bilgic, M., Getoor, L., Gallagher, B., Eliassi-Rad, T.: Collective classification in network data. In: The AI Magazine (2008)
14. Settles, B., Craven, M.W.: An analysis of active learning strategies for sequence labeling tasks. In: Conference on Empirical Methods in Natural Language Processing (2008)
15. Wu, Y., Xu, Y., Singh, A., Yang, Y., Dubrawski, A.W.: Active learning for graph neural networks via node feature propagation. arXiv Preprint: arXiv:1910.07567 (2019)

16. Wu, Z., Pan, S., Chen, F., Long, G., Zhang, C., Yu, P.S.: A comprehensive survey on graph neural networks. IEEE Trans. Neural Netw. Learn. Syst. **32**, 4–24 (2019). https://doi.org/10.1109/TNNLS.2020.2978386
17. Xu, X., Yuruk, N., Feng, Z., Schweiger, T.A.J.: SCAN: a structural clustering algorithm for networks. In: Proceedings of the 13th ACM SIGKDD International Conference on Knowledge Discovery and Data Mining, KDD '07, pp. 824-833. Association for Computing Machinery, New York (2007). https://doi.org/10.1145/1281192.1281280

Multi-armed Bandits with Generalized Temporally-Partitioned Rewards

Ronald C. van den Broek, Rik Litjens, Tobias Sagis, Nina Verbeeke,
and Pratik Gajane[✉]

Eindhoven University of Technology, Eindhoven, Netherlands
{r.c.v.d.broek,r.litjens,t.g.m.sagis,n.c.verbeeke}@student.tue.nl,
p.gajane@tue.nl

Abstract. Decision-making problems of sequential nature, where decisions made in the past may have an impact on the future, are used to model many practically important applications. In some real-world applications, feedback about a decision is delayed and may arrive via partial rewards that are observed with different delays. Motivated by such scenarios, we propose a novel problem formulation called multi-armed bandits with generalized temporally-partitioned rewards. To formalize how feedback about a decision is partitioned across several time steps, we introduce β-*spread property*. We derive a lower bound on the performance of any uniformly efficient algorithm for the considered problem. Moreover, we provide an algorithm called `TP-UCB-FR-G` and prove an upper bound on its performance measure. In some scenarios, our upper bound improves upon the state of the art. We provide experimental results validating the proposed algorithm and our theoretical results.

Keywords: Multi-armed bandits · Delayed rewards · Temporally-partitioned rewards

1 Introduction

The classical multi-armed bandit (MAB, or simply bandit) problem is a framework to model sequential decision-making [5]. In a MAB problem, the learning agent is faced with a finite set of K decisions or *arms*, and a decision taken by the agent is symbolized by pulling an arm. Feedback about the decisions taken is available to the agent via numerical rewards. Multi-arm bandit literature typically focuses on scenarios where rewards are assumed to arrive immediately after pulling an arm. In contrast, the works on delayed-feedback bandits (e.g., [8,10]) assume a delay between pulling an arm and the observation of its corresponding reward. In those studies, the reward is assumed to be concentrated in a single round that is delayed. This setting can be extended by allowing the reward to be partitioned into partial rewards that are observed with different delays. This type of bandit problem, known as MAB with Temporally-Partitioned Rewards (TP-MAB), was introduced by [11].

In the TP-MAB setting, an agent receives subsets of the reward over multiple rounds. The complete reward of an arm is the sum of the partial rewards

I. Miliou et al. (Eds.): IDA 2024, LNCS 14641, pp. 41–52, 2024.
https://doi.org/10.1007/978-3-031-58547-0_4

obtained by pulling the arm. [11] present α-smoothness to characterize the reward structure. The α-smoothness property states that the maximum reward in a group of consecutive partial rewards cannot exceed a fraction of the maximum reward (precise definition given in Definition 2). However, the assumption of α-smoothness does not fit well if the cumulative reward is not uniformly spread. In this article, we introduce a more generalized way of formulating how an arm's delayed cumulative reward is spread across several rounds.

As a motivating application, consider websites (e.g., Coursera, Khan Academy, edX) that provide Massive Open Online Courses (MOOCs). Such websites aim to provide users with useful recommendations for courses. This problem can be modeled as a TP-MAP problem. A course, which consists of a series of video lectures, might be thought of as an arm. A course can be recommended to a user by an agent, which corresponds to pulling an arm. When the student follows a course, the agent can observe partial rewards (e.g., by checking the watch time retention). In this setting, α-smoothness rarely captures the actual cumulative reward distribution. Many students watch the video lectures at the beginning of a course but never finish the last few lectures, making the spread of partial rewards non-uniform. As a result, the existing work on delayed-feedback bandits and the algorithms proposed by [11] may fail to recommend courses that are relevant for the user. Motivated by such scenarios, we investigate a more generalized way of formulating the reward structure.

Our Contributions

1. We introduce a novel MAB formulation with a generalized way of describing how an arm's delayed cumulative reward is distributed across rounds.
2. We prove a lower bound on the performance measure of any uniformly efficient algorithm for the considered problem.
3. We devise an algorithm `TP-UCB-FR-G` and prove an upper bound on its performance measure. The proven upper bounds are tighter than the state of the art in some scenarios.
4. We provide experimental results that validate the correctness of our theoretical results and the effectiveness of our proposed algorithm.

2   Background and Related Work

The *non-anonymous* delayed feedback bandit problem was considered in [8], where it is assumed that knowledge of which action resulted in a specific delayed reward is available. Recently, a variety of delayed-feedback scenarios were studied in MAB settings different from ours, such as linear and contextual bandits [1,13, 15], non-stationary bandits [14].

The majority of past research on the delayed MAB setting assumes that the entire reward of an arm is observed at once, either after some bounded delay [8,10] or after random delays from an unbounded distribution with finite expectation [7,12]. Our article studies the setting in which the reward for an arm is spread over an interval with a finite maximum delay value. This is consistent with the applications that we aim to model, such as MOOC providers mentioned in Sect. 1. To the best of our knowledge, this setting was first analyzed in [11]. They introduced the Multi-Armed Bandit with Temporally-Partitioned Rewards (TP-MAB) setting. In the TP-MAB setting, a stochastic reward that is received by pulling an arm is partitioned over partial rewards observed during a finite number of rounds followed by the pull. In [11], it is assumed that the arm rewards follow α-smoothness property (precise definition given in Definition 2).

While the study by [11] provides promising results, it is based on the strong assumption that the α-smoothness property holds. As a result, their proposed solutions are not suitable for a broader variety of scenarios where rewards are partitioned non-uniformly. As a remedy, we propose to use general distributions that can more accurately characterize how the received reward is partitioned. Consider a scenario (e.g., as described Sect. 1) in which additional information is available about how the cumulative reward is spread over the rounds. By generalizing the reward structure, our approach is able to handle partitioned rewards in which the maximum reward per round is not partitioned uniformly across rounds, such as those shown on the right side of Fig. 1.

Two novel algorithms, TP-UCB-EW and TP-UCB-FR, leveraging α-smoothness property are introduced in [11]. The setup of TP-UCB-FR is most suitable for leveraging assumed distribution in a generalized setting. Subsequently, we use TP-UCB-FR as a baseline for our proposed algorithm.

3 Problem Formulation

Consider a MAB problem with K arms over a time horizon of T rounds, where $K, T \in \mathbb{N}$. At every round $t \in \{1, 2, \ldots, T\}$ an arm from the set of arms $\{1, 2, \ldots, K\}$ is pulled. The reward for an arm i is drawn from an unknown distribution on $[0, 1]$ with mean μ_i. The performance of an algorithm $\mathfrak{A}$ after T time steps is measured using expected *regret* denoted as $\mathcal{R}_T(\mathfrak{A})$.

Definition 1 (Regret). *The regret of an algorithm $\mathfrak{A}$ after T time steps is* $\mathcal{R}_T(\mathfrak{A}) := \mu^* T - \sum_{i=1}^{K} \mu_i \cdot \mathbb{E}[N_i(T)]$, *where* $\mu^* := \max_{1 \le i \le K} \mu_i$ *and* $N_i(T) = num$-*ber of times an arm i is selected till time t.*

The total reward is temporally partitioned over a set of rounds $T' = \{1, 2, \ldots, \tau_{\max}\}$. Let $x_{t,m}^i (m \in T')$ denote the partitioned reward that the learner receives at round m, after pulling the arm i at round t. It is known to the agent which arm pull produced this reward. The cumulative reward is completely collected by the learner after a delay of at most $\tau_{\max}$. Each per-round reward $x_{t,m}^i$ is the realization of a random variable $X_{t,m}^i$ with support in $[0, \overline{X}_m^i]$. The cumulative reward collected by the learner from pulling arm i at round t is denoted by r_t^i and it is the realization of a random variable R_t^i such that $R_t^i := \sum_{n=1}^{\tau_{\max}} X_{t,n}^i$ with support $[0, \overline{R}^i]$. Straightforwardly, we observe that $\overline{R}^i := \sum_{n=1}^{\tau_{\max}} \overline{X}_n^i$.

It is shown in [11] that, in practice, per-round rewards for an arm provide information on the cumulative reward of the arm. They introduce α-smoothness

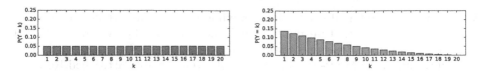

Fig. 1. α-smoothness reward distribution for the MOOC setting (left) and a near-perfect approximation of the reward distribution using β-spread (right)

property (defined in Definition 2) that partitions the rewards such that each partition corresponds to the sum of a set of consecutive per-round rewards. Formally, let $\alpha \in T'$ be such that α is a factor of $\tau_{\max}$. The cardinality of each partition, where we refer to partition as 'z-group', is denoted by $\phi := \frac{\tau_{\max}}{\alpha}$ with $\phi \in \mathbf{N}$. We can now define each z-group $z_{t,k}^i, k \in \{1, 2, ..., \alpha\}$ as the realization of a random variable $Z_{t,k}^i$, with support $[0, \overline{Z}_{\alpha,k}^i]$, such that for every k:

$$Z_{t,k}^i := \sum_{n=t+(k-1)\phi}^{t+k\phi-1} X_{t,n}^i \tag{1}$$

Definition 2 (α-smoothness). *For $\alpha \in \{1, ..., \tau_{max}\}$, the reward is α-smooth iff $\frac{\tau_{max}}{\alpha} \in \mathbb{N}$ and for each $i \in \{1, ..., K\}$ and $k \in \{1, 2, ..., \alpha\}$ the random variables $Z_{t,k}^i$ are independent and s.t. $\overline{Z}_{\alpha,k}^i = \overline{Z}_\alpha^i = \frac{\overline{R}^i}{\alpha}$.*

The α-smoothness property ensures that all temporally-partitioned rewards contribute towards bounding the values of future rewards within the same window. If the α-smoothness holds, then the maximum cumulative reward in a z-group $\overline{Z}_{\alpha,k}^i$ is equal for all z-groups $k \in \{1, 2, ..., \alpha\}$. Therefore, we can say that $\forall_k \in \{1, 2, ..., \alpha\}, \overline{Z}_{\alpha,k}^i = \overline{Z}_\alpha^i$. The assumption of α-smoothness is unsuitable for scenarios in which the cumulative reward is not uniformly partitioned across rounds. The goal of this article is to generalize the spread of the rewards across z-groups. To that end, one has to eliminate the assumption that every z-group has an equal probability of attaining a partial reward. To accomplish this, we replace α-smoothness with β-*spread property* that allows for modeling scenarios in which the cumulative reward is distributed non-uniformly across rounds i.e., a property that allows $\overline{Z}_{\alpha,k}^i$ to differ across z-groups.

Our Solution approach: β-spread property

Definition 3 (β-spread). *For $\alpha \in \{1, ..., \tau_{max}\}$, the reward is β-smooth if and only if*

1. *$\frac{\tau_{max}}{\alpha} = \phi$ with $\phi \in \mathbb{N}$,*
2. *the reward distribution can be described by a distribution $\mathcal{D}$ on a finite integer domain $\{1, 2, ..., \alpha\}$ with probability mass function $P_\mathcal{D}(k)$, and*

3. *for each* $i \in \{1, ..., K\}$ *and* $k \in \{1, 2, ..., \alpha\}$ *the random variables* $Z_{t,k}^i$ *are independent and s.t.* $\overline{Z}_{\alpha,k}^i = P_{\mathcal{D}}(k) \cdot \overline{R}^i$.

Based on prior information about how the cumulative reward is distributed over the rounds, the actual reward distribution can be approximated by a distribution $\widehat{\mathcal{D}}$ with corresponding probability mass function $P_{\widehat{\mathcal{D}}}(k)$, as long as it adheres to the definition of β-spread. We specify this, because the true reward distribution might not be known exactly in all cases. However, our solution approach requires at least some knowledge of the reward distribution.

As an example, consider the MOOC setting described at the end of Sect. 1. Consider the case where the watch time retention is considered a partial reward, and reduces linearly over time. The reward distribution under α-smoothness over the z-groups is illustrated in Fig. 1 (left). Since we expect the partial reward to reduce linearly over time, the distribution of rewards under α-smoothness is inappropriate. Rather, we closely approximate the linear reduction with a Beta-binomial distribution with parameters $\alpha = 1$ and $\beta = 3$ (right), which will result in lower cumulative regret. In this article, we use Beta-binomial distributions frequently due to its capability to describe a wide variety of distributions.

4   Lower Bound on Regret

Using the β-spread property, we can derive the following lower bound for a uniformly efficient policy i.e., any policy with regret in $\mathcal{O}(T^x)$ with $x < 1$.

Theorem 1. *The regret of any uniformly efficient policy $\mathfrak{U}$ applied to a TP-MAB problem with the β-spread property after T time steps is lower bounded as*

$$\lim_{T \to +\infty} \inf \frac{\mathcal{R}_T(\mathfrak{U})}{\ln T} \geq \sum_{i:\mu_i < \mu^*} \frac{2}{(\alpha+1)} \sum_{k=1}^{\alpha} kP_{\mathcal{D}}(k) \cdot \alpha \sum_{k=1}^{\alpha} (P_{\mathcal{D}}(k))^2 \frac{\Delta_i}{\alpha \mathcal{KL}\left(\frac{\mu_i}{R_{\max}}, \frac{\mu^*}{R_{\max}}\right)}$$

where $\Delta_i := \mu^* - \mu_i$ *and* $\mathcal{KL}(p, q) := $ *Kullback-Leibler divergence between Bernoulli random variables with means* p *and* q *[9].*

Comparison with the Lower Bound given by [11] By assuming α-smoothness, the following lower bound for TP-MAB was proved in [11]:

$$\lim_{T \to +\infty} \inf \frac{\mathcal{R}_T(\mathfrak{U})}{\ln T} \geq \sum_{i:\mu_i < \mu^*} \frac{\Delta_i}{\alpha \mathcal{KL}\left(\frac{\mu_i}{R_{\max}}, \frac{\mu^*}{R_{\max}}\right)}. \tag{2}$$

Notice that the difference between the lower bound with the β-spread property and the lower bound derived in [11] lies in two factors. The first factor, $\frac{2}{(\alpha+1)} \sum_{k=1}^{\alpha} kP_{\mathcal{D}}(k)$ is equal to the normalized expected value of our assumed reward spread distribution. $\sum_{k=1}^{\alpha} kP_{\mathcal{D}}(k)$ calculates the expected value for the chosen discrete distribution. $\frac{(\alpha+1)}{2}$ is the expected value when the chosen distribution is the uniform distribution. Hence, its inverse can be seen as a normalization term. The second factor, $\alpha \sum_{k=1}^{\alpha} (P_{\mathcal{D}}(k))^2$, can be seen as a normalized approximation of the *index of coincidence* [6] between rewards. The index of coincidence, $\sum_{k=1}^{\alpha} (P_{\mathcal{D}}(k))^2$ determines the probability of two reward points being observed in the same z-group. Its minimal value equals $\frac{1}{\alpha}$ and occurs when the α-smoothness property holds (uniform distribution). The value is maximal and equal to 1 if all rewards fall into one z-group. Multiplying the index of coincidence with α gives this factor more weight in the lower bound and extends the domain from $[\frac{1}{\alpha}, 1]$ to $[1, \alpha]$. This essentially means that it is 'harder' for algorithms to perform well when the rewards come in bulk, rather than over the course of multiple rounds.

The lower bound given in Theorem 1 resolves to the lower bound given by [11] in Eq.(2) in case of α-smoothness. However, our lower bound for the considered problem setting is tighter when $\frac{2}{(\alpha+1)} \sum_{k=1}^{\alpha} k P_{\mathcal{D}}(k) \cdot \alpha \sum_{k=1}^{\alpha} (P_{\mathcal{D}}(k))^2 > 1$. This means that rewards that are expected to be observed late, or rewards that come all at the same time, contribute negatively to the performance of an algorithm in the described setting. Rewards that are expected to be observed early or are more spread out contribute positively.

Proof Sketch for Theorem 1 We start by constructing two MAB problem instances that call for different behaviors from the algorithm attempting to solve them. Then, we use the change-of-distribution argument to show that any uniformly efficient algorithm cannot efficiently distinguish between these instances. Please consult the extended version of this article given in [3] for the complete proof of Theorem 1.

5   Proposed Algorithm and Regret Upper Bound

In this section, we propose an algorithm that makes use of the β-spread property in the TP-MAB setting and prove an upper bound on its regret.

5.1   Proposed Algorithm: TP-UCB-FR-G

Our proposed algorithm TP-UCB-FR-G given below is a non-trivial extension of TP-UCB-FR [11]. In TP-UCB-FR-G, the most significant modification is the confidence interval c_{t-1}^i which is rigorously built to suit the β-spread property.

As input, the algorithm takes a smoothness constant $\alpha \in [\tau_{max}]$, a maximum delay τ_{max} and a probability mass function $P_{\widehat{D}}$. The algorithm uses $P_{\widehat{D}}$ to be able to give a proper judgment of an arm before all the delayed partial rewards are observed. This is realized by replacing the not yet received partial rewards with fictitious realizations, or in other words, the expected estimated rewards. At round t, the fictitious reward vectors are associated with each arm pulled in the span $H := \{t - \tau_{max} + 1, ..., t - 1\}$. These fictitious rewards are denoted by $\tilde{\boldsymbol{x}}_h^i = \left[\tilde{x}_{h,1}^i, ..., \tilde{x}_{h,\tau_{max}}^i \right]$ with $h \in H$, where $\tilde{x}_{h,j}^i := x_{h,j}^i$, if $h + j \leq t$ (the reward

---

Algorithm 1: TP-UCB-FR-G

1: **Input:** $\alpha \in [\tau_{max}], \tau_{max} \in \mathbb{N}^*, P_{\widehat{\mathcal{D}}}$
2: **for** $t \in \{1, ..., K\}$ **do**
3:     Pull an arm $i_t \leftarrow t$
4: **for** $t \in \{K+1, ..., T\}$ **do**
5:     **for** $i \in \{1, ..., K\}$ **do**
6:         Compute $\hat{R}^i_{t-1}$ and c^i_{t-1} as in (3) and (4)
7:         $u^i_{t-1} \leftarrow \hat{R}^i_{t-1} + c^i_{t-1}$
8:     Pull arm $i_t \leftarrow z = \text{argmax}_{i \in [K]} u^i_{t-1}$
9:     Observe $x^{i_h}_{h,t-h+1}$ for $h \in \{t - \tau_{max} + 1, ..., t\}$

---

has already been seen), and $\tilde{x}^i_{h,j} = 0$, if $h + j > t$ (the reward will be seen in the future). The corresponding fictitious cumulative reward is $\tilde{r}^i_h := \sum_{j=1}^{\tau_{max}} \tilde{x}^i_{h,j}$. In the initialization phase of the algorithm (lines 2-3), each arm is pulled once. Later, at each time step t, the upper confidence bounds u^i_{t-1} are determined for each arm i by computing the estimated expected reward $\hat{R}^i_{t-1}$ and confidence interval c^i_{t-1} using Eqs. (3) and (4) respectively.

$$\hat{R}^i_{t-1} := \frac{1}{N_i(t-1)} \left(\sum_{h=1}^{t-\tau_{max}} r^i_h \mathbb{1}_{\{i_h = i\}} + \sum_{h \in H} \tilde{r}^i_h \mathbb{1}_{\{i_h = i\}} \right), \tag{3}$$

$$c^i_{t-1} := \frac{\phi \bar{R}^i}{N_i(t-1)} \sum_{k=1}^{\alpha} k P_{\widehat{\mathcal{D}}(k)} + \bar{R}^i \sqrt{\frac{2 \ln(t-1) \sum_{k=1}^{\alpha} \left(P_{\widehat{\mathcal{D}}(k)} \right)^2}{N_i(t-1)}}, \tag{4}$$

where $N_i(t-1)$ is the number of times arm i has been pulled up to $t-1$ and i_h represents the arm that was pulled at time h. The algorithm then pulls the arm i with the highest upper confidence bound u^i_{t-1} and observes its rewards.

5.2   Regret Upper Bound of TP-UCB-FR-G

Theorem 2. *In the TP-MAB setting with β-spread reward, the regret of TP-UCB-FR-G after T time steps with $P_{\widehat{\mathcal{D}}}(k)$ matching $P_{\mathcal{D}}(k)$ is upper bounded as*

$$\mathcal{R}_T(\textit{TP-UCB-FR-G})$$

$$\leq \sum_{i:\mu_i < \mu^*} \frac{4 \ln T \, (\bar{R}^i)^2 \sum_{k=1}^{\alpha} (P_{\mathcal{D}}(k))^2}{\Delta_i} \cdot \left(1 + \sqrt{1 + \frac{\Delta_i \phi \sum_{k=1}^{\alpha} k P_{\mathcal{D}}(k)}{\bar{R}^i \ln T \sum_{k=1}^{\alpha} (P_{\mathcal{D}}(k))^2}} \right)$$

$$+ 2\phi \sum_{k=1}^{\alpha} k P_{\mathcal{D}}(k) \sum_{i:\mu_i < \mu^*} \bar{R}^i + \left(1 + \frac{\pi^2}{3} \right) \sum_{i:\mu_i < \mu^*} \Delta_i.$$

Observe that $\sum_{k=1}^{\alpha} kP_{\mathcal{D}}(k)$ is equal to the expected value of our assumed reward spread distribution, similar to the factor in the lower bound but not normalized. The other factor is the index of coincidence $\sum_{k=1}^{\alpha} (P_{\mathcal{D}}(k))^2$, which also occurs in the lower bound but is not weighted for the upper bound.

Comparison with the Upper Bound given in [11]. Let us compare our upper bound with the upper bound given in [11]. For the latter bound to hold, the α estimate given as input to their algorithm has to match the α of the real reward distribution as well. Note that $\sum_{k=1}^{\alpha} kP_{\mathcal{D}}(k) = \frac{\alpha+1}{2}$ in case of α-smoothness. For other assumed distributions with $\sum_{k=1}^{\alpha} kP_{\mathcal{D}}(k) < \frac{\alpha+1}{2}$ our upper bound on the regret is lower. Furthermore, choosing a β-spread distribution as input with a low mean and a low index of coincidence will result in a better upper bound, by Theorem 2, compared to choosing $\widehat{\mathcal{D}}$ with rewards centered towards the end (high mean) and not spread out (high index of coincidence).

Proof Sketch of Theorem 2. Here we provide a proof sketch for Theorem 2. Please refer to the extended version of the article given in [3] for the complete proof. The approach can be divided into three steps. Firstly, we show that the probability that an optimal arm is estimated significantly lower than its mean is bounded by t^{-4}. Secondly, we show the probability of a suboptimal arm being estimated significantly higher than its mean is bounded by t^{-4}. Finally, we assess the algorithm's ability to differentiate between optimal and suboptimal arms.

6   Experimental Results

We compare our proposed algorithm `TP-UCB-FR-G` with `TP-UCB-FR` [11], `UCB1` [2], and `Delayed-UCB1` [8]. We use two experimental settings – a synthetically generated environment and a real-world playlist recommendation scenario. In these settings, we inherit learners used in the provided experiments in [11], and create new learner configurations using `TP-UCB-FR-G`. As input distributions for the new learners, we use Beta-Binomial distributions with unique parameter values for each learner. The Beta-Binomial distribution gives us the opportunity to model extreme scenarios, which should result in more insightful experimental results. We observe that other distributions do not grant the flexibility of a Beta-Binomial distribution, as demonstrated in experiments deferred to the appendix given with the extended version of this article [3]. In the plots under this section, we use the notation `TP-UCB-FR-G`(α, `dist_name`) to denote a learner for our algorithm, where `dist_name` is the name of the Beta-Binomial distribution for which the exact parameters are shown in Table 1. Further details about the used Beta-Binomial distributions and experimental settings are given in the appendix of the extended version of this article [3].

6.1   Setting 1: Synthetic Environment

The distribution of rewards in this setting are s.t. $Z_{t,k}^i \sim \frac{\overline{R}^i}{\alpha} Beta[a_k^i, b_k^i]$ where $Beta$ is a Beta distribution with a, b s.t. rewards are distributed according

Table 1. Parameter values for Beta-Binomial distributions

Distribution name	α	β	Distribution name	α	β	Distribution name	α	β
`extreme_begin`	1	100	`begin_middle`	2	4	`end`	8	2
`very_begin`	1	16	`middle`	5	5	`very_end`	16	1
`begin`	2	8	`middle_end`	4	2			

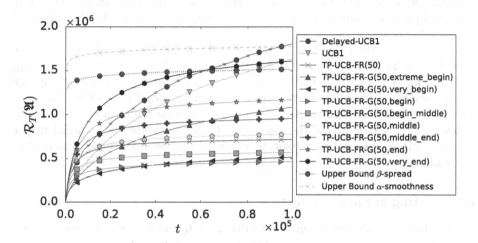

Fig. 2. Regret against time for Setting 1.2 with $\tau_{\max} = 100$ and $\alpha_{est} = 50$

to the spread of the corresponding setting. Again, we model $K = 10$ arms, an α-smoothness constant of $\alpha = 20$ and a maximum reward s.t. convergence to an optimal arm takes longer. That is, $\overline{R}^i = 100\zeta^i$ with $\zeta \in \{1, 3, 6, 9, 12, 15, 18, 21, 22, 23\}$. However, there is a difference in the $\tau_{\max}$, α_{est} and the parameters used for the assumed Beta distribution by the learners. The exact configurations can be found in the appendix given with the extended version of this article [3]. In general, there are 8 combinations consisting of 4 configurations with 2 scenarios each. The configurations differ in $\tau_{\max}$ and α_{est}, whereas the scenarios differ in distribution parameters. Generally, there is one where the rewards are observed late after the pull (Setting 1.1), and one where the results are observed just after the pull (Setting 1.2). We use learners with the distributions given in Table 1).

Results Let us denote $\Delta(s_1, s_2)$ for $s_1, s_2 \in \{$Setting 1.1, Setting 1.2$\}$ as the absolute difference in cumulative regret between Settings s_1 and s_2. The pairwise differences in cumulative regret observed between settings are marginal. As an example, $\Delta($Setting 1.1, Setting 1.2$) \approx 4.8 \times 10^3$ for TP-UCB-FR-G(50, `very_end`) which is the highest difference in average regret observed across all compared settings. Since the regret of TP-UCB-FR-G(50, `very_end`) averaged over T is $\approx 1.61 \times 10^6$, the observed change of $\approx 0.3\%$ is negligible. Furthermore, the same experiment performed with different values for both $\tau_{\max}$ and α_{est} seems to confirm the same marginal change.

For example, Setting 1 for $\tau_{\max} = 200$ and $\alpha_{est} = 20$ results in a maximum change in average regret of only $\approx 0.5\%$. These findings indicate that the performance of TP-UCB-FR-G learners in a uniformly distributed aggregate rewards setting is indistinguishable from that in a non-uniformly distributed aggregate rewards setting. Therefore, we can state that TP-UCB-FR-G$(\alpha, \texttt{begin})$ delivers a significant performance increase compared to the learner proposed by [11]. The gain that we observe for the mentioned settings is as high as $\approx 48.2\%$. An extensive performance summary is deferred to the appendix given with the extended version of this article [3].

In Fig. 2, the theoretical upper bound of TP-UCB-FR-G as well as the upper bound of the TP-UCB-FR algorithm is plotted on top of the results for the Setting 1.2. The figure shows that the upper bound proposed in this article is tighter in this setting. Note that the theoretical upper bounds for TP-UCB-FR-G and TP-UCB-FR only hold for specific learners that assume the data generating distribution precisely and that the 'very end' learner exceeds the β-spread upper bound. This shows another reason to estimate the assumed distribution optimistically.

6.2 Setting 2: Spotify Playlists

We evaluate our algorithm on real-world data by addressing the user recommendation problem introduced by [11], using the Spotify dataset [4]. We select the $K = 6$ most played playlists as the arms to be recommended. Each time a playlist i is selected, the corresponding rewards x_t^i for the first $N = 20$ songs are sampled from the dataset. In this setting, the α-smoothness is $\alpha = 20$, the maximum delay $\tau_{\max} = 4N = 80$ and the results are averaged over 100 independent runs.

Results In Fig. 3, we observe that optimistic learners significantly outperform the baseline TP-UCB-FR(20). We focus on TP-UCB-FR-G(20, begin), since it is by far the best-performing learner. This learner achieves a decrease of $\approx 26.3\%$ in regret averaged over time horizon T. Table 2 summarizes the performance gains of TP-UCB-FR-G learners in the Spotify setting.

Table 2. TP-UCB-FR-G learners and their decrease in regret

Learner TP-UCB-FR-G$(\alpha_{est} = 20)$	Regret $(\times 10^4)$	Decrease $(\%)$
extreme_begin	4.40	≈ -72.5
very_begin	2.40	≈ 5.9
begin	**1.88**	$\approx$ **26.3**
begin_middle	2.11	≈ 17.2

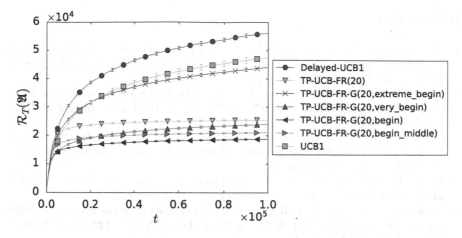

Fig. 3. Regret against time for begin-oriented learners in the Spotify setting

We also observe that overly optimistic learners perform worse than TP-UCB-FR(20). However, TP-UCB-FR-G(20, begin) outperforms TP-UCB-FR(20) for larger t, making it a better option for playlist recommendations.

7   Concluding Remarks and Future Work

In this paper, we model sequential decision-making problems with delayed feedback using a novel formulation called multi-armed bandits with generalized temporally-partitioned rewards. To generalize delayed reward distributions, we introduce the β-spread property. We establish a tighter lower bound for the TP-MAB setting with the β-spread property compared to the TP-MAB setting with the α-smoothness property. We also introduce the TP-UCB-FR-G algorithm, which exploits the β-spread property. We demonstrate that in certain scenarios, the upper bound of this algorithm can be lower than that of the TP-UCB-FR algorithm, thus surpassing the upper bounds of the classical UCB1 and Delayed-UCB1 algorithms as well. Finally, we demonstrate that our algorithm outperforms TP-UCB-FR and other UCB algorithms in diverse experiments using synthetic and real-world data, achieving a remarkable 26.3% decrease in regret compared to the state-of-the-art TP-UCB-FR algorithm.

A possible future research direction is to explore removing the restriction of the β-spread property to discrete probability distributions bounded by a finite domain of size α. This can enhance the algorithm's flexibility and broaden its practical applications. Additionally, a valuable extension involves considering scenarios where arms are treated as subsets, each assigned distinct α-values and distributions. Moreover, an intriguing area of exploration involves studying scenarios where the partitioned reward time span, denoted as $\tau_{\max}$, varies.

Acknowledgements. This work is supported by the Dutch Research Council (NWO) in the framework of the TEPAIV research project (project number 612.001.752).

References

1. Arya, S., Yang, Y.: Randomized allocation with nonparametric estimation for contextual multi-armed bandits with delayed rewards. Stat. Probab. Lett. **164**, 108818 (2020). https://doi.org/10.1016/j.spl.2020.108818
2. Auer, P., Cesa-Bianchi, N., Fischer, P.: Finite-time analysis of the multiarmed bandit problem. Mach. Learn. 2002 **47**(2), 235–256 (2002). https://doi.org/10.1023/A:1013689704352
3. van den Broek, R.C., Litjens, R., Sagis, T., Siecker, L., Verbeeke, N., Gajane, P.: Multi-armed bandits with generalized temporally-partitioned rewards (2023). https://arxiv.org/abs/2303.00620
4. Brost, B., Mehrotra, R., Jehan, T.: The music streaming sessions dataset. CoRR **abs/1901.09851** (2019). http://arxiv.org/abs/1901.09851
5. Bubeck, S., Cesa-Bianchi, N.: Regret analysis of stochastic and nonstochastic multi-armed bandit problems. Found. Trends Mach. Learn. **5**(1), 1–122 (2012). https://doi.org/10.1561/2200000024
6. Friedman, W.F.: The index of coincidence and its applications in cryptanalysis, vol. 49. Aegean Park Press California (1987)
7. Gael, M.A., Vernade, C., Carpentier, A., Valko, M.: Stochastic bandits with arm-dependent delays. In: Proceedings of the 37th International Conference on Machine Learning, vol. 119, pp. 3348–3356 (2020)
8. Joulani, P., György, A., Szepesvári, C.: Online learning under delayed feedback. In: Proceedings of the 30th International Conference on International Conference on Machine Learning, vol. 28. ICML'13 (2013)
9. Kullback, S., Leibler, R.A.: On information and sufficiency. Ann. Math. Stat. **22**(1), 79–86 (1951)
10. Mandel, T., Liu, Y.E., Brunskill, E., Popović, Z.: The queue method: handling delay, heuristics, prior data, and evaluation in bandits. In: Twenty-Ninth AAAI Conference on Artificial Intelligence (2015)
11. Romano, G., Agostini, A., Trovò, F., Gatti, N., Restelli, M.: Multi-armed bandit problem with temporally-partitioned rewards: When partial feedback counts. In: Proceedings of the Thirty-First International Joint Conference on Artificial Intelligence (2022). https://doi.org/10.24963/ijcai.2022/472
12. Vernade, C., Cappé, O., Perchet, V.: Stochastic bandit models for delayed conversions. In: Conference on Uncertainty in Artificial Intelligence (2017). https://hal.science/hal-01545667
13. Vernade, C., Carpentier, A., Lattimore, T., Zappella, G., Ermis, B., Brückner, M.: Linear bandits with stochastic delayed feedback. In: Proceedings of the 37th International Conference on Machine Learning, vol. 119, pp. 9712–9721 (2020). https://proceedings.mlr.press/v119/vernade20a.html
14. Vernade, C., György, A., Mann, T.A.: Non-stationary delayed bandits with intermediate observations. In: Proceedings of the 37th International Conference on Machine Learning. ICML'20 (2020)
15. Zhou, Z., Xu, R., Blanchet4, J.: Learning in Generalized Linear Contextual Bandits with Stochastic Delays. Curran Associates Inc. (2019)

GloNets: Globally Connected Neural Networks

Antonio Di Cecco[1], Carlo Metta[2], Marco Fantozzi[3],
Francesco Morandin[3], and Maurizio Parton[1(✉)]

[1] University of Chieti-Pescara, Pescara, Italy
maurizio.parton@gmail.com
[2] ISTI-CNR, Pisa, Italy
[3] University of Parma, Parma, Italy

Abstract. Deep learning architectures suffer from depth-related performance degradation, limiting the effective depth of neural networks. Approaches like ResNet are able to mitigate this, but they do not completely eliminate the problem. We introduce Global Neural Networks (GloNet), a novel architecture overcoming depth-related issues, designed to be superimposed on any model, enhancing its depth without increasing complexity or reducing performance. With GloNet, the network's head uniformly receives information from all parts of the network, regardless of their level of abstraction. This enables GloNet to self-regulate information flow during training, reducing the influence of less effective deeper layers, and allowing for stable training irrespective of network depth. This paper details GloNet's design, its theoretical basis, and a comparison with existing similar architectures. Experiments show GloNet's capability to self-regulate, and its resilience to depth-related learning challenges, such as performance degradation. Our findings position GloNet as a viable alternative to traditional architectures like ResNets.

Keywords: Neural Networks · Deep Learning · Skip Connections

1 Introduction

Deep learning's success in AI is largely due to its hierarchical representation of data, with initial layers learning simple features and deeper ones learning more complex, nonlinear transformations of these features [4]. Increasing depth should enhance learning, but sometimes it leads to performance issues [5,10]. Techniques like normalized initialization [10,11,16,24] and normalization layers [1,15] enable up to 30-layer deep networks, but performance degradation persists

A. Di Cecco—National PhD in AI, XXXVIII cycle, health and life sciences, UCBM.
C. Metta—EU Horizon 2020: G.A. 871042 SoBig-Data++, NextGenEU - PNRR-PEAI (M4C2, investment 1.3) FAIR and "SoBigData.it".
F. Morandin and M. Parton—Funded by INdAM groups GNAMPA and GNSAGA.
A. Di Cecco, C. Metta, M. Fantozzi, F. Morandin, M. Parton—Computational resources provided by CLAI laboratory, Chieti-Pescara and Italy.

I. Miliou et al. (Eds.): IDA 2024, LNCS 14641, pp. 53–64, 2024.
https://doi.org/10.1007/978-3-031-58547-0_5

at greater depths without skip connections. This issue, detailed in the original ResNet paper [12], stems from the fact that learning identity maps is not easy for a deeply nonlinear layer. ResNet idea is to focus on learning nonlinear "residual" information, with a backbone carrying the identity map. This brilliant solution has been key in training extremely deep networks that, when weight-sharing and batch normalization are used, can scale up to thousands of layers.

Deeper neural networks should not experience performance degradation. Theoretically, a deeper network could match the performance of an n-layer network by similarly learning features $\mathcal{G}_1, \ldots, \mathcal{G}_n$ in its initial layers, then minimizing the impact of additional layers. With this ability to *self-regulate*, such a network could effectively be "infinitely deep". However, even with ResNet architectures, performance degradation persists beyond a certain depth, see Fig. 5b or [8]. This issue can be due to various factors, see for instance [10, 18, 26], and may partly arise from the inability of modern architectures to self-regulate their depth. Our paper introduces a novel technique to enable self-regulation in neural network architectures, overcoming these depth-related performance challenges.

Novel Contributions. The main contribution of this paper is introducing and testing GloNet, an explainable-by-design layer that can be superimposed on any neural network architecture, see Sect. 2. GloNet's key feature is its capacity to self-regulate information flow during training. It achieves this by reducing the influence of the deepest layers to a negligible level, thereby making the training more stable, preventing issues like vanishing gradients, and making the network trainable irrespective of its depth, see Sect. 5.

This self-regulation capabilities of GloNet lead to several significant benefits:

1. **Faster training:** GloNet trains in half the ResNet time while achieving comparable performance. Beyond the depth threshold where ResNet begins to degrade, GloNet trains in less than half the time and outperforms ResNet.
2. **ResNet alternative:** The inability of ResNet-based architectures to self-regulate depth makes GloNet a preferable option, particularly for applications requiring very deep architectures.
3. **No Network Architecture Search needed:** GloNet networks inherently find their effective depth, eliminating the need for computationally expensive Network Architecture Search methods to determine optimal network depth.
4. **More controllable efficiency/performance trade-off:** Layers can be selectively discarded to boost efficiency, allowing a controlled trade-off between efficiency and performance, optimizing the network for specific requirements.

2   Notation and Model Definition

A feedforward neural network is described iteratively by a sequence of L blocks:

$$\mathbf{x}_{l+1} = \mathcal{G}_l(\mathbf{x}_l), \quad l = 0, \ldots, L-1, \tag{1}$$

where $\mathbf{x}_0$ denotes the input vector, and $\mathbf{x}_{l+1}$ is the output from the l-th block. In this context, a "block" is a modular network unit, representing a broader concept than a traditional "layer". Each block function $\mathcal{G}_l$ typically merges a non-linearity, such as ReLU, with an affine transformation, and may embody more complex structures, like the residual blocks in ResNet.

At the end of the sequence (1), a classification or regression head $\mathcal{H}$ is applied to $\mathbf{x}_L$. For instance, a convolutional architecture could use a head with average pooling and a fully connected classifier. The fundamental principle in deep learning is that $\mathcal{G}_0, \ldots, \mathcal{G}_{l-1}$ hierarchically extract meaningful features from the input $\mathbf{x}_0$, that can then be leveraged by computing the output $\mathcal{H}(\mathbf{x}_L)$ of the network.

When the blocks in (1) are simple layers like an affine map followed by a non-linearity (this description comprises, for instance, fully connected and convolutional neural networks), all features extracted at different depths are exposed to the head by a single feature vector $\mathbf{x}_L$ *that has gone through several non-linearities*. This fact leads to several well-known drawbacks, like vanishing gradients or difficulty in learning when the task requires more direct access to low-level features. When using ReLU and shared biases, some low-level information could actually be destroyed. Several excellent solutions have been proposed to these drawbacks, like for instance residual networks [12,13], DenseNets [14], and preactivated units with non-shared biases [18].

We propose an alternative solution: a modification to (1), consisting of a simple layer between the feature-extraction sequence and the head, computing the sum of every feature vector. The architecture is designed to receive information uniformly from all parts of the network, regardless of their level of abstraction:

$$\begin{cases} \mathbf{x}_{l+1} = \mathcal{G}_l(\mathbf{x}_l), \quad l = 0, \ldots, L-1 \\ \mathbf{x}_{L+1} = \sum_{l=1}^{L} \mathbf{x}_l = \sum_{l=0}^{L-1} \mathcal{G}_l(\mathbf{x}_l) \\ \text{output} = \mathcal{H}(\mathbf{x}_{L+1}) \end{cases} \tag{2}$$

We refer to the additional layer in (2) as a *GloNet layer*, because all the features $\mathbf{x}_l$ that without GloNet would be preserved only "locally" up to the next $\mathbf{x}_{l+1} = \mathcal{G}_l(\mathbf{x}_l)$, appear now in the "global" feature vector $\mathbf{x}_{L+1} = \sum_{l=1}^{L} \mathbf{x}_l$ as summands. When feature vectors have different dimensions, adaptation to a common dimension is required before the sum, see Sect. 6 for details.

Remark 1 (Depth self-regulation). GloNet provides skip connections directly to the head, without learning a residual map, see (2) and Fig. 1. This direct *and simultaneous* backpropagation from each block promotes uniform information spread to the head. SGD-like training favors shorter paths [17,29,30], leading *GloNet to accumulate information mostly in the initial blocks rather than the latter ones*, by diminishing the impact of deeper layers. Thus, GloNet self-regulates its depth during training, resembling an "infinitely deep" architecture. Section 5 provides empirical support for this claim.

Remark 2 (Explainability-by-design). The global feature vector $\mathbf{x}_{L+1}$ combines blocks' outputs, transforming the neural network into a sub-networks ensemble, each learning features at different levels of detail, from basic in early blocks to more complex in later ones. The linear aggregation in the GloNet layer allows for

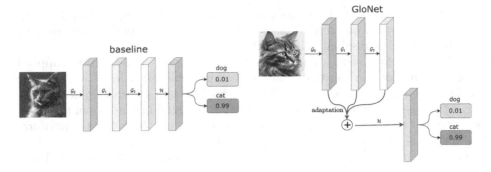

Fig. 1. On the left, a 3-blocks neural network followed by a classification head. On the right, the same architecture with GloNet modification in red. (Color figure online)

block-level feature interpretability. For input-level interpretability, saliency maps can be utilized, with the GloNet layer's additivity making their interpretation straightforward. By integrating these explainability techniques, GloNet offers a detailed insight into the prediction process and critical input features, thus serving as an 'explainable-by-design' tool [9].

3   Related Work

In a ResNet with an activation-free backbone, also known as ResNetv2 [13], blocks $\mathcal{G}_l$ are defined as $\mathrm{Id} + \mathcal{F}_l$, where Id is the skip connection and $\mathcal{F}_l$ is the "residual block", computing two times Conv $\circ$ ReLU $\circ$ BN, where BN is batch normalization. Unrolling the ResNetv2 equation $\mathbf{x}_{l+1} = \mathbf{x}_l + \mathcal{F}_l(\mathbf{x}_l)$ from any block output $\mathbf{x}_l$ (see [13, Equation 4]) gives:

$$\mathbf{x}_L = \mathbf{x}_l + \sum_{i=l}^{L-1} \mathcal{F}_i(\mathbf{x}_i) \qquad (3)$$

This equation shows that ResNet, much like GloNet, passes the output $\mathbf{x}_l$ of each block directly to the head. However, a key distinction lies in how the head accesses these outputs: ResNet requires distinct pathways for simultaneous access to different outputs, whereas GloNet's head achieves simultaneous access to each output via the GloNet layer. This unique capability of GloNet may contribute to its additional features compared to ResNetv2, as explored in Sect. 5.

Moreover, GloNet is faster than ResNet (not requiring batch normalization), and can be seen as an ensemble computing the sum of models of increasing complexity, giving an explainable-by-design model (differently from ResNet).

Unlike DenseNet [14], aggregating each block with all subsequent blocks through concatenation, GloNet connects only to the last block, with summation for aggregation. This approach avoids the parameter explosion given by concatenation in DenseNet, and maintains the original complexity of the model.

Finally, GloNet can be viewed as a network with early exits at every block, adapted and aggregated before the head. See [21,23,27] for the early exit idea.

4 Implementing GloNet

Implementing GloNet in a specific architecture is not always straightforward, see Sect. 2. Here we explain how to implement GloNet in common situations.

GloNet and Skip Connections. Existing skip connections, like those in ResNet or DenseNet architectures, should be replaced with GloNet connections. Otherwise, training would not converge as we would be adding the identity to the output multiple times. Note that GloNet provides only skip connections to the GloNet layer, and does not ask the blocks to learn a residual map.

GloNet and Batch Normalization. GloNet modifies block outputs based on their contribution to the task, reducing the impact of deeper blocks, which goes against the goal of batch normalization to maintain consistent input statistics. However, removing batch (or layer) normalization before GloNet's aggregation layer is not an issue, because GloNet provides an alternative form of regularization that, as shown in Sect. 5, achieves similar performance.

GloNet into Residual Networks. Residual architectures use skip connections and normalization. Once those are removed from a ResNetv2 block computing Id + affine map ∘ ReLU ∘ BN ∘ affine map ∘ ReLU ∘ BN, one is left with two simpler blocks affine map ∘ ReLU, and each of those blocks can potentially be aggregated into the GloNet layer. In this case, one ResNetv2 block corresponds to two simpler blocks. This is what we do in this paper, and for this reason when GloNet has n blocks, its equivalent ResNetv2 architecture has $n/2$ blocks.

GloNet into Vision Transformers. When using more complex architectures like transformers, several different choices can be made, each one potentially affecting the final performance of the GloNet-enhanced model. In this initial exploration of GloNet, we propose a straightforward integration with a Vision Transformer (ViT) [7] adapted to CIFAR-10. The image is segmented into 4×4 patches, its class encoded, and then concatenated with the patch encoding and a positional embedding. This series is then fed into a cascade of n encoders with 4 attention heads each, which are accumulated into a GloNet layer, and passed to a classification head. In our experiment, we compared $n = 4, 5$ and 6.

5 Experiments

In this section we provide experiments supporting the core claims of our paper, as stated in the Introduction. In particular, we focus on showing that GloNet trains much faster than ResNet, that GloNet performances are on par with ResNet's ones, that GloNet can self-regulate its depth, that GloNet does not need batch normalization, and that GloNet is virtually immune to depth-related problems. All the experiments can be reproduced using the source code provided at [6].

SGEMM Fully Connected Regression. We experimented with a regression task from the UCI repository [2,3], focused on predicting the execution time of matrix multiplication on an SGEMM GPU kernel. See [2,3] for details.

Table 1. Average epoch's training time in seconds at different depths, for GloNet and its equivalent ResNetv2 baseline, on SGEMM regression task.

architecture	depth						
	10	24	50	100	200	600	1000
ResNetv2	42	65	96	147	231	410	675
Glonet	27	39	55	86	121	175	289

Since GloNet has skip connections, to obtain a fair comparison we used a ResNetv2-like baseline. For comparison, we used also a vanilla baseline, identical to the ResNetv2 baseline but without skip connections. Moreover, since GloNet does not use batch normalization (in fact, GloNet self-regulation capabilities can be tampered by normalization, see Sect. 4), we also experimented with a vanilla and a ResNetv2 baseline with the BN layer removed. GloNet and the corresponding baselines (denoted by vanilla, ResNetv2, vanilla-no-BN, and ResNetv2-no-BN in figures) have a similar amount of parameters, the only difference being given by the trainable BN parameters. All blocks have 16 units. All models starts with a linear layer mapping the 14-dimensional input to $\mathbb{R}^{16}$, and ends with the head, a linear layer with 1 unit. GloNet models have an additional GloNet layer before the head *with no additional parameters*. The number of blocks ranges in [10, 24, 50, 100, 200], respectively (halved for the ResNets because every block is twice the layers of the corresponding non-ResNet model).

For training, we used MSE loss, L^2-regularization with a coefficient of 10^{-5} (10^{-4} showed no improvements), Adam optimizer with batch size 1024, learning rate set to 0.01, He normal initializer for weights, and zero initializer for biases. We trained all models for 200 epochs, the point at which baselines plateaued, potentially favoring them over GloNet. The first thing to notice is that with GloNet *training takes almost half or less than half the time of ResNet*, see Table 1. This is because GloNet, differently from ResNet, does not need batch normalization.

After 200 epochs, we compared the best test errors and learning curves, see Fig. 2 and Table 2. Across the different block configurations tested, GloNet and ResNet errors were similar (but note that in particular with 200 blocks ResNet shows potential for further improvement). However, the training shapes in Fig. 2 suggested a unique aspect of GloNet not present in the ResNetv2 baseline: GloNet's training was unaffected by the increasing depth, as shown by the shape of the learning curve that remained consistent whether the network had 10, 24, 50, 100, or 200 blocks. On the contrary, ResNet learning curve became flatter for increasing depth. See the caption of Fig. 2 for more details.

To further explore this feature, GloNet was tested with even deeper models (600 and 1000 blocks), and compared against the corresponding ResNetv2 baseline. Even at these substantial depths, GloNet's learning curve maintained its shape, as shown in Fig. 3. Moreover, GloNet's performance remained stable across these varying depths, maintaining a best test error of around

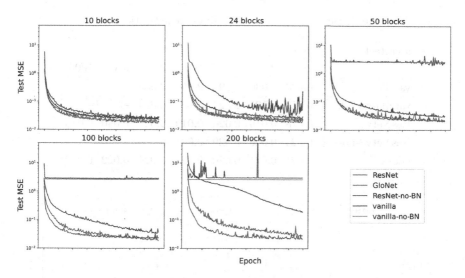

Fig. 2. Test errors while training for 200 epochs GloNet and four corresponding baselines at different depths on SGEMM. Starting at 24 blocks, GloNet and ResNetv2 outperform vanilla, as expected due to skip connections. At 50 and 100 blocks, both ResNetv2 and GloNet show depth resilience, with ResNetv2 slightly leading in best test error. ResNetv2-no-BN exhibits significantly higher error, highlighting ResNetv2's reliance on batch normalization. At 100 blocks GloNet curve shows a markedly higher curvature than ResNetv2. At 200 blocks, GloNet outperforms ResNetv2 in best test error and curve shape. However, ResNetv2's shape at 200 epochs indicates potential for further improvement, see Sect. 6.

0.02 regardless of the number of blocks $(10, 24, 50, 100, 200, 600,$ or $1000)$, see Table 2. In contrast, ResNet's showed a clear decline as the network depth increased. While its best test error at 200 epochs remained around 0.02 up to 200 blocks, this error increased to 0.04 and 0.05 at 600 and 1000 blocks, respectively, see again Table 2. As happened with 100 and 200 blocks, ResNet learning curve was flatter, diverging from GloNet's more consistent curve shape as depth increased.

GloNet accumulates information in the first few blocks and uses only the required capacity for a specific task and architecture, leading to minimal output from subsequent blocks, see Remark 1. This is not observed in baseline models with or without batch normalization, see Fig. 4, and likely contributes to GloNet's stable performance as network depth increases, in contrast to the degradation observed in the baseline models under similar depth conditions.

Remark 3. Figure 2 shows that even shallow models are over-parametrized for the task at hand, as indicated by the stable performance across varying depths. To understand GloNet behaviour in the under-parametrization regime, we experimented with 48 blocks, 4 units per block. This setup was chosen to achieve approximately one-quarter of the optimal parameter count, which is estimated

Table 2. Best test errors across different network depths.

architecture	depth						
	10	24	50	100	200	600	1000
vanilla	0.023	0.030	1.997	2.750	2.985	–	–
vanilla-no-BN	0.018	0.021	2.624	2.624	2.624	–	–
ResNetv2	0.018	0.019	0.020	0.019	0.027	0.040	0.048
ResNetv2-no-BN	0.024	0.026	0.029	0.033	0.189	–	–
Glonet	0.021	0.021	0.020	0.022	0.021	0.022	0.022

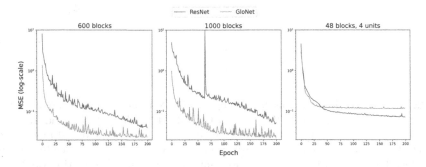

Fig. 3. Test MSE learning curves of GloNet (orange) versus ResNetv2 (blue) at 600 (left) and 1000 (center) blocks, when compared with Fig. 2, show a significant drop of ResNet's performance at these depths, unlike GloNet's stable performance. The learning curves on the right, with 48 blocks with 4 units each, show that GloNet behaves as expected even in the under-parametrization regime, achieving faster convergence but higher error compared to ResNet. This is consistent with GloNet behavior in the over-parametrization regime from Fig. 2. (Color figure online)

to be 12 blocks with 16 units each, see Fig. 4. Maintaining units while reducing the depth from 10 to 4 blocks was not feasible, because GloNet and ResNet would become nearly theoretically identical. In this under-parametrization scenario, we observed a training shape as in Fig. 2, and a decline in performance: GloNet achieved a best MSE of 0.116, while ResNet recorded a best MSE of 0.072. See Fig. 3.

MNIST Fully Connected Classification. To verify that GloNet's automatic selection of optimal depth and resilience to depth during training were not specific to the SGEMM regression task, we used the same architecture on MNIST image classification. Despite fully connected architectures not being ideal for image classification, this task allowed us to greatly expand the number of input features, which theoretically could pose a greater challenge to shallower models.

The only difference from the architecture used in SGEMM is the head, which in this case is a fully connected layer followed by a SoftMax layer on 10 classes. We tested GloNet and a convolutional baseline across 6, 10, 24, 50, 60, 80, 100,

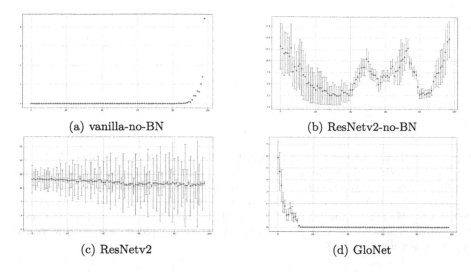

(a) vanilla-no-BN

(b) ResNetv2-no-BN

(c) ResNetv2

(d) GloNet

Fig. 4. L1 norm of block outputs, 100 blocks, 200 epochs, SGEMM, mean and standard deviation from 5000 samples. GloNet (d) uses only the first 12 blocks and achieves the least variance. In ResNetv2 (c) all blocks contribute, as expected with BN. ResNetv2-no-BN (b) uses outputs in different ways, due to the residual maps mixed with the skip connections. The baseline vanilla-no-BN (a) shows the opposite behavior to GloNet, emphasizing late outputs and diminishing earlier ones, indicating shorter gradient paths to the last blocks.

and 200 blocks, with halved numbers for ResNet. Figure 5a shows that GloNet automatically chooses the optimal number of blocks. However, notice that in this case, unlike Fig. 4c, ResNetv2 outputs show a decreasing shape, probably because the trainable parameters of the batch normalization are in this case able to force a small mean and variance on the last blocks. This indicates that also with batch normalization the network struggles to self-regulate its depth. Figure 5b confirms GloNet resilience to depth, and shows a severe performance degradation of ResNetv2 when depth goes above 50 blocks.

CIFAR10 Convolutional Classification. We further experimented with a ResNet20 on CIFAR10. ResNet20 is a ResNetv2 with 3 stages of 3 residual blocks each, described in [12]. We compared a ResNet20 architecture with its GloNet version, obtained by removing the backbone and adding a GloNet layer before the classification head, as detailed in Sect. 4. We trained for 200 epochs. Learning curves are completely overlapping, with best test errors 91.12% and 91.08% for ResNet and GloNet respectively. This experiment shows that also with convolutional architectures, GloNet performs on par with the traditional ResNet architecture, despite taking half the time for training.

GloNet for Vision Transformer Classification. A Visual Transformer (ViT) is a transformer applied to sequences of feature vectors extracted from image patches [7]. Feeding the n encoders outputs into a GloNet layer before the classification head gives GloNet-ViT, see Sect. 4. Here we compare ViT and

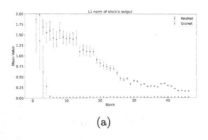 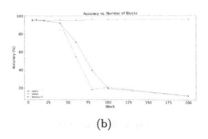

(a)   (b)

Fig. 5. (a) Block's outputs L1 norm, GloNet (orange), and ResNetv2 (blue), 50 blocks, 200 epochs, MNIST. Mean and standard deviation from 5000 samples. GloNet uses only the first 4 blocks. In ResNetv2, all blocks give a significant contribution, as expected because of BN. (b) Accuracies of GloNet (orange), ResNetv2 (blue), and vanilla (green), 200 epochs, MNIST, against 6, 10, 24, 50, 60, 80, 100, and 200 blocks. GloNet is consistent across depths, while vanilla and ResNetv2 show a rapid decline with increased depth. (Color figure online)

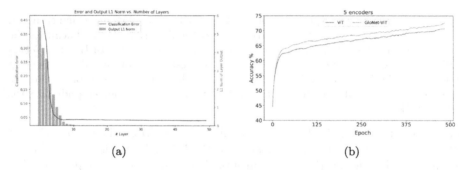

(a)   (b)

Fig. 6. (a) Block's outputs L1 norm of GloNet (red), smaller networks performance (blue), 50 blocks, 200 epochs, MNIST. Performance plateaus after block eight. (b) Accuracy of ViT (blue), and GloNet-ViT (orange) with 5 encoders. Best accuracies 0.709, 0.727 for ViT, GloNet-ViT, respectively. (Color figure online)

GloNet-ViT trained up to plateau on CIFAR-10, with 4, 5 and 6 encoders. With 4 and 6 encoders, ViT and GloNet-ViT accuracies overlap. With 5 encoders, GloNet-ViT improves over ViT, see Fig. 6b. ViT best accuracies are 0.707, 0.709, and 0.725, and GloNet-ViT best accuracies are 0.709, 0.727, and 0.729, for 4, 5, and 6 encoders, respectively. This is a proof-of-concept experiment, without any model-finetuning, aiming only at showcasing the robustness and versatility of GloNet for complex architectures like transformers.

Controllable Efficiency/Performance Trade-Off. To show GloNet's ability to optimize efficiency versus performance, we trained a 50-block GloNet fully connected architecture on MNIST for 200 epochs. We then progressively removed the last block, adjusted the GloNet layer for fewer blocks, and evaluated the shallower model without retraining. As Fig. 6a shows, removing up to 42 blocks barely affected accuracy, demonstrating how GloNet's can be used to balance efficiency and performance.

6   Conclusions and Future Works

GloNet is a technique enhancing existing models, making them resilient to depth-related learning issues, without increasing complexity or compromising performance. Compared to ResNet, GloNet offers significant advantages without any drawbacks, at least in tested scenarios. In particular, it achieves similar training outcomes in nearly half the time at depths where ResNet remains stable, and maintains consistent performance at greater depths where ResNet falters.

This paper is a proof-of-concept requiring further investigation, as briefly outlined hereafter. Firstly, our experiments used same shape intermediate block-wise representations. When this is not the case, adaptation to a common shape using embeddings to preserve information or projections to align output dimensions can be used, as in ResNet [12]. Secondly, our focus was on tasks not demanding deep architectures. GloNet would certainly benefit from a more in-depth analysis on tasks that benefit from very deep architectures, for instance reinforcement learning settings like the ones described in [19,20,22,25,28]). Moreover, future investigations should aim for robust performance assessments through multiple runs, instead of relying on single-runs, and for dedicated hyperparameter tuning. For example, by separately optimizing GloNet and ResNet in the over-parametrization regime, we could specifically investigate whether ResNet's peak accuracy surpasses that of GloNet, and by what margin, as suggested by the trends observed in Fig. 2. Thirdly, GloNet's application in explainability presents an interesting avenue for exploration, see Remark 2. Lastly, as shown in Fig. 6a, GloNet offers a post-training method to balance energy efficiency and performance, potentially impacting sectors like IoT.

References

1. Ba, L.J., Kiros, J.R., Hinton, G.E.: Layer normalization. CoRR **abs/1607.06450** (2016). http://arxiv.org/abs/1607.06450
2. Ballester-Ripoll, R., Paredes, E.G.: SGEMM GPU Kernel Performance dataset (2018). https://archive.ics.uci.edu/ml/datasets/SGEMM+GPU+kernel+performance
3. Ballester-Ripoll, R., Paredes, E.G., Pajarola, R.: Sobol tensor trains for global sensitivity analysis. RE & SS **183**, 311–322 (2019)
4. Bengio, Y., Courville, A., Vincent, P.: Representation learning: a review and new perspectives. IEEE TPAMI **35**(8), 1798–1828 (2013)
5. Bengio, Y., Simard, P., Frasconi, P.: Learning long-term dependencies with gradient descent is difficult. IEEE TNN **5**(2), 157–166 (1994)
6. Di Cecco, A.: GloNet repository (2024). https://github.com/AntonioDiCecco/GloNet
7. Dosovitskiy, A., et al.: An image is worth 16x16 words: transformers for image recognition at scale. In: ICLR (2021). https://openreview.net/forum?id=YicbFdNTTy
8. Ebrahimi, M.S., Abadi, H.K.: Study of residual networks for image recognition. In: LNNS. vol. 284, pp. 754–763 (2021)

9. Gianfagna, L., Di Cecco, A.: Explainable AI with Python. Springer, Cham (2021). https://doi.org/10.1007/978-3-030-68640-6
10. Glorot, X., Bengio, Y.: Understanding the difficulty of training deep feedforward neural networks. In: JMLR. JMLR, vol. 9, pp. 249–256 (2010)
11. He, K., Zhang, X., Ren, S., Sun, J.: Delving deep into rectifiers: surpassing human-level performance on imagenet classification. In: ICCV, pp. 1026–1034 (2015)
12. He, K., Zhang, X., Ren, S., Sun, J.: Deep residual learning for image recognition. In: IEEE CVPR, pp. 770–778 (2016)
13. He, K., Zhang, X., Ren, S., Sun, J.: Identity mappings in deep residual networks. In: Leibe, B., Matas, J., Sebe, N., Welling, M. (eds.) ECCV 2016. LNCS, vol. 9908, pp. 630–645. Springer, Cham (2016). https://doi.org/10.1007/978-3-319-46493-0_38
14. Huang, G., Liu, Z., Maaten, L.V.D., Weinberger, K.Q.: Densely connected convolutional networks. In: CVPR, pp. 2261–2269 (2017)
15. Ioffe, S., Szegedy, C.: Batch normalization: accelerating deep network training by reducing internal covariate shift. In: ICML, pp. 448-456 (2015)
16. LeCun, Y.A., Bottou, L., Orr, G.B., Müller, K.-R.: Efficient BackProp. In: Montavon, G., Orr, G.B., Müller, K.-R. (eds.) Neural Networks: Tricks of the Trade. LNCS, vol. 7700, pp. 9–48. Springer, Heidelberg (2012). https://doi.org/10.1007/978-3-642-35289-8_3
17. Martin, C.H., Mahoney, M.W.: Implicit self-regularization in deep neural networks. JMLR **22**, 1–165 (2021). http://jmlr.org/papers/v22/20-410.html
18. Metta, C., et al.: Increasing biases can be more efficient than increasing weights. In: IEEE/CVF WACV (2024)
19. Morandin, F., Amato, G., Fantozzi, M., Gini, R., Metta, C., Parton, M.: SAI: a sensible artificial intelligence that plays with handicap and targets high scores in 9×9 go. In: ECAI, pp. 403–410 (2020). https://doi.org/10.3233/FAIA200119
20. Morandin, F., Amato, G., Gini, R., Metta, C., Parton, M., Pascutto, G.: SAI: a sensible Artificial Intelligence that plays Go. In: IJCNN, pp. 1–8 (2019)
21. Panda, P., Sengupta, A., Roy, K.: Conditional deep learning for energy-efficient and enhanced pattern recognition. In: DATE, pp. 475-480 (2016)
22. Pasqualini, L., et al.: Score vs. winrate in score-based games: which reward for reinforcement learning? In: ICMLA, pp. 573–578 (2022)
23. Rajesh, G.: A benchmark repository of Early Exit Neural Networks in .onnx format (2021). https://github.com/gorakraj/earlyexit_onnx
24. Saxe, A., McClelland, J., Ganguli, S.: Exact solutions to the nonlinear dynamics of learning in deep linear neural networks. In: ICLR (2014)
25. Silver, D., et al.: Mastering the game of go with deep neural networks and tree search. Nature **529**(7587), 484–489 (2016)
26. Srivastava, R.K., Greff, K., Schmidhuber, J.: Training very deep networks. In: NeurIPS, pp. 2377–2385 (2015)
27. Teerapittayanon, S., McDanel, B., Kung, H.T.: Branchynet: fast inference via early exiting from deep neural networks. In: ICPR (2016). https://doi.org/10.1109/ICPR.2016.7900006
28. Wu, D.J.: Accelerating Self-Play Learning in Go. AAAI20-RLG workshop (2020). https://arxiv.org/abs/1902.10565
29. Zhang, C., Bengio, S., Hardt, M., Recht, B., Vinyals, O.: Understanding deep learning requires rethinking generalization. In: ICLR (2017)
30. Zhang, C., Bengio, S., Hardt, M., Recht, B., Vinyals, O.: Understanding deep learning (still) requires rethinking generalization. Commun. ACM **64**(3), 107–115 (2021)

Mind the Data, Measuring the Performance Gap Between Tree Ensembles and Deep Learning on Tabular Data

Axel Karlsson[1], Tianze Wang[1,3], Slawomir Nowaczyk[2],
Sepideh Pashami[2], and Sahar Asadi[1(✉)]

[1] King, Stockholm, Sweden
{axel.karlsson,tianze.wang,sahar.asadi}@king.com
[2] Center for Applied Intelligent Systems Research, Halmstad University,
Halmstad, Sweden
{slawomir.nowaczyk,sepideh.pashami}@hh.se
[3] KTH Royal Institute of Technology, Stockholm, Sweden
tianzew@kth.se

Abstract. Recent machine learning studies on tabular data show that ensembles of decision tree models are more efficient and performant than deep learning models such as Tabular Transformer models. However, as we demonstrate, these studies are limited in scope and do not paint the full picture. In this work, we focus on how two dataset properties, namely dataset size and feature complexity, affect the empirical performance comparison between tree ensembles and Tabular Transformer models. Specifically, we employ a hypothesis-driven approach and identify situations where Tabular Transformer models are expected to outperform tree ensemble models. Through empirical evaluation, we demonstrate that given large enough datasets, deep learning models perform better than tree models. This gets more pronounced when complex feature interactions exist in the given task and dataset, suggesting that one must pay careful attention to dataset properties when selecting a model for tabular data in machine learning – especially in an industrial setting, where larger and larger datasets with less and less carefully engineered features are becoming routinely available.

Keywords: Tabular data · Gradient boosting · Tabular Transformers

1   Introduction

Transformer models [24], and Deep Learning (DL) more generally, have demonstrated exceptional success across various domains [8,19,20,23]. They are particularly outstanding for homogeneous and unstructured data, including audio, computer vision, and natural language processing. Given their weaker inherent

---

The authors share the first authorship.

I. Miliou et al. (Eds.): IDA 2024, LNCS 14641, pp. 65–76, 2024.
https://doi.org/10.1007/978-3-031-58547-0_6

bias, they can learn rich representations provided sufficient amounts of data. Conversely, it is structured data that is most prominent in many industrial settings due to its interpretability and ease of processing in "traditional" ways, without employing Machine Learning (ML) and DL. Surprisingly, ubiquitous tabular data in real-world applications is yet to be "conquered" by DL models. The common preconception in the field is that "classical" ML techniques, primarily tree-based algorithms, yield the best results. Analysis of Kaggle competitions [17] exemplifies this.

Tabular datasets are usually small to medium-scale compared to datasets generally used for learning from text and images. Moreover, tabular datasets are usually sparse [4] (features with very few non-default values) and typically contain missing values [4]. In addition, tabular data are heterogeneous in nature, which presents further challenges to DL methods that are well-designed to work with homogenous data such as images. TabNet [1] is a well-known first attempt to approach learning on tabular data with Transformers, and many other works such as [9] show promising results of Transformer models on tabular datasets. However, recent studies [10,22] demonstrate the advantage of the well-established Gradient-bossted Decision Trees (GBDT), e.g., XGBoost [5], in both training efficiency and test performance.

This paper aims to understand the factors influencing the relatively weaker performance of DL models on tabular data. It builds on recent analysis, such as [10,22], and extends these studies along two dimensions: first is the dataset size since it is expected that DL models are more data-hungry. While many public tabular datasets are small, large ones are gaining popularity and are likely to become more common in the future. Second is the complexity of the features since many tabular datasets result from extensive feature engineering and contain "well-behaving" features. However, the amount of preprocessing can be prohibitive and time-consuming in many applications, and one can expect tabular data with more raw features to appear.

This work focuses on the performance comparison between GBDT and Transformer models with a hypothesis-driven approach. We underline two potential contributing factors for the performance gap between GBDT and Transformer models reported in previous studies [9–11], and construct experiments for empirical comparison of their relevance. Besides the well-known impact of the dataset size, we also analyze feature complexity and dependency. We assume one reason why GBDT works well on many datasets is that tabular features are often well-designed with a lot of manual effort and expert knowledge. In the future, tabular datasets will likely not have this benefit due to quantity, and, as a consequence, one can expect the feature interactions to be more complex. We hypothesize that DL models can deal with such challenges better than tree-based models can, as DL models are less reliant on feature engineering [13].

Our contributions can be summarized as (1) an empirical comparison between the Tabular Transformer and XGBoost models on small to large-scale real-world datasets, (2) a study comparing how input feature complexity affects the performance of these models, and (3) outline the potential reasons for the differences that open future research directions.

2   Preliminaries

In this section, we introduce the preliminaries for our work. First, we discuss key properties and characteristics of tabular data. Then, we give an overview of the ML approaches that will be the focus of this paper.

2.1   Tabular Data

Tabular data are ubiquitous in real-world applications with practical applications in many domains, e.g., medicine, finance, manufacturing, etc. [4,22]. The supervised learning task can be formulated as finding an approximation of a function that maps the input features to the target: $\hat{f} \approx f : \mathcal{X} \rightarrow \mathbf{Y}$, where $\mathcal{X} = \{\mathbf{x_1}, \mathbf{x_2}, \dots, \mathbf{x_n}\}$ represents the input feature space, and each $\mathbf{x_i}$ is a vector of k features describing an instance. The key characteristic of tabular datasets is that each feature vector $\mathbf{x_i}$ can be represented as $\mathbf{x_i} = (x_i^1, x_i^2, \dots, x_i^k)$, where k is the number of features in the dataset. These feature vectors $\mathbf{f_j}$ correspond to columns in matrix $\mathcal{X}$ and typically represent semantically meaningful attributes of data instances. Historically, a lot of effort was put into creating "useful" features, and they were expected to correspond to the output in a rather straightforward manner. Often, independence or conditional independence of features was implicitly assumed. Nowadays, however, as we are moving toward collecting raw events or data on a large scale automatically, these assumptions are being violated more often, for example, in industrial IoT settings, where data is collected routinely and without detailed human oversight.

The output space $\mathbf{Y} = \{y_1, y_2, ..., y_n\}$ is typically a categorical label for classification tasks and a continuous value for regression tasks. The prediction for an unseen instance $\mathbf{x'}$ is given by $\hat{f}(\mathbf{x'})$. The learning goal is for prediction $\hat{f}(\mathbf{x'})$ to be as close as possible to the real value $f(\mathbf{x'})$.

2.2   Tree Ensembles

Tree-based ML techniques have evolved significantly since their invention, from learning individual Decision Trees (DTs) to state-of-the-art ensemble methods. DTs, known for their simplicity and interpretability, are the foundation for more complex models. The advent of ensemble methods, combining multiple DTs to improve predictive accuracy and mitigate overfitting, essentially began with Random Forests. The biggest step forward was the introduction of XGBoost (Extreme Gradient Boosting), building upon the concept of gradient boosting and iteratively correcting the errors of previous trees [5].

The GBDT implementation, XGBoost, is a *de facto* method of choice for most practitioners due to its high efficiency and performance, particularly in dealing with large and complex datasets. XGBoost handles sparse data, which is common in many real-world scenarios, notably well. A regularization term in the objective function helps to reduce overfitting.

The algorithm constructs a series of DTs sequentially, where each subsequent tree attempts to correct the errors made by the previous ones. One key innovation in XGBoost is its efficient handling of computational challenges by optimizing both the tree structure and leaf weights at once. Variants of XGBoost, Light-GBM [12] and CatBoost [7], offer specific enhancements like faster training and improved handling of categorical data.

Despite its numerous strengths, XGBoost also has limitations. A notable one is the relatively high computational cost for larger datasets. This is partly due to the sequential nature of boosting, which is more time-consuming compared to methods that build trees in parallel, like Random Forests. Additionally, XGBoost can be prone to overfitting if not properly tuned, especially in cases where data is noisy or the number of training examples is limited.

XGBoost's popularity and success in handling tabular data can be attributed to its superiority in dealing with well-structured data. Tabular datasets often contain a mix of continuous and categorical variables. The robustness of XGBoost in handling different data types, along with its ability to efficiently manage missing values and sparseness, makes it particularly well-suited for these scenarios. Its ability to capture complex non-linear relationships through a combination of simple decision trees, while maintaining interpretability and reducing overfitting through regularization, renders it highly effective for a wide range of tabular data applications.

2.3   Deep Learning

Deep learning (DL) is increasingly becoming an important player in the realm of tabular data, promising distinct benefits similar to the ones realized in other areas, e.g., image and speech recognition. One of the primary strengths of DL in handling tabular data lies in its ability to automatically learn feature interactions without manual intervention. Deep neural networks, especially those with multiple hidden layers, are adept at uncovering complex, non-linear relationships within data. This capability is particularly beneficial for tabular datasets with intricate underlying patterns that simpler models might struggle to capture. Additionally, DL models can be trained end-to-end, allowing them to simultaneously handle raw data and its transformation, deal with multimodal data, or extend to settings like transfer, multi-task, or federated learning.

However, DL also presents several challenges when applied to tabular data. One notable weakness is the need for extensive training data and high computational training cost. Another issue is the lack of interpretability as DL models are often considered "black boxes," making it difficult to understand how specific feature(s) influence predictions. On the other hand, DL models excel in capturing complex, hierarchical patterns in data, a trait that becomes increasingly valuable as the data volume and complexity grow. However, this complexity often requires more computational resources, regularization, and hyperparameter-tuning. The following ML approaches are evaluated in this paper

Multilayer Perceptron (MLP) is a feedforward artificial neural network consisting of at least three layers (an input layer, multiple hidden layers, and an output layer), where each node uses a nonlinear activation function, allowing

MLPs to distinguish data that is not linearly separable. It has been shown [10] that MLPs are particularly good at capturing smooth decision boundaries.

TabNet includes an automatic feature extraction technique for tabular data, which achieves state-of-the-art performance for several classification tasks [1]. TabNet is among the first attempts to apply Transformer models on tabular data, and it achieves comparable or better results compared to GBDT methods.

FT-Transformer is a Transformer-based state-of-the-art architecture that has been shown to outperform other DL models on tabular data benchmarks [9]. TF-Transformer encodes all types of features, including both categorical and numerical, into embeddings using a novel Feature Tokenizer architecture. These embeddings are then processed through a stack of Transformer layers. Each Transformer layer operates at the feature level of an individual object, ensuring a fine-grained and nuanced analysis of the data. What sets the FT-Transformer apart is its utilization of the final representation of the special "[CLS]" token for making predictions. This approach allows for a more comprehensive attention mechanism across features that is particularly suitable for dealing with the heterogeneous nature of tabular data.

3   Related Work

Learning on tabular data is challenging; compared to image and natural language data, tabular data are heterogeneous by nature as they can come from many different sources, which usually leads to dense numerical and sparse categorical features [4]. Furthermore, tabular datasets often exhibit missing values [21] and outliers [6]. The outstanding performance of tree-based ensemble ML models, including XGBoost [5], remains valid for several benchmark data sets [22] despite the dominating success of DL models for unstructured data.

A recent benchmark [10] compared DL models and tree-based models on tabular data, showing that GBDTs remain superior on medium-sized datasets from the accuracy and speed perspectives. At the same time, an ensemble of deep models and GBDT outperformed GBDT alone [22], indicating the possibility that DL methods capture complementary information.

Recently, many works used DL models [1,9,11] to learn from tabular data. The differentiability offers the possibility to easily integrate with other models to provide an end-to-end gradient descent trainable pipeline. A further question revolves around whether the recent advancements in DL, particularly Transformer models [24] across various domains of DL, e.g., audio, vision, and natural language, could offer alternative solutions for handling tabular data in large datasets. TabNet [1] is among the first to apply Transformer models on tabular data, and it achieved comparable or better results compared to GBDT methods in certain tasks. Following that trend, many other works [4,9] also report superior performance of Transformer models on tabular data. However, some recent empirical comparison works [10,14,16] found out that while Transformer models perform well on some tabular datasets, the overall performance still lagged behind GBDT models when benchmarked with a large variety of datasets.

Table 1. Summary of datasets. The last column represents the important feature of each dataset selected using the correlation value of the feature to the target.

Dataset Name	#Samples	#Features		Downstream Task	Important Feature
		Num	Cat		
Forest Cover [3]	581,012	10	44	Multi-class classification	Elevation
Higgs [10]	940,160	24	0	Binary classification	m_bb
Taxi [10]	581,835	9	0	Regression	total_amount

4   Methodology and Design of Experiments

In this work, we empirically compare and try to explain the performance gap between GBDTs and Tabular Transformer models. In this comparison, for GBDTs, we use the XGBoost implementation [5]. We use FT-Transformer implementation of Transformer-based models since it has shown promising results across different tasks on recent tabular data benchmarks [9]. We also include TabNet [1], which has been used conventionally as a deep model on tabular data due to its performance and interpretability. We also include a simple two-layer MLP as a baseline in the comparisons.

Prior empirical comparisons [4,10,16,22] offer important insights on when GBDT models outperform Tabular Transformers. However, most of the available comparisons are limited to small- to medium-scale tabular datasets. In [10], the datasets are limited to 10,000 samples and a partial study of up to 50,000 samples. In [22], half of the datasets are smaller than 15,000 samples, which heavily influences the overall conclusions in the paper. While our work does not evaluate a similarly large number of datasets, we focus on how the landscape of data affects the performance gap of Tabular Transformer models and GBDT in an effort to understand this gap. In particular, we study the impact of dataset size and feature dependencies on model performance. Three large-scale datasets are used: Forest Cover [3], Higgs [10], and Taxi [10]. Table 1 summarizes the datasets used in our empirical evaluation and their key properties.

Dataset Size. Modern DL models are data-hungry, and they require a lot of data to learn from. Tree-based ensemble algorithms, on the other hand, are typically more data efficient, which usually translates to better performance on datasets of small sample sizes.

Previous work did not fully explore the limits of the learning capability of the two competing approaches, which led us to question how the comparison looks when the dataset size is not a limitation. Our hypothesis is that, in the comparison using small datasets, the available samples fail to provide sufficient training signal due to DL's inherently weaker inductive bias. Learning from a larger dataset would allow DL models to catch up with GBDTs in terms of performance and possibly surpass them. In this work, we empirically compare the performance of tree-based models and tabular Transformer models on large datasets and try to capture the impact on performance as we train the model on the following subsets: small (1,000 samples), medium (10,000 samples), and large (100,000 samples and the full dataset).

Feature Dependency. What we learned from other domains of DL, e.g., vision, is that DL models can learn to synthesize higher-level features. For example, in computer vision, a CNN model would learn from neighboring pixels to represent lines, circles, eyes, and noses in cat and dog images, which will later contribute to the task of image classification. However, such higher-level features are really hard to hand-engineer or to learn with simpler models like regression.

Intuitively, DL models could replace, in a similar fashion, some of the feature engineering efforts that are today required for tabular data – learn better features by iterating through the data samples during training. The learning of simple and complex features could significantly ease feature engineering as the task can be optimized end to end during the model training instead of the resource-intensive manual process based on domain knowledge.

Tabular datasets are often designed and collected with downstream tasks in mind. This implies that the most relevant features (often with clear decision boundaries) are likely already present in the dataset, e.g., certain thresholds on monthly salary are good indicators for predicting if a loan will be faulty or not. Tree-based models can easily capture such downstream tasks by recreating the thresholds. However, in cases where the critical information is not directly presented in the dataset, it poses a larger challenge for GBDTs. For example, they can potentially suffer when aggregating critical information across multiple features, e.g., weighted summation of two features. DL models, on the other hand, are more robust to such scenarios due to their ability to synthesize new features from existing features.

To enforce a setting where critical information for downstream prediction tasks is not directly presented in the dataset as a single feature, we employ feature decomposition. We consider two types of such "reverse feature engineering" operations: (1) summation and (2) multiplication. Decomposing a selected feature x_a yields n sub-features $x_a^1, \cdots, x_a^n$. In the summation case, we create these n sub-features such that $x_a^i = r_i * x_a$ ($i \in \{1, ..., n\}$), with the constraint $\sum_{i=1}^n r_i = 1$, and therefore $x_a = \sum_{i=1}^n x_a^i$. In the multiplication case, the constraint is $\prod_{i=1}^n r_i = 1$, and therefore $x_a = \prod_{i=1}^n x_a^i$. The values of r_i are randomly chosen for each sample independently, preventing the model from reconstructing the decomposed feature globally, and instead forcing it to learn the relationship. In this work, we empirically evaluate feature decomposition with $n = 2$ and $n = 3$ sub-features as a first step toward understanding how feature dependency affects the performance of tabular models.

Model Selection. For XGBoost and FT-Transformer, the default hyperparameters were used [5,9]. The hyperparameters for the MLP were batch size: 32, hidden layer size: 32, and learning rate 0.01 using Adam optimizer. After initial experimentation with TabNet, we found that it underperforms with default hyperparameters. After tuning the hyperparameters on the datasets, the following were used: batch size: 512, N_{steps}: 1, γ: 1, λ_{sparse}: 0.001, and learning rate: 0.02. N_a, N_d: 256 for Forest Cover, 64 for Higgs, 128 for Taxi, and the rest were set to default as in [1]. We split datasets into train, validation, and test datasets. For Forest Cover, we use the test set in [18]. For the other datasets, we keep the same fixed-size test dataset that contains 20% of samples in the original dataset for a fair comparison between different models and feature-splitting methods. The size of the training dataset varies depending on the experiment setup, i.e., 1,000, 10,000, 100,000, and all samples available for training. The validation dataset size also varies and is kept as 50% of the corresponding training dataset.

Evaluation Metrics. To test our hypothesis, we report the performance using Accuracy for the tasks in Forest Cover and Higgs and R2 for Taxi. Results are presented as the average performance of each model over three trials. To evaluate the impact of the feature splitting on different models, we additionally compare the feature importance for each model, using mean absolute SHAP values [15].

5 Results

In this section, we present the results and the analysis of our empirical comparison of Tree-based models and Tabular Transformer models.

5.1 Impact of Training Dataset Size

Figure 1 shows the impact of the training dataset size on the performance of different models across the three datasets. We observe that as the number of samples in the training dataset increases, the performance consistently increases, for all models. Furthermore, while XGBoost outperforms TabNet and

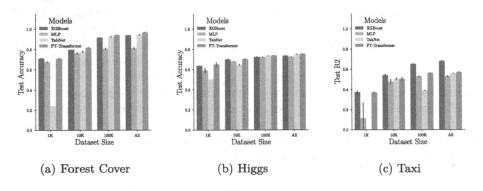

|(a) Forest Cover | (b) Higgs | (c) Taxi |

Fig. 1. Performance improvements as dataset size increases.

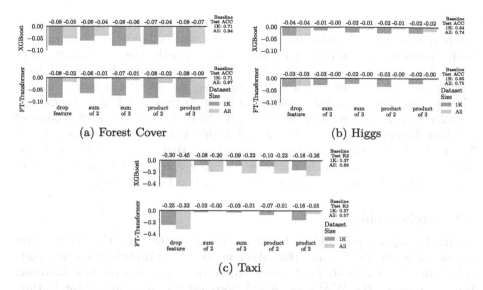

Fig. 2. Performance drop for different feature splitting methods: "drop feature" excludes the important feature; "sum of 2" splits the important feature into two sub-features whose summation equals the original feature, etc.

FT-Transformer on the smallest training datasets, the performance gap becomes smaller as the size of the training dataset increases. In the case of the Forest Cover and Higgs datasets, TabNet and FT-Transformer end up outperforming XGBoost. For the Taxi dataset, the performance of TabNet and FT-Transformer also increases with the number of samples in the training dataset, even if, for all the available data, the gap in performance between the models remains. As an indication of the difference in computational costs of models, we report the training time until convergence on the full Forest Cover dataset. Using only the CPU for training, XGBoost takes 328 s. Using one NVIDIA Tesla T4 GPU, MLP, TabNet, and FT-Transformer take 822, 735, and 5721 s, respectively.

5.2 Feature Complexity

Figure 2 shows the performance drop for the evaluated models when removing the most important feature and using the different feature splitting methods, relative to models trained without feature splitting. Figure 2a shows the drops in test accuracy for the Forest Cover dataset. For small training datasets with 1000 samples, removing the Elevation feature, the test accuracy drops around 7 or 8 percent for all models compared to not including the feature. Test accuracy drops more when splitting a feature using the product method than using the sum method, which is intuitive as the product of two numbers is harder to recover for both XGBoost and FT-Transformer. Also, the performance drops more when we split the important feature into three sub-features rather than two, given that the information is spread across more features.

While XGBoost and FT-Transformer experience similar performance drops when trained on one thousand instances, the comparison landscape changes when both models are trained on larger datasets. For example, the performance drops of FT-Transformer (around 1 or 2%) are much lower than those of XGBoost models (around 4 or 6%), with feature splitting methods "sum of 2", "sum of 3", and "product of 2". For "product of 3", even though FT-Transformer has a larger drop (9%) compared to XGBoost (7%), the absolute performance of FT-Transformer ($97 - 9 = 88\%$) is still higher than that of XGBoost ($94 - 7 = 87\%$). These consistent lower drops indicate that, compared to tree-based models, Tabular Transformer could better recover critical information spread across multiple features, especially when trained on large datasets.

5.3 Explainability

Figure 3 compares pre- and post-split feature importance for the Forest Cover dataset. The two sub-features (Elevation_sub01, Elevation_sub02) created post-split have, for XGBoost, much lower importance than the original Elevation feature. Instead, the Wilderness_Area4 becomes the most important feature, and a significant reshuffling of the ranking takes place. In contrast, for the FT-Transformer, the two sub-features retain their original importance and remain the most important ones. There are also no other changes in the ranking. This, in addition to the performance drops observed in Fig. 2, indicates that FT-Transformer is capable of significantly better reconstruction of the original feature – as opposed to XGBoost which suffers from a larger decrease in performance after feature splitting and fails to capture the importance of the original feature.

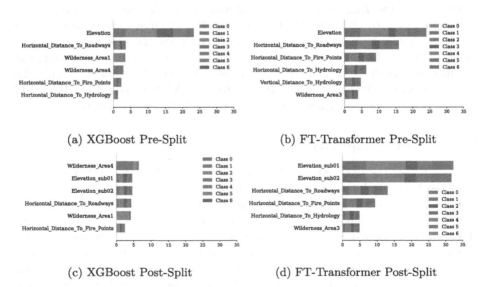

(a) XGBoost Pre-Split   (b) FT-Transformer Pre-Split

(c) XGBoost Post-Split   (d) FT-Transformer Post-Split

Fig. 3. SHAP feature importance pre/post-split of Elevation for Forest Cover.

6   Conclusions and Future Work

In this work, we exemplify situations where Transformer-based models close the gap to tree-based ensembles on tabular data, just as they have for natural language processing or computer vision. This is done by focusing on experimental settings where the strengths of these models have the opportunity to play a significant role. Based on our results, we can conclude that provided that the dataset size is sufficiently large (100K+ samples), Tabular Transformer models are likely to be preferable to XGBoost with regard to model performance. In this setting, Transformer-based methods can handle more complex feature interactions and dependencies better than XGBoost regarding model performance.

Transformer-based models have not matched their success in tabular data as in other areas. This might be because tabular data, unlike images or text, does not form a manifold. Each tabular data sample usually has less information compared to an image or a sentence. Additionally, features in these fields are spatially correlated, while tabular features are often more independent. Another potential factor as to why GBDTs remain so successful in this domain could be the impact of feature engineering. If one spends less time on feature engineering, Transformer-based models could hypothetically compensate. In our work, feature splitting is a simple attempt at demonstrating this concept in order to provide an initial understanding. We leave extensions of this idea of "reverse" feature engineering to future work.

In future work, we plan to expand the evaluation beyond the current, rather limited, dataset scope to include different numbers and types of features, data sample sizes, ratios between the number of features and samples, the number of missing values, and class proportions.

Another promising future direction is understanding task complexity. As can be noticed, XGBoost outperforms DL models on the Taxi dataset. We hypothesize that this is due to the relatively simple linear relation between the most important feature (total_amount) and the target (tip_amount), which is easy for the XGBoost model to capture. Besides studying diverse datasets, we propose exploring data (possibly synthetic) where controlling task complexity is viable (for example, during data generation). Another advantage of synthetic data is the access to ground truth feature importance for both SHAP approximations and inspection of decision boundaries [2].

Acknowledgements. This work was partially carried out with support from the Knowledge Foundation and from Vinnova (Sweden's innovation agency) through the Vehicle Strategic Research and Innovation program, FFI.

References

1. Arik, S.Ö., Pfister, T.: TabNet: attentive interpretable tabular learning. In: AAAI, vol. 35, pp. 6679–6687 (2021)
2. Barr, B., et al.: Towards ground truth explainability on tabular data (2020)

3. Blackard, J., Dean, D.: Comparative accuracies of artificial neural networks and discriminant analysis in predicting forest cover types from cartographic variables. Comput. Electron. Agric. **24**, 131–151 (1999)
4. Borisov, V., Leemann, T., Seßler, K., Haug, J., Pawelczyk, M., Kasneci, G.: Deep neural networks and tabular data: a survey. IEEE Trans. Neural Netw. Learn. Syst. (2022)
5. Chen, T., Guestrin, C.: XGBoost: a scalable tree boosting system. In: Proceedings of the 22nd ACM SIGKDD, pp. 785–794 (2016)
6. Dixon, W.: Processing data for outliers. Biometrics **9**(1), 74–89 (1953)
7. Dorogush, A.V., Ershov, V., Gulin, A.: CatBoost: gradient boosting with categorical features support. ArXiv **abs/1810.11363** (2018)
8. Dosovitskiy, A., Beyer, L., Kolesnikov, A., et al.: An image is worth 16×16 words: transformers for image recognition at scale. preprint arXiv:2010.11929 (2020)
9. Gorishniy, Y., Rubachev, I., Khrulkov, V., Babenko, A.: Revisiting deep learning models for tabular data. Adv. Neural. Inf. Process. Syst. **34**, 18932–18943 (2021)
10. Grinsztajn, L., Oyallon, E., Varoquaux, G.: Why do tree-based models still outperform deep learning on typical tabular data? Adv. Neural. Inf. Process. Syst. **35**, 507–520 (2022)
11. Kadra, A., Lindauer, M., Hutter, F., Grabocka, J.: Well-tuned simple nets excel on tabular datasets. NeurIPS **34**, 23928–23941 (2021)
12. Ke, G., et al.: LightGBM: a highly efficient gradient boosting decision tree. In: Proceedings of the 31st International Conference on Neural Information Processing Systems, pp. 3149-3157. NIPS'17, Curran Associates Inc. (2017)
13. Krizhevsky, A., Sutskever, I., Hinton, G.E.: ImageNet classification with deep convolutional neural networks. In: Advances in Neural Information Processing Systems, vol. 25 (2012)
14. Levin, R., et al.: Transfer learning with deep tabular models. In: The Eleventh International Conference on Learning Representations (2023)
15. Lundberg, S.M., Lee, S.I.: A unified approach to interpreting model predictions. In: Advances in Neural Information Processing Systems, vol. 30 (2017)
16. McElfresh, D., et al.: When do neural nets outperform boosted trees on tabular data? (2023)
17. Miller, C., Hao, L., Fu, C.: Gradient boosting machines and careful pre-processing work best: ASHRAE great energy predictor iii lessons learned (2022)
18. Mitchell, R., Adinets, A., Rao, T., Frank, E.: XGBoost: scalable GPU accelerated learning. CoRR **abs/1806.11248** (2018)
19. Radford, A., et al.: Robust speech recognition via large-scale weak supervision. In: International Conference on Machine Learning, pp. 28492–28518. PMLR (2023)
20. Rampášek, L., et al.: Recipe for a general, powerful, scalable graph transformer. Adv. Neural. Inf. Process. Syst. **35**, 14501–14515 (2022)
21. Sánchez-Morales, A., Sancho-Gómez, J.L., Martínez-García, J.A., Figueiras-Vidal, A.R.: Improving deep learning performance with missing values via deletion and compensation. Neural Comput. Appl. **32**, 13233–13244 (2020)
22. Shwartz-Ziv, R., Armon, A.: Tabular Data: deep learning is not all you need. Inf. Fusion **81**, 84–90 (2022)
23. Touvron, H., et al.: Llama 2: Open foundation and fine-tuned chat models. arXiv preprint arXiv:2307.09288 (2023)
24. Vaswani, A., et al.: Attention is all you need. In: NeurIPS, vol. 30 (2017)

A Remark on Concept Drift for Dependent Data

Fabian Hinder[(✉)] , Valerie Vaquet , and Barbara Hammer

CITEC, Bielefeld University, Bielefeld, Germany
{fhinder,vvaquet,bhammer}@techfak.uni-bielefeld.de

Abstract. Concept drift, i.e., the change of the data generating distribution, can render machine learning models inaccurate. Several works address the phenomenon of concept drift in the streaming context usually assuming that consecutive data points are independent of each other. To generalize to dependent data, many authors link the notion of concept drift to time series. In this work, we show that the temporal dependencies are strongly influencing the sampling process. Thus, the used definitions need major modifications. In particular, we show that the notion of stationarity is not suited for this setup and discuss an alternative we refer to as consistency. We demonstrate that consistency better describes the observable learning behavior in numerical experiments.

Keywords: Concept Drift · Dependent Data · Concept Drift Detection

1 Introduction

The world surrounding us is subject to constant change, which affects the increasing amount of data collected over time. As those changes, referred to as *concept drift* or drift for shorthand [21], constitute a major issue in many applications, considerable research is focusing on this phenomenon. While in some applications, the goal is keeping an accurate model in the presence of drifts (*stream or online leaning*) [6], in *monitoring setups* the goal is detecting and gathering information about the drift to inform appropriate reactions [15].

Frequently, the general assumption is that independent observations are collected over time even if they arrive in direct succession [6,15]. However, in many real-world scenarios, temporal dependencies in the data have to be expected. This is for example the case when considering *time series* [3,20,22]. In this setting, the absence of change across multiple time series is described by the concept of *stationarity*. This notion is relevant if the goal is to obtain knowledge about the time series-generating process.

In this work, we demonstrate that this does not align with the goal and setup of machine learning and monitoring, where we want to keep an up-to-date model

Funding in the frame of the ERC Synergy Grant "Water-Futures" No. 951424 is gratefully acknowledged.

I. Miliou et al. (Eds.): IDA 2024, LNCS 14641, pp. 77–89, 2024.
https://doi.org/10.1007/978-3-031-58547-0_7

based on our observations or detect anomalous behavior therein. In particular, we typically observe only one stream rather than multiple instantiations thereof. To substantiate this claim, we analyze the connection of drift and stationarity in settings relevant to machine learning applications. We propose the notion of *consistency* when dealing with streaming data that stems from one time series. Moreover, we demonstrate in experiments, that this formalization captures the intuitive notion of drift for possibly dependent data streams.

This paper is organized as follows: In Sect. 2 we recap the concept of drift and stochastic processes used later on. We categorize the possible setups and analyze them according to their convergence properties (Sect. 2.3). In Sect. 3, we provide examples showcasing that stationarity and drift do not align in settings of interest for machine learning (Sect. 3.1). Thus, we propose the notion of consistency when dealing with a data stream possibly subject to time dependencies (Sect. 3.2). We derive a first method checking for this concept (Sect. 3.3). Finally, we confirm the theoretical arguments with numerical experiments (Sect. 4) and conclude the work (Sect. 5).

2   Problem Setup

In this work, we consider the problem of stream learning, i.e., at every given time point τ only the data (X_i, T_i) observed until τ, i.e., $T_i \leq \tau$, is available. The observations X_i can be independent of each other, dependent on the past (time series), or the time point T_i of observation (drift). Depending on the specific setup, learning faces different problems, e.g., model adaption. In the following, we first give a precise formal definition of the notion of drift (of independent data) and stochastic processes as a model for time series. We then discuss the challenges of defining drift for dependent data.

2.1   A Probability Theoretical Framework for Concept Drift

In the classical setup, machine learning assumes a time-invariant distribution $\mathcal{D}$ on the data space $\mathcal{X}$ and data points as i.i.d. samples drawn from it. This assumption is violated in many real-world applications, in particular, when learning on data streams. As a consequence, machine learning models can become inaccurate over time. As a first step to tackle this issue, we incorporate time into our considerations. To do so, let $\mathcal{T}$ be an index set representing time and a (potentially different) distribution $\mathcal{D}_t$ for every $t \in \mathcal{T}$. Drift refers to the phenomenon that those distributions differ for different points in time [21]. It is possible to extend this setup to a general statistical interdependence of data and time via a distribution $\mathcal{D}$ on $\mathcal{T} \times \mathcal{X}$ which decomposes into a distribution P_T on $\mathcal{T}$ describing the likelihood of observing a data point at a given time and the conditional distributions $\mathcal{D}_t$ on $\mathcal{X}$ describing the data at time t [11,15]. One of the key findings of [11] is a unique characterization of the presence of drift by the property of statistical dependency of observation time T and data point X.

Definition 1. *Let $\mathcal{X}, \mathcal{T}$ be countably generated measurable spaces. Let $(\mathcal{D}_t, P_T)$ be a drift process [11] during $\mathcal{T}$ on $\mathcal{X}$, i.e., a distribution P_T on $\mathcal{T}$ and Markov kernel $\mathcal{D}_t$ from $\mathcal{T}$ to $\mathcal{X}$. A time window $W \subset \mathcal{T}$ is a P_T non-null set. A sample (of size N) is an $\mathcal{T} \times \mathcal{X}$ N-tuple $S = ((T_1, X_1), \ldots, (T_N, X_N))$ with $T_i \sim P_T$, $X_i \mid [T_i = t] \sim \mathcal{D}_t$ i.i.d. We say that $\mathcal{D}_t$ has drift if the probability of observing two different distributions in one sample is larger than 0, i.e., $\mathbb{P}_{T,S \sim P_T}[\mathcal{D}_T \neq \mathcal{D}_S] > 0$.*

Considerable work focuses on independent data streams with drift. Many authors consider the model accuracy as a proxy for drift and update the model if a decline in the accuracy is observed [6]. As the connection between model loss and drift in the data generating distribution is rather lose [12,13], relying on distribution-based drift detection schemes instead of model loss-based schemes enables reliable monitoring [15]. Assuming independent data points X_i, allows applying tools from classical statistics or learning theory [11,12]. In contrast, given temporal dependencies in the observed data, these strategies cannot be used successfully.

2.2 Stochastic Processes

Instead of drift processes, dependent data streams or time series can be modeled by stochastic processes [4]:

Definition 2. *Let $(\Omega, \mathcal{A}, \mathbb{P})$ be a probability space, $(\mathcal{X}, \Sigma_\mathcal{X})$ a measurable space, and $\mathcal{T}$ an index-set. A $\mathcal{X}$-valued stochastic process over $\mathcal{T}$ is a map $X_\bullet : \mathcal{T} \times \Omega \to \mathcal{X}$ such that $X_t : \Omega \to \mathcal{X}$ is a $\mathcal{X}$-valued random variable for all $t \in \mathcal{T}$. We call the maps $t \mapsto X_t(\omega)$ the paths of X_t. The process X_t is called stationary iff the distribution of X_t is time invariant, i.e., $\mathbb{P}[X_t \in A] = \mathbb{P}[X_s \in A]$ for all $s, t \in \mathcal{T}, A \subset \mathcal{X}$. For $\mathcal{T} \subset \mathbb{R}$ we call $p \in \mathcal{T}$ a change point iff $t \to \mathbb{P}[X_t \in A]$ is not constant for all neighborhoods of p. Assuming $(\mathcal{T}, \Sigma_\mathcal{T})$ is a measurable space, we call X_t measurable iff $X_\bullet$ is measurable with respect to $\mathcal{A} \otimes \Sigma_\mathcal{T}$.*

A measurable stochastic process is a random variable that takes values in the space of measurable functions from $\mathcal{T}$ to $\mathcal{X}$. In particular, measurability is required if we want to integrate along the path as done in ergodic processes.

The main difference between stochastic and drift processes is that a stochastic process has one value for every point in time whereas a drift process yields a distribution. For example: Measuring the temperature of an object over time is a stochastic process because we read exactly one value. In contrast, a stream of ballots should be considered a drift process because the distribution is more interesting than a single vote.

2.3 A Taxonomy of Change Detection in Data Streams

Different setups can be considered in the context of concept drift and dependent data. Intuitively, the absence of stationarity is the same as the presence of drift.

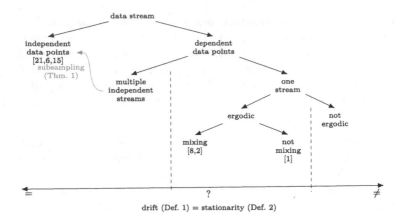

Fig. 1. Taxonomy of change detection setups in data streams ordered by agreement of drift and stationarity.

We investigate this relationship in more detail: We first propose a categorization of the possible setups as visualized in Fig. 1. In the remainder of this section, we describe the different setups.

Independent Data Points. The classical setup of concept drift states that we have several independent data points $X_1, \ldots, X_N, \ldots$ that are sampled from potentially different distributions. This can be investigated for general machine learning [6,12,21] or for drift detection [7,9–11,15]. As we can characterize a drift process solely by the values of all functions $f : \mathcal{X} \to \mathbb{R}$ integrated over all time windows W [11,13] we can test for drift as follows: instead of only tracking the data points we additionally consider the time of observation, $T_1, \ldots, T_N, \ldots$. Using this information we can compute the value

$$\frac{\sum_{i=1}^{N} f(X_i(\omega))\mathbf{1}[T_i(\omega) \in W]}{\sum_{i=1}^{N} \mathbf{1}[T_i(\omega) \in W]} \xrightarrow{N \to \infty} \iint f(x)\mathrm{d}\mathcal{D}_t(x)\mathrm{d}P_T(t \mid W) \quad \mathbb{P}\text{-a.s.} \quad (1)$$
$$= \mathbb{E}[f(X) \mid T \in W],$$

where ω is the random-event. In particular, the limit always applies, and the value on the right-hand side does not depend on ω only on the drift process $(\mathcal{D}_t, P_T)$ (and f and W). This justifies defining drift as $\mathcal{D}_{T_1} \neq \mathcal{D}_{T_2}$ being observed with a probability larger than zero. Thus, in an independent data stream, the presence of drift is the same as the absence of stationarity (up to a P_T null set).

Multiple (Independent) Streams. In real-world applications, data points frequently are not independent. For example, this includes weather, ecosystem, stock market, or general sensory data [21]. Instead of observing an independent data stream, we observe one or multiple stochastic processes. Let us consider the case of multiple ones first: Let $X_t^{(1)}, \ldots, X_t^{(M)}, \ldots$ be sampled independently from the same distribution and each modeling the temporal dependency of consecutive observations. This setup can be reduced to the independent data streams as above:

Theorem 1. *Let $\mathcal{T} = [0,1]$, and $\mathcal{X} = \mathbb{R}^d$ or more generally both countably generated. Let $X_t^{(1)}, X_t^{(2)}, \dots$ be i.i.d. measurable stochastic processes and P_T any probability measure on $\mathcal{T}$. Let $T_1, T_2, \dots \sim P_T$ be i.i.d. and independent of the $X_t^{(i)}$. Denote by $X_i = X_{T_i}^{(i)}$. Then all (T_i, X_i) are i.i.d. and $X_i \mid [T_i = t] \sim \mathbb{P}_{X_t}$ for all $t \in \mathcal{T}$, i.e., they form a sample of the drift process $(\mathbb{P}_{X_t}, P_T)$.*

Furthermore, $(\mathbb{P}_{X_t}, P_T)$ has no drift (in the sense of Definition 1) if and only if there exists a P_T null-set N such that X_t restricted onto $\mathcal{T} \setminus N$ is stationary.

Proof. All proofs can be found in the ArXiv version [16].

Thus, we consider the case of a single dependent path which is the standard setting for time series analysis [3,23] in the remainder of this paper.

Ergodicity and Mixing. Considering only one dependent process we first encounter a phenomenon that we refer to as the *uncertainty principle of drift learning*. To understand this assume we want to estimate the value of $\mathbb{E}[f(X) \mid T = t]$. As the probability of making an observation exactly at time t is zero, we replace it with a small time window W. In the independent setup, we simply sample more and more data points and gain better and better estimation (see Eq. (1)) and then successively shrink W towards t. However, in the single-path setup, we can only sample one path at several time points. As a consequence, the inner integral on the right-hand side of Eq. (1) becomes a *time-average*:

$$\frac{\sum_{i=1}^{N} f(X_{T_i(\omega)}(\omega)) \mathbf{1}[T_i(\omega) \in W]}{\sum_{i=1}^{N} \mathbf{1}[T_i(\omega) \in W]} \xrightarrow{N \to \infty} \int f(X_t(\omega)) \mathrm{d}P_T(t \mid W).$$

As can be seen, the right-hand side now does depend on ω and does not necessarily converge to the mean of the underlying distribution during W also known as *space average*, i.e., $\iint f(x) \mathrm{d}\mathcal{D}_t(x) \mathrm{d}P_T(t \mid W)$. Intuitively speaking, dependency limits the amount of local information and due to drift, we cannot obtain more information by considering larger time windows. The special cases in which the convergence holds are summarized in Fig. 2 and discussed in the following paragraph.

If the time average converge to the space average the process is called *ergodic* [4]. Such processes allow precise estimations and learning under the assumption of stationarity [1]. Yet, they have arbitrarily slow convergence [18]. To counteract this issue many authors consider *mixing processes* that impose a rate on "becoming independent". This leads to uniform convergence similar to the independent case and has been used by several authors to prove statements on learnability in the stationary [17,24] and non-stationary setup [2,8].

Yet, both ergodic and mixing processes usually require infinite temporal horizons to achieve convergence. As we are usually concerned with limited time the results might be interesting from a theoretical point of view but have little

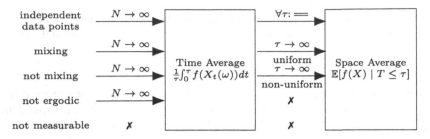

Fig. 2. Two stages of converges for different types of streams. Left to right: Empirical mean, path integral/time-average, mean of the underlying distribution.

practical relevance. Also, due to the uncertainty principle we introduced before most make additional requirements that limit the non-stationarity in one way or another. In particular, the relation to drift detection in the monitoring setup is not clear.

Not-Ergodic Processes. Ergodicity usually describes that the time average equals the space average for infinitely large time windows. We will call processes for which there are either no time windows for which equality holds or for which this cannot be predicted *not-ergodic*.

It is obvious that stationarity is not a reasonable tool to discuss not-ergodic processes as there is no utilizable connection to the underlying distributions. This implies that there cannot be any formal guarantees regarding generalization properties in the usual sense.

Research Questions. Based on the presented analysis we notice that there do exist theoretical results on detecting drift in dependent data but their relevance for practical setups is questionable. This leads us to the following questions that we address in the remainder of this paper:

1. Is non-stationarity a suitable definition for the notion of concept drift for dependent data?
2. Does there exist an alternative notion that is more suitable for practical application (this includes testability on finite temporal horizons)?

3   Consistency Property

In this section, we investigate the suitability of stationarity to model the equivalent of drift as used for independent data streams on time series. First, we provide (counter-)examples that show that this is not the case. Based thereon, we derive the concept of temporal consistency to formalize drift for time series. Finally, we derive a method to test consistency.

3.1   Drift is not Non-Stationarity

That the presence of drift is not the same as the absence of stationarity for dependent data streams can be shown by the following, simple counterintuitive examples:

Example 1. Let $\mathcal{X} = \{-1, 1\}, \mathcal{T} = [0, 1]$ and B and C be independent random variables with $\mathbb{P}[B = -1] = 1 - \mathbb{P}[B = 1] = p$ for $p \in [0, 1]$, and $C \sim \mathcal{U}(\mathcal{T})$. Consider $X_t = (1 - 2 \cdot \mathbf{1}[C > t]) \cdot B$ the process whose paths take values in $\{-1, 1\}$ and switch sign exactly once (see Fig. 3). Then we have $\mathbb{P}[X_t = -1] = t(1-p) + (1-t)p$ and therefore X_t is stationary if and only if $p = \frac{1}{2}$ and otherwise every $t \in \mathcal{T}$ is a change point. Yet, every path has exactly one jump at time C.

Example 2. Consider a teacher student scenario: Let $\mathcal{X} = \mathbb{R}^d \times \{0, 1\}$, $\mathcal{T} = \mathbb{Z}$, $X_t \sim \mathcal{U}([-1, 1]^d), X_t \perp\!\!\!\perp X_s \forall t \neq s$, and $y_t = \mathbf{1}[w_t^\top X_t > 0]$ where w_t is the teacher weight vector which is independently resampled every $N > 1$ time steps, i.e., $w_{iN+j} = w_{iN} \forall i \in \mathbb{Z}, j = 0, \dots, N-1$, $w_{iN} \sim \mathcal{N}(0, 1)$, $w_t \perp\!\!\!\perp w_s \forall |t - s| > N$. Then (X_t, y_t) is a labeled data stream that is ψ-mixing and stationary. Yet, every N time steps the teacher is changed in a completely random way.

Both examples show a common theme: If we consider one path only then there are obvious changes, but if we average over several paths then those changes cancel out (in Example 1 for $p = 1/2$). Thus, we can conclude that in contrast to stationarity, drift is a path-wise notion. A naïve way to solve this problem is to apply the definition of stationarity path-wise, i.e., $\mathbb{P}[X_S = X_T] = 1$. However, such processes are essentially constant:

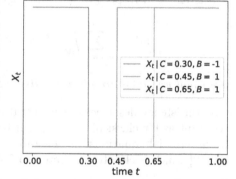

Fig. 3. Sample paths from Example 1.

Proposition 1. *Let $\mathcal{T} = [0, 1]$, and $\mathcal{X} = \mathbb{R}^d$ or more generally both countably generated with $\mathcal{X}$ having a measurable diagonal. Let X_t be a measurable stochastic process and P_T a measure on $\mathcal{T}$ with $S, T \sim P_T$ independent of each other and X_t. It holds $X_S = X_T$ $\mathbb{P}$-a.s. if and only if X_t is constant, i.e., $X_t(\omega) = C(\omega)$ for P_T-a.e. t and $\mathbb{P}$-a.e. ω.*

Thus, equality is far too restrictive for dependent data and hence of little interest. However, the definition of drift as "a change of the distribution over time" stems from the batch setup with independent data where we do not expect any change and can make arbitrarily precise measurements as discussed in Sect. 2.3. If we expect some kind of temporal dynamics, like seasonality, it might be better to rephrase this as "the distribution changes in an unexpected way" where "unexpected" means "in a way our model cannot cope with without retraining". This is a more suitable description as shown in the following example:

Example 3. Let $T, X = \mathbb{R}$. For $\varepsilon \in [0, \pi]$ consider the family of stochastic processes $X_t^{(\varepsilon)}(\omega) = \sin(t + U^{(\varepsilon)}(\omega))$ with $U^{(\varepsilon)} \sim \mathcal{U}([0, 2\pi - \varepsilon])$ uniformly distributed. Then $X_t^{(0)}$ is the only stationary process in this family. Yet, given two points $S, T \sim \mathcal{U}([0, 2\pi])$ we can estimate U from $X_T^{(\varepsilon)}, X_S^{(\varepsilon)}$ with probability 1.

In the next section, we explore the idea of "unexpected change" in more detail, leading to our substitute for the notion of stationarity.

3.2   Temporal Consistency

In the last section, we discussed that stationarity is a sub-optimal choice to describe the notion of drift in a dependent setup, rather we should look for "unexpected changes of a single stream". We will now formalize the idea, referring to the new notion as *consistency*. Formally, we define a collection of functions that assign a loss to the paths. Consistency then refers to time-invariant losses.

Definition 3. *Let P_T be a probability measure on T. Let $\mathcal{F}$ be a collection of measurable functions from $T \times X$ to $\overline{\mathbb{R}}$. A measurable path $x_t : T \to X$ is consistent with respect to $\mathcal{F}$ iff $\inf_{f^* \in \mathcal{F}} \int f^*(x_t, t) \mathrm{d}P_T(t) < \infty$ and*

$$\inf_{\substack{f_1^*, \dots, f_n^* \in \mathcal{F} \\ (W_1, \dots, W_n) \in \mathcal{W}}} \sum_{i=1}^{n} \int_{W_i} f_i^*(x_t, t) \mathrm{d}P_T(t) = \inf_{f^* \in \mathcal{F}} \int f^*(x_t, t) \mathrm{d}P_T(t)$$

are equal where $\mathcal{W}$ are all finite disjoint coverings of T by time windows.

Consistency describes our expectations of how a system evolves. This can be constant as for classical drift, x_t can belong to a function class $\mathcal{C}$, e.g., defined by differential equations, or generalizes the notion of model drift as considered by [13] which is closely connected to $\mathcal{H}$-model drift [12]:

Proposition 2. *Let $X \subset \mathbb{R}^d$ and $T = [0, 1]$. Let $\mathcal{C}$ be a collection of measurable functions from T to $\mathbb{R}^d$ and $\mathcal{H}$ a hypothesis class with loss ℓ on X. It holds:*

1. *Consider $\mathcal{F} = \{f(x, t) = \mathbb{1}[x \neq g(t)] \cdot \infty \mid g \in \mathcal{C}\}$ then x_t is $\mathcal{F}$-consistent if and only if $x_t \in \mathcal{C}$ in the L^0-sense.*
2. *For continuous $\mathcal{C}$ that approximate X at all t. Consider $\mathcal{F} = \{f(x, t) = \|x - g(t)\|_2 \mid g \in \mathcal{C}\}$ then x_t is $\mathcal{F}$-consistent if and only if $x_t \in \overline{\mathcal{C}}^{L^p}$.*
3. *Consider $\mathcal{F} = \{f(x, t) = \ell(h, x) \mid h \in \mathcal{H}\}$ then x_t is not $\mathcal{F}$-consistent implies that the drift process (δ_{x_t}, P_T) has $\mathcal{H}$-model drift [12].*

In the next section, we derive an algorithm that checks for consistency in the simple case when we reference a class of functions.

3.3   Measuring Consistency of a Noisy Stochastic Processes

As a proof of concept, we showcase the workings of consistency using a simple example similar to Proposition 2.2 to check that a stochastic process is contained in a function class $\mathcal{C}$. We consider a noisy (measured) version of the function f, i.e., $X_t = f(t) + \varepsilon_t$ where ε_t is an independent noise process, and we want to check whether $f \in \mathcal{C}$. To do so we use a newly designed loss that can compensate for the noise. We assume that $\mathcal{C}$ specifies global phenomena, e.g., linear, polynomial, or exponential trends, periodicity, etc. Thus, $\mathcal{C}$ contains no local phenomena. An algorithmic solution is then given by the following: We fit a model $\hat{f} \in \mathcal{C}$ to X_t and then check whether $f - \hat{f}$ does contain any structure using a local model like k-NN. If this is not the case, i.e., the local model is not better than random chance, then $f \in \mathcal{C}$. In case the paths are equidistantly sampled, the k-NN simplifies to the following loss which has the desired properties as a simple computation shows.

Lemma 1. *Let $\mathcal{T} = [0,1]$ with sampling time points $t_i^{(n)} = i/n$. Let $k : \mathbb{N} \to \mathbb{N}$ such that $k(n) \leq n$ and odd, denote by $k_0(n) = (k(n) - 1)/2$. Let $f, \hat{f} : \mathcal{T} \to \mathcal{X}$ be measurable maps and $X_t = f(t) + \varepsilon_t$ with $\varepsilon_t \perp\!\!\!\perp \varepsilon_s \forall t \neq s, \mathbb{E}[\varepsilon_t] = 0$ and denote by $X_t^{(n)}$ the subsample according to $t_i^{(n)}$. We define the k-local MSE as:*

$$\mathrm{lMSE}_k(\hat{f} \mid X_t^{(n)}) = \frac{1}{n - k(n)} \sum_{i=k_0(n)}^{n-k_0(n)} \left(\frac{1}{k(n)} \sum_{j=-k_0(n)}^{k_0(n)} X_{t_{i+j}^{(n)}}^{(n)} - \hat{f}(t_{i+j}^{(n)}) \right)^2.$$

If $|f(t)|, |\hat{f}(t)| < C$, $k(n) < n$, and $\mathrm{Var}(\varepsilon_t), \mathbb{E}[\varepsilon_t^4]$ are bounded, then it holds

$$\left| \mathbb{E}[\mathrm{lMSE}_k(\hat{f} \mid X_t^{(n)})] - \mathrm{MSE}(\hat{f} \mid f) \right| \leq \max \left\{ \frac{\sum_{i=1}^{n} \mathrm{Var}\left(\varepsilon_{t_i^{(n)}}\right)}{k(n)(n - k(n))}, \frac{4C^2 \cdot k(n)}{n - k(n)} \right\}.$$

If also $k(n) \to \infty$ and $k(n)/n \to 0$ as $n \to \infty$ then the k-local MSE of the empirical process converges in probability to the MSE of the true signal as $n \to \infty$.

The k-local MSE is essentially the usual MSE except that smoothing is applied to the training data and the model. Indeed, for linear models with non-linear preprocessing, this allows for efficient training. Yet, in our experiments, training with the usual MSE usually leads to better results, and the local MSE is only used for testing. In the next section, we empirically evaluate the different methods to probe consistency. We also check the validity of the arguments presented in Sect. 3.1 using numerical evidence.

4   Numerical Evaluation

We perform numerical evaluations on artificial data to confirm the presented theory experimentally.[1]

---

[1] The experimental code as well as all datasets can be found at https://github.com/ FabianHinder/Remark-on-Dependent-Drift.

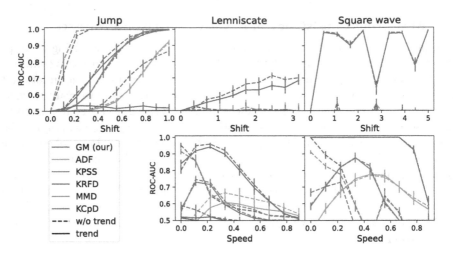

Fig. 4. Mean (and standard deviation) of results over 10×100 runs. x-axis shows intensity of time shift/slow down and y-axis ROC-AUC-Score.

4.1   Testing Stationarity

We start by verifying that stream learning algorithms indeed operate path-wise and thus are not well described using the notion of stationary. To do so we generate paths according to the distribution in Example 1 and Example 3 in a stationary and non-stationary case. We then apply the Kernel Change-point Detector [9] (KCpD) to the resulting paths. In the case of Example 1 the detector determines the value of C (the moment of the jump) with less than 2 samples offset each time, independent of whether the distribution was stationary or not. Similarly, for Example 3 every point is considered a change point. Furthermore, using the sub-sampling process described in Theorem 1 the detector alerts a drift if and only if the initial distribution is not stationary, as predicted by the theorem. We also applied the Augmented Dickey-Fuller [5] (ADF), and the Kwiatkowski-Phillips-Schmidt-Shin [19] (KPSS) test which are used to test for stationarity in time series to the data. The resulting p-values do not differ for the stationary and non-stationary cases for both tests. We thus conclude that the methods do not check for the stationarity of the underlying distribution or rather fail to do so. Instead, they rather consider the properties of the single path. Conducting the same experiment on Example 2 also including supervised drift detectors including ADWIN and DDM [21] lead the same result.

4.2   Evaluation of Method

To evaluate how well existing methods can deal with consistency we consider the following synthetic datasets: "Jump" similar to Example 1, "Lemniscate" $f(t) = C_1 \cos(t) + C_2 \sin(2t)/2$, and "Square wave" where f is given by a square wave. In all cases, we added normal distributed noise and considered a version with an additive linear trend (which is not considered as drift). Drift is induced in the middle of the path. In the case of Jump a jump in the mean value is introduced, for Lemniscate and Square wave we either added a shift in time, i.e., $f(t)$ for $t < 0$, $f(t + \Delta), \Delta > 0$ for $t \geq 0$, or slowed down time, i.e., $f(t)$ for $t < 0$, $f(ct), c \in [0, 1)$ for $t \geq 0$. All with different intensities. We considered ADF, KPSS, and KCpD as well as MMD [7] and KFRD [10] as methods from the literature, where the latter two are provided with the correct change point candidate. ADF and KPSS performed trend corrections. For KCpD we used the number of found change points, for all other methods the returned p-value.[2] As suggested by Lemma 1, we use the lMSE as an indicator for non-consistency using a linear model with (trig-)polynomial (up to degree 15) as preprocessing (GM). For every setup, we performed 1,000 independent runs and evaluated the obtained scores using the ROC-AUC to measure how well the methods separate the drifting and non-drifting paths [14]. The results are presented in Fig. 4. As can be seen, the classical drift detectors KCpD, MMD, and KFRD only work for Jump without trend and in case of an extreme slow down, although KCpD and KFRD were explicitly designed for time series data [9, 10]. The other methods are less affected by the trend. ADF and KPSS perform quite well on Jump and in the cases of slow down. However, none of the methods from the literature can deal with a time shift on any dataset. Our method performs quite well on all datasets. In particular, it is the only one that can deal with time shifts.

5 Conclusion

In this work, we argued that drift and stationarity do not align in many setups realistic for online learning on and monitoring of dependent data. Next to providing theoretical counter-examples, we ran a numerical evaluation confirming our considerations. Besides, we proposed the concept of consistency which is more suitable to the tasks at hand. Yet, it is still unclear how problematic the discrepancy between stationarity and drift is. Also, besides the notion of (first order) stationarity we studied in this paper, the notion of wide-sens, nth-order, and strict stationarity might offer a well-suited choice for some cases. An in-depth investigation of this seems interesting. Furthermore, so far we have only considered the notion of consistency for stochastic processes. A description of the phenomenon of dependent data streams using "dependent drift process" and a suited generalization of the notion of consistency are still outstanding. Additionally, the notion of consistency is very similar to the notion of model drift studied in [13]. There it was shown that such approaches are not suited for monitoring tasks in the independent setting. Though we believe that this does not pose a problem as the proof is heavily founded on the independence assumption, considering this problem in detail appears to be very relevant.

---

[2] Autoregressive models based on Ridge and k-NN regression are considered but only work for Jump. This also rules out supervised detectors that process their losses.

References

1. Adams, T.M., Nobel, A.B.: Uniform convergence of vapnik-chervonenkis classes under ergodic sampling. Ann. Probab. **38**(4), 1345–1367 (2010)
2. Agarwal, A., Duchi, J.C.: The generalization ability of online algorithms for dependent data. IEEE Trans. Inf. Theory **59**(1), 573–587 (2012)
3. Aminikhanghahi, S., Cook, D.J.: A survey of methods for time series change point detection. Knowl. Inf. Syst. **51**(2), 339–367 (2017)
4. Borodin, A.N.: Stochastic Processes. Springer (2017). https://doi.org/10.1007/978-3-319-62310-8
5. Dickey, D., Fuller, W.: Distribution of the estimators for autoregressive time series with a unit root. JASA. J. Am. Stat. Assoc. **74**, 427–431 (1979)
6. Ditzler, G., Roveri, M., Alippi, C., Polikar, R.: Learning in nonstationary environments: a survey. IEEE Comp. Int. Mag. **10**(4), 12–25 (2015)
7. Gretton, A., Borgwardt, K.M., Rasch, M.J., Schölkopf, B., Smola, A.J.: A kernel method for the two-sample-problem. In: NIPS, pp. 513–520 (2006)
8. Hanneke, S., Yang, L.: Statistical learning under nonstationary mixing processes. In: The 22nd International Conference on Artificial Intelligence and Statistics, pp. 1678–1686. PMLR (2019)
9. Harchaoui, Z., Cappé, O.: Retrospective mutiple change-point estimation with kernels. In: 2007 IEEE/SP 14th Workshop on Statistical Signal Processing, pp. 768–772. IEEE (2007)
10. Harchaoui, Z., Moulines, E., Bach, F.: Kernel change-point analysis. In: Koller, D., Schuurmans, D., Bengio, Y., Bottou, L. (eds.) NIPS, vol. 21 (2008)
11. Hinder, F., Artelt, A., Hammer, B.: Towards non-parametric drift detection via dynamic adapting window independence drift detection (DAWIDD). In: International Conference on Machine Learning, pp. 4249–4259. PMLR (2020)
12. Hinder, F., Vaquet, V., Brinkrolf, J., Hammer, B.: On the change of decision boundary and loss in learning with concept drift. In: Crémilleux, B., Hess, S., Nijssen, S. (eds) International Symposium on Intelligent Data Analysis, pp. 182–194. Springer, Cham (2023). https://doi.org/10.1007/978-3-031-30047-9_15
13. Hinder, F., Vaquet, V., Brinkrolf, J., Hammer, B.: On the hardness and necessity of supervised concept drift detection. In: 2th International Conference on Pattern Recognition Applications and Methods (2023)
14. Hinder, F., Vaquet, V., Hammer, B.: Suitability of different metric choices for concept drift detection. In: Bouadi, T., Fromont, E., Hüllermeier, E. (eds.) IDA 2022. LNCS, vol. 13205, pp. 157–170. Springer, Cham (2022). https://doi.org/10.1007/978-3-031-01333-1_13
15. Hinder, F., Vaquet, V., Hammer, B.: One or two things we know about concept drift – a survey on monitoring evolving environments. arXiv preprint arXiv:2310.15826 (2023)
16. Hinder, F., Vaquet, V., Hammer, B.: A remark on concept drift for dependent data. arXiv preprint arXiv:2312.10212 (2023)
17. Kontorovich, L.: Measure Concentration of Strongly Mixing Processes with Applications. Carnegie Mellon University (2007)
18. Krengel, U.: On the speed of convergence in the ergodic theorem. Monatshefte für Mathematik **86**(1), 3–6 (1978)
19. Kwiatkowski, D., Phillips, P.C., Schmidt, P., Shin, Y.: Testing the null hypothesis of stationarity against the alternative of a unit root: how sure are we that economic time series have a unit root? J. Econometrics **54**(1–3), 159–178 (1992)

20. Lim, B., Zohren, S.: Time-series forecasting with deep learning: a survey. Phil. Trans. R. Soc. A **379**(2194), 20200209 (2021)
21. Lu, J., Liu, A., Dong, F., Gu, F., Gama, J., Zhang, G.: Learning under concept drift: a review. IEEE Trans. Knowl. Data Eng. **31**(12), 2346–2363 (2018)
22. Shalev-Shwartz, S., Ben-David, S.: Understanding Machine Learning: From Theory to Algorithms. Cambridge university press (2014)
23. Truong, C., Oudre, L., Vayatis, N.: Selective review of offline change point detection methods. Signal Process. **167**, 107299 (2020)
24. Yu, B.: Rates of convergence for empirical processes of stationary mixing sequences. Ann. Probab. **22**(1), 94–116 (1994)

Representation Learning

Variational Perspective on Fair Edge Prediction

Antoine Gourru[1]([envelope]) [iD], Charlotte Laclau[1,2] [iD], Manvi Choudhary[1],
and Christine Largeron[1] [iD]

[1] Laboratoire Hubert Curien, UMR CNRS 5516, Saint-Etienne, France
{antoine.gourru,charlotte.laclau,manvi.choudhary,
christine.largeron}@univ-st-etienne.fr
[2] Télécom Paris, Institut Polytechnique de Paris, Paris, France
charlotte.laclau@telecom-paris.fr

Abstract. Algorithmic fairness has been of great interest in the machine learning community and more recently in the graph context. In this paper, we address the problem of dyadic fairness where the task at hand is edge prediction, and the population of interest (nodes) is divided into a protected and a non-protected group, e.g. men and women. The goal is then to ensure that there should be no statistically significant difference in the prediction outcomes between the two groups, after accounting for any relevant factors that may impact the outcome. To proceed, we design a novel loss based on the variational information bottleneck principle to learn individual node representation while controlling a given level of dyadic fairness. The optimization of the loss is done with a Graph Neural Network. Experiments carried out on several real-world datasets confirmed the capacity of the proposed method, to maintain high accuracy on the edge prediction task while significantly reducing potential bias.

Keywords: Node embedding · Edge prediction · Fairness

1 Introduction

Machine learning (ML) models are created to optimize a certain notion of performance, modeled as a metric such as accuracy. This measure is intended to represent the overall usefulness for users. However, ML models make decisions without taking into account the side effects or negative impacts they may have on specific groups of users. Due to these biased decisions, certain users are subject to discrimination because of their race, gender, or social background. This bias is rooted in the fact that the models rely heavily on sensitive information that affects the structure of the training data. The field of algorithmic fairness aims to address this bias in ML by either enforcing the independence of ML model predictions conditionally on some specific variables (group fairness) or ensuring that similar users receive similar predictions (individual fairness). Despite the

This work was partially funded by the French National Research Agency (ANR) in the context of the FAMOUS project.

I. Miliou et al. (Eds.): IDA 2024, LNCS 14641, pp. 93–104, 2024.
https://doi.org/10.1007/978-3-031-58547-0_8

popularity of algorithmic fairness in ML, most contributions have focused on tabular data [23], and have overlooked more complex data such as graphs.

In this paper, we address the problem of group fairness when the input data is a graph. We assume that each node is equipped with one sensitive attribute that should bear no consequences on the quality and output of an edge prediction model. This type of fairness, referred to as dyadic fairness [20], entails that the fairness of a prediction regarding two nodes should not be influenced by their sensitive attributes. Let us consider the example of a social network of bloggers, where a node represents a blogger and two nodes are connected if bloggers follow each other. Political inclination can be considered as a protected attribute and such networks present a risk of partisan or political bias. Indeed, an edge prediction model might tend to promote only links between people having the same political ideology, since they are more likely to be connected in the network, which leads to the formation of online bubbles and online polarization. The interest for this theme has increased over the last years, with contributions that propose to either mitigate bias by altering (rewiring) the edge distribution [10,19,20,32,34] or by adding constraints on the node embeddings [5,16,28]. For this latter, we can differentiate between models that are intrinsically modifying Graph Neural Networks (GNN) modules or propagation rules [2,27,35,36] from models that are using adversarial techniques [4,9,12,22] to enforce dyadic fairness and regularization based approaches [6,26]. See [7,11] for a more exhaustive list of contributions.

While these contributions have all proven efficient on numerous benchmark graphs, they suffer from the following shortcomings. First, except for [6,19], existing models do not allow for explicit control of the level of fairness desired in the model. This is of particular importance as legislation generally judges the bias of a model by a minimum fairness score that is deemed acceptable (e.g. the 80% rule from Title VII in the US). Second, contributions dedicated to edge prediction are mostly trained in two independent steps: (1) learning a vector representation of nodes, i.e. the node embeddings and (2) learning an edge prediction model that takes these trained embeddings as input. Fairness is usually injected upstream of the prediction - on the graph directly or in the embeddings. As a result, one can expect that this decoupling will not completely prevent the recurrence of bias in the prediction.

In this paper, we propose a new approach to tackle the aforementioned issues. We design an end-to-end model called LEarning FAir Variational Embedding (LEAVE[1]) for inferring fair node representations trained specifically for the edge prediction task. The proposed objective function allows to control the trade-off between edge prediction accuracy and bias reduction, making our model more resilient to diverse graphs. In summary, the main contributions of this paper are:

- We derive a model from the Variational Information Bottleneck principle (VIB) [3,25] to tackle the problem of edge prediction on graphs. This model is based on a contrastive loss that simultaneously learns accurate node embeddings for edge prediction and mitigates potential bias.

---

[1] see https://github.com/AntoineGourru/leave for code, and experimental details..

- Fairness in our case is defined in two different ways within the same objective function. On the one hand, we aim to minimize an *adversarial* term corresponding to a classifier whose goal is to predict the sensitive attribute, and on the other hand, we explore the potential of learning stochastic representations to mitigate bias through uncertainty.
- We implement this model on two different GNN architectures and demonstrate its utility on several real-world graphs.

The remaining sections of the paper are structured as follows. In Sect. 2, we review related prior works on fairness for graphs. We introduce our model in Sect. 3. Section 4 presents the evaluation results of our model on various datasets. Finally, we conclude this paper in Sect. 5.

2 Related Work

Fairness in the context of graph data is a fast-growing research field [7,11,31]. The first contributions in this field were node embedding models [5,16,28] where the fairness objective was formulated as a problem of learning node embeddings are independent of the sensitive attribute. For instance, FairWalk [28] and Cross-Walk [16] extend the popular Node2vec [13] by proposing fair random walks along the graph. DeBayes [5] on the other hand, extends Conditional Network Embedding (CNE) [15], a Bayesian approach where the sensitive information is modeled as a prior distribution. Since then, other approaches have been proposed with the same idea of altering node embedding learning models, but with a more generic perspective, through regularisation [6] or adversarial filtering like in CFC (Compositional Fairness Constraints) [4]. A key advantage of these latter models is that they remove the constraint of choosing one specific node embedding model. However, all these approaches are task-agnostic, and while they guarantee fairness at the node level (i.e. for node classification), they do not account for fairness when considering pairs of nodes (i.e. dyadic fairness).

A second wave of contributions focused on embedding-agnostic procedures [8,19–21], where the goal is to *repair* or rewire the original graph structure. However, these methods fall into the category of pre-processing methods which guarantee that the input is unbiased but they are not able to control potential bias which can appear while learning the embeddings or solving the task at hand.

Most recent contributions focused on backing up fairness constraints with Graph Neural Networks (GNNs). In [27], the authors perform linear debiasing of node embeddings at training time with a particular neural network component termed, MONET unit. As a result, the embeddings learned with MONET are only fair w.r.t. linear models for the edge prediction task, meaning that any non-linear model used for solving the edge prediction task is able to retrieve part of the bias from the embeddings. Our contribution relies on similar general principles to FairGNN [9], FIPR [6]. FairGNN is an adversarial approach based on Graph Neural Networks (GNN) focusing on the task of fair node classification, therefore dealing with fairness at the individual level. While we also propose to solve an objective function that exhibits similarities with adversarial-type of models, ours LEAVE is dedicated to fairness when one considers tuples of nodes (edge prediction) and we propose to strengthen fairness through uncertainty. In addition, FairGNN is limited to binary sensitive attributes while LEAVE can consider the multi-class case as well. Moreover, in LEAVE, a hyper-parameter allows us to have explicit control over the amount of fairness induced in the model. To the best of our knowledge, only FIPR [6], a fair regularizer based on the I-projection of a graph for edge prediction is comparable with our proposal in that matter and, thus will be considered as the strongest baseline in the experimental evaluation (see Sect. 4).

3   Variational Fairness-Aware Node Embedding

3.1   Problem Set-Up

We consider a network represented by an undirected and unweighted graph $G = (V, E)$, where V is the set of n vertices and E, is the set of m observed edges, e.g. users of a social media and friendship relation. y is a link indicator function defined for each pair of nodes such that $y_{u,v} = 1$ if there is a link between node u and node v and $y_{u,v} = 0$ otherwise. Moreover, we assume the existence of A, a categorical sensitive attribute that assigns a value belonging to $\{0, \cdots, \ell\}$ to each node of the graph. For example, in a social network where the edges represent interactions between individuals, A can be the gender or the continent of origin of each individual represented by a node. We define s, a sensitive attribute indicator for each pair of nodes (u, v) such that $s_{u,v} = 1$ if u and v have similar sensitive attribute values, $s_{u,v} = 0$ otherwise.

3.2   Definition of the Loss

In this subsection, we define our loss function. We rely on the Variational Information Bottleneck principle [3], as defined for dyadic tasks by [25]. We extend the formulation to include a fairness regularization.

We focus on edge prediction, which consists of predicting the existence of an edge between two nodes. Formally, given a pair of nodes (u, v) we aim to predict their link probability expressed as $p(y_{u,v}|u, v)$. Modern machine learning-based methods, e.g. GNN, rely on learning latent representations of nodes that we note z, e.g. $z_u \in \mathbb{R}^d$ is the representation of node u, and Z is the set of all latent representations. Hence, we utilize the latent representations to compute the link probability, $p(y_{u,v}|u, v) = q_\theta(y_{u,v}|z_u, z_v)$, where q_θ is some trainable function. From a utility perspective, this means that we want Z to preserve as much information as possible about the graph structure. As shown by [25], treating z as a random variable following $p(z|x)$ leads to the following likelihood :

$$\mathcal{L}_{util} = \mathbb{E}_{z_u \sim p(z_u|x_u), z_v \sim p(z_v|x_v)} \left[\log q_\theta(y_{u,v}|z_u, z_v) \right] \tag{1}$$

where $q_\theta(y_{u,v}|z_u, z_v)$ is the likelihood score for a pair of nodes parametrized by θ and x_u and x_v are some initial representations of the nodes in the graph (see

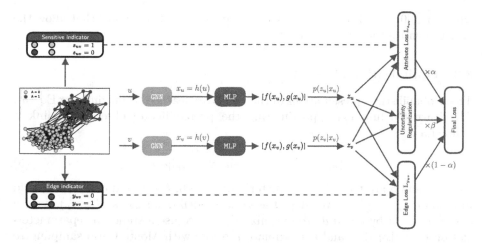

Fig. 1. Architecture of **LEAVE** where the graph encoder is a GNN.

the last paragraph of the Sect. 3.3). We define $p(z|x)$ as a Gaussian of dimension d with diagonal variance, as was done in previous works [25].

$$z|x \sim \mathcal{N}(f(x), g(x)), \tag{2}$$

where f and g are multilayer perceptrons (MLP). We also include a Kullbach-Liebler regularizer

$$\mathcal{L}_{reg} = KL(p(z|x)\|r(z)) \tag{3}$$

where $p(z|x)$ is the latent conditional distribution of z whose parameters are learned, and $r(z)$ is a marginal term fixed to a unit Gaussian distribution $\mathcal{N}(0, I)$ ensuring that the representation learned are compact. It allows the representations to be independent, or at least less dependent from the original representation x. This is of high interest when the initial representations of the nodes are themselves biased. Additionally, it allows to optimize a lower bound of the real likelihood (see [25])

We eventually include a fairness regularizer that quantifies the dependence between the node representations Z and the sensitive attribute s, which we try to minimize. Let $q_{\theta'}(s_{u,v}|z_u, z_v)$ denote the function that approximates the probability of two nodes sharing the same sensitive attribute $(p(s_{u,v}|u,v))$. The resultant fair loss is then determined as follows:

$$\mathcal{L}_{fair} = \mathbb{E}_{z_u \sim p(z_u|x_u), z_v \sim p(z_v|x_v)} \left[\log q_{\theta'}(s_{u,v}|z_u, z_v)\right]. \tag{4}$$

The full fair likelihood becomes:

$$\mathcal{L}_{VIB} = (1 - \alpha)\mathcal{L}_{util} - \alpha\mathcal{L}_{fair} - \beta\mathcal{L}_{reg}, \tag{5}$$

where $\alpha \in [0, 1]$ and $\beta \geq 0$ are two hyperparameters of the model that allow the user to control the level of fairness.

3.3   Optimization of LEAVE

The probabilistic representations of the nodes are learned by maximizing Eq. (5). The learnable function approximating the probability of observing a link is defined as:

$$q_\theta \left(y_{uv} | z_u, z_v \right) := \sigma \left(-a || z_u - z_v ||_2 + b \right) ; \tag{6}$$

where a and b are trainable parameters, s.t. $a > 0$, $b \in \mathbb{R}$ and σ is the sigmoid function $\sigma(t) = \frac{1}{1+e^{-t}}$. We adopt the same modeling for $q_{\theta'} \left(s_{uv} | z_u, z_v \right)$, with a and b replaced by c and d. To optimize $\mathcal{L}_{VIB}$, we use a standard reparameterization of $p(z|x)$ [17], and approximate the loss with Monte Carlo sampling to obtain a smooth gradient flow. We draw $2k$ samples in pairs and epochs.

Fair Negative Sampling. The majority of node embedding approaches are designed to be trained in differentiating between positive pairs (nodes connected to each other) and negative pairs (nodes not connected to each other) within the graph. Additionally, in our case, one needs to account for the sensitive group information to define a negative pair. Indeed, node embeddings must not only distinguish connected pairs from unconnected ones but also keep nodes belonging to the same protected group apart. To proceed, we propose the following method that is well suited to the problem of fairness. Considering a source node, the set of candidate nodes to build a negative pair are (1) not connected to the source node, and (2) have the same sensitive attributes. The target node is drawn from this set, and we add the edge between the source and the random target in the negative example set.

Definition of x LEAVE relies mainly on x, the initial representation of the nodes. These can be hand-created features, adjacency vectors, or pre-trained representations. In this work, we propose to use a third function $x_u = h(u, G)$ whose parameters are trained *simultaneously*. Using this modeling choice makes the model general and independent of the initial modeling of the nodes. In our experiments, we used the GAT architecure [33] as illustrated in Fig. 1 describing the overall architecture of LEAVE.

4   Experiments and Results

We evaluate LEAVE on several benchmark datasets described in Table 1.

In Polblogs [1], nodes represent blogs and edges the hyperlinks between two blogs. We consider the political leaning of blogs to be a sensitive attribute (binary). For LastFM [30], nodes are users and the edges indicate mutual follower relationships. The sensitive attribute is the country of the user. In Facebook

Table 1. Statistics of the graphs. #groups: number of modalities for the sensitive attribute; *dens.*: density of the graph; r: *fair* mixing coefficient.

Dataset	$|V|$	$|E|$	r	*dens.*	S	#groups
Polblogs	1,490	19,090	0.81	$2e^{-2}$	party	2
LastFM	7,624	27,806	0.86	$1e^{-3}$	country	16
Facebook	22,470	171,002	0.82	$7e^{-4}$	page-type	4

[29], nodes correspond to Facebook pages and an edge to mutual likes between the pages. The page type is treated as the sensitive attribute. We also propose to take a closer look at the mixing coefficient with respect to the protected attribute. Mixing in social network analysis allows us to evaluate the tendency of the nodes to connect with others having similar attributes [24]. Assortative mixing is related to homophily and this coefficient has been shown to be a strong indicator of segregation in online and offline networks [14]. We adapt it in the context of dyadic fairness (r in Table 1). The coefficient r lies in $[-1, 1]$, where 1 corresponds to the perfectly assortative case, i.e. nodes with the same value for A exclusively connect with each other; -1 corresponds to the dissortative case. For all three considered benchmarks, we note that r is high, LastFM being the more biased structure in that regard. A direct implication of this remark is that one can expect that in these cases, imposing fairness constraints in the learning process will result in a significant drop in accuracy for the edge prediction.

4.1   Edge Prediction Protocol

We train the models on 70% of the observed edges and used the remaining 10% and 20% for hyperparameter tuning and testing, respectively. For each edge connecting two nodes, we draw one negative example with the procedure described in Sect. 3.

4.2   Evaluation Metrics

We use three common metrics for edge prediction and fairness evaluation so as to cover three distinct aspects. The representation bias focuses on the bias at the node embedding level, the AUC measures the edge prediction performance, and the disparate impact assesses the dyadic fairness at the task level.

Representation Bias (RB). Originally proposed by [4], and then formalized by [5], RB evaluates the bias in the node embeddings, by considering A as the target variable. RB $\in [0, 1]$ and is ideally close to 0.5 meaning that the classifier learned from the node embeddings makes random predictions for the sensitive attribute. RB alone is not sufficient to evaluate fair edge prediction as it focuses on the bias at the node representation level. In our experiments, we use a one-vs-all logistic regression with l2 regularization.

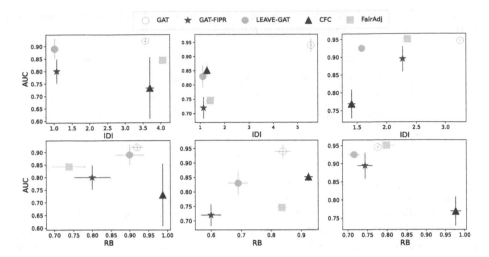

Fig. 2. Results on Polblogs (left column), LastFM (middle), and Facebook (right): AUC vs.IDI on first row, AUC vs. RB. on second row. Horizontal and vertical lines represent the standard deviation along both dimensions. When these lines are not visible, it means that the standard deviation is close to zero. AUC, IDI, and RB are optimal at values of 1, 1, and 0.5 respectively.

Disparate Impact (DI) evaluates the bias at the edge prediction level [19] and is computed as the ratio between the probabilities of predicting an edge between two nodes conditional on the fact that these nodes belong to the same or to different sensitive groups. An ideal DI is equal to 1. For ease of reading, we present the inverse of this value in our experiments, which we note as IDI. The higher the IDI value, the higher the probability of predicting a link between nodes with the same attribute compared to nodes with different attributes (which is the situation most often observed).

Area Under the ROC Curve (AUC) measures the performance of the task at hand, and lies in $[0, 1]$, 1 being the optimal value.

4.3   Baselines for Edge Prediction

We implement our model with a Graph Attention Network [33] as encoding function $h(.)$. We evaluate **LEAVE** and perform an ablation study to highlight the impact of the different terms composing the loss. **GAT** corresponds to the case where α and β are both fixed at 0, and $z = x$ is the output of the encoder without probabilistic aspect. These models focus on link prediction performance (high AUC). **LEAVE-**$_{woVIB}$ is a version in which $\alpha \neq 0$ but $z = x$ (z deterministic and $\beta = 0$). These models focus on finding a trade-off between link prediction performance and fairness (low IDI). **LEAVE** includes both the dyadic fairness regularizer ($\alpha \neq 0$) and the uncertainty term ($\beta \neq 0$). We hope here to strengthen the fairness of both the representations (RB) and the link prediction (IDI).

We experiment with three additional competitors. FairAdj [20], that learns a fair adjacency matrix, and incorporates a Variational Graph Autoencoder [18] for link prediction. We also use the adversarial approach proposed by [4] referred to as Compositional Fairness Constraints (CFC) in the following. Both these methods were implemented in Python[2] by authors of [20]. Finally, we compare our approach with the Fair I-Projection Regularizer FIPRmethod [6] that relies on a regularization term that encourages dyadic fairness. This regularization can be added to any probabilistic network model. This approach was shown to outperform every existing work such as [4, 5, 28]. We have chosen these baselines because they more specifically address the problem of fair link prediction and because they are flexible in the choice of the embedding method. We select each method hyperparameters using grid search on a validation set. Details concerning the hyperparameter tuning are provided in the github[3].

4.4 Analysis of Results

Comparison with Competitors. Figure 2 shows the results obtained on all datasets. We recall that the optimal model should be located in the upper left corner (high AUC (near 1) and low IDI (near 1)/RB (Near 0.5)).

First, and not surprisingly, on Polblogs and LastFM, GAT without fairness awareness obtains better AUCs, but the worst IDIs and RBs. The two objectives, fairness and performance, are therefore antagonists, as shown by the high mixing coefficient.

Second, our model achieves higher or comparable AUCs than its competitors for all three datasets while getting better IDI in most cases. For these two metrics, on LastFM our results are similar to CFC, however, we strongly outperform this latter in terms of RB. It means that even if CFC debias the model to obtain fair edge prediction, using these representations in another pipeline might reproduce or even amplify bias (higher RB than GAT). On the other hand, LEAVE does not suffer from this shortcoming.

Finally, FIPR is our closest and strongest competitor. For IDI, we always outperform or obtain comparable results with this latter while getting a significantly better AUC. For RB, the results are more mixed: on two out of three datasets, FIPR obtains RB closer to the optimal value but the observed difference is proportional to its loss in AUC (around 10%). On Facebook, we obtain both better AUC and more optimal RB.

Impact of VIB. Table 2 shows the differences between LEAVE and LEAVE$_{wov_{IB}}$. By making the representation probabilistic in LEAVE versus LEAVE$_{wov_{IB}}$, we improved fairness while maintaining, or even improving, AUC. This clearly

---

[2] https://github.com/brandeis-machine-learning/FairAdj.
[3] https://github.com/AntoineGourru/leave.

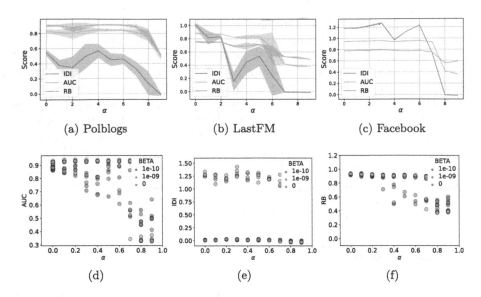

(a) Polblogs   (b) LastFM   (c) Facebook

(d)   (e)   (f)

Fig. 3. First row: impact of α with LEAVE$_{wo_{VIB}}$ for all three benchmark (a to c); second row: impact of the combination of α and β for LEAVE on Polblogs in terms of AUC (d), IDI (e) and RB (f). Each point corresponds to the results obtained with a different random seed. To improve the visualization, IDI is in log-scale, therefore the optimal score is 0.

demonstrates the usefulness of applying this framework to the task. The introduction of a probabilistic dimension enabled us to predict "unexpected" but likely links.

Impact of α and β Finally, Fig. 3 shows the impact of the α alone and of the combination of α and β together. We present the logarithm of the IDI, meaning that the model is fair if it is close to 0. As expected, when α increases, AUC, RB, and IDI decrease, confirming an improvement of fairness at the expense of link prediction performance. Nevertheless, the evolution of IDI is not monotonous, and we observe regions of increase of this measure, but also regions where IDI decreases without disturbing AUC (the ideal solution). Finally, RB and AUC evolve jointly, suggesting that the performances are, on these data and with these encoders, a function of the separability of the representations from the sensitive attribute. A stronger constraint on the representations is, therefore, necessary to debias these architectures. Regarding the impact of β, we observe once again a similar pattern for AUC and RB, where the addition of the uncertainty term attached to β allows us to obtain a smoother and more interesting trade-off between an accurate and a fair representation. For DI, we observe that adding the second regularization term β is actually sufficient to achieve fairness from this point of view, whatever the value of α, as log of IDI is close to 0 when $\alpha = 0$.

Table 2. Impact of using a probabilistic encoder. Average scores, $mean_{std}$, for AUC and IDI. Best results are in bold, and significant difference according to a signed-rank Wilcoxon test (e.g. p-value lower than 0.01) are indicated by a star.

	Polblogs		Lastfm		Facebook	
	AUC	IDI	AUC	IDI	AUC	IDI
LEAVE	**0.89**$_{.04}$	**1.02**$^{*}_{.08}$	**0.83**$_{.04}$	**1.11**$^{*}_{.08}$	0.92$_{.02}$	**1.58**$^{*}_{.02}$
LEAVE$_{woVIB}$	0.87$_{.02}$	1.64$_{.07}$	**0.83**$_{.03}$	1.47$_{.06}$	**0.93**$_{.00}$	2.08$_{.01}$

5 Conclusion

We addressed the problem of fair edge prediction with LEAVE, an end-to-end model optimizing a function that simultaneously learns node embeddings reflecting node similarities in the graph while imposing dyadic-fairness constraint through two regularisation terms. These two terms allow to strengthen fairness while preserving a good prediction performance. In addition, we can explicitly control the trade-off between capturing the relational structure of the graph and reducing the potential bias for edge prediction. Our experimental results confirm the good behavior of LEAVE. Further research perspectives are many, including an extension of our model to the case of continuous sensitive attributes.

References

1. Adamic, L.A., Glance, N.: The political blogosphere and the 2004 us election: divided they blog. In: International Workshop on Link Discovery, pp. 36–43 (2005)
2. Agarwal, C., Lakkaraju, H., Zitnik, M.: Towards a unified framework for fair and stable graph representation learning. In: UAI, pp. 2114–2124. PMLR (2021)
3. Alemi, A., Fischer, I., Dillon, J., Murphy, K.: Deep variational information bottleneck. In: International Conference on Learning Representations (2017)
4. Bose, A., Hamilton, W.: Compositional fairness constraints for graph embeddings. In: ICML, pp. 715–724 (2019)
5. Buyl, M., Bie, T.D.: Debayes: a bayesian method for debiasing network embeddings. In: International Conference on Machine Learning, pp. 2537–2546 (2020)
6. Buyl, M., Bie, T.D.: The kl-divergence between a graph model and its fair i-projection as a fairness regularizer. In: ECML-PKDD, pp. 351–366 (2021)
7. Choudhary, M., Laclau, C., Largeron, C.: A survey on fairness for machine learning on graphs (2022). https://arxiv.org/abs/2205.05396
8. Current, S., He, Y., Gurukar, S., Parthasarathy, S.: Fairegm: fair link prediction and recommendation via emulated graph modification. In: EAAMO (2022)
9. Dai, E., Wang, S.: Say no to the discrimination: learning fair graph neural networks with limited sensitive attribute information. In: WSDM, pp. 680–688 (2021)
10. Dong, Y., Liu, N., Jalaian, B., Li, J.: EDITS: modeling and mitigating data bias for graph neural networks. In: Web Conference, pp. 1259–1269 (2022)
11. Dong, Y., Ma, J., Wang, S., Chen, C., Li, J.: Fairness in graph mining: a survey. IEEE Trans. Knowl. Data Eng. **35**(10), 10583–10602 (2023)

12. Fisher, J., Mittal, A., Palfrey, D., Christodoulopoulos, C.: Debiasing knowledge graph embeddings. In: Proceedings of EMNLP, pp. 7332–7345 (2020)
13. Grover, A., Leskovec, J.: Node2vec: scalable feature learning for networks. In: Proceedings of ACM SIGKDD KDD, pp. 855–864 (2016)
14. Hofstra, B., Corten, R., Van Tubergen, F., Ellison, N.B.: Sources of segregation in social networks: a novel approach using facebook. Am. Sociol. Rev. **82**(3), 625–656 (2017)
15. Kang, B., Lijffijt, J., De Bie, T.: Conditional network embeddings. In: ICLR (2019)
16. Khajehnejad, A., Khajehnejad, M., Babaei, M., Gummadi, K.P., Weller, A., Mirza-soleiman, B.: Crosswalk: fairness-enhanced node representation learning. AAAI **36**(11), 11963–11970 (2022)
17. Kingma, D.P., Welling, M.: Auto-encoding variational bayes. Proceedings of the International Conference on Learning Representations (ICLR) (2014)
18. Kipf, T.N., Welling, M.: Variational graph auto-encoders. arXiv preprint arXiv:1611.07308 (2016)
19. Laclau, C., Redko, I., Choudhary, M., Largeron, C.: All of the fairness for edge prediction with optimal transport. In: AISTATS, pp. 1774–1782. PMLR (2021)
20. Li, P., Wang, Y., Zhao, H., Hong, P., Liu, H.: On dyadic fairness: Exploring and mitigating bias in graph connections. In: ICLR (2021)
21. Li, Y., Wang, X., Ning, Y., Wang, H.: FairLP: towards fair link prediction on social network graphs. Proc. Int. AAAI Conf. Web Soc. Media **16**, 628–639 (2022)
22. Masrour, F., Wilson, T., Yan, H., Tan, P., Esfahanian, A.: Bursting the filter bubble: fairness-aware network link prediction. AAAI **34**(01), 841–848 (2020)
23. Mehrabi, N., Morstatter, F., Saxena, N., Lerman, K., Galstyan, A.: A survey on bias and fairness in machine learning. ACM Comput. Surv. **54**(6), 1–35 (2021)
24. Newman, M.E.J.: Mixing patterns in networks. Phys. Rev. E **67**(2) (2003). https://doi.org/10.1103/physreve.67.026126
25. Oh, S.J., Murphy, K.P., Pan, J., Roth, J., Schroff, F., Gallagher, A.C.: Modeling uncertainty with hedged instance embeddings. In: ICLR (2018)
26. Oneto, L., Navarin, N., Donini, M.: Learning deep fair graph neural networks. In: European Symposium on Artificial Neural Networks, pp. 31–36 (2020)
27. Palowitch, J., Perozzi, B.: Monet: debiasing graph embeddings via the metadata-orthogonal training unit. In: ASONAM (2020)
28. Rahman, T.A., Surma, B., Backes, M., Zhang, Y.: Fairwalk: Towards fair graph embedding. In: IJCAI, pp. 3289–3295 (2019)
29. Rozemberczki, B., Allen, C., Sarkar, R.: Multi-scale attributed node embedding. J. Complex Netw. **9**(2), cnab014 (2021)
30. Rozemberczki, B., Sarkar, R.: Characteristic functions on graphs: birds of a feather, from statistical descriptors to parametric models. In: CIKM, pp. 1325–1334 (2020)
31. Saxena, A., Fletcher, G., Pechenizkiy, M.: Fairsna: algorithmic fairness in social network analysis (2022). https://arxiv.org/abs/2209.01678
32. Spinelli, I., Scardapane, S., Hussain, A., Uncini, A.: Fairdrop: biased edge dropout for enhancing fairness in graph representation learning. In: TAI (2021)
33. Veličković, P., Cucurull, G., Casanova, A., Romero, A., Liò, P., Bengio, Y.: Graph attention networks. In: ICLR (2018)
34. Wang, N., Lin, L., Li, J., Wang, H.: Unbiased graph embedding with biased graph observations. In: Web Conference, pp. 1423–1433 (2022)
35. Wang, Y., Zhao, Y., Dong, Y., Chen, H., Li, J., Derr, T.: Improving fairness in graph neural networks via mitigating sensitive attribute leakage. In: KDD (2022)
36. Zhang, T., et al.: Fairness in graph-based semi-supervised learning. Knowl. Inf. Syst. **65**(2), 543–570 (2022)

Node Classification in Random Trees

Wouter W. L. Nuijten$^{(\boxtimes)}$ ⓘ and Vlado Menkovski ⓘ

Eindhoven University of Technology, Eindhoven, Netherlands
w.w.l.nuijten@tue.nl

Abstract. We propose a method for the classification of objects that are structured as random trees. Our aim is to model a distribution over the node label assignments in settings where the tree data structure is associated with node attributes (typically high dimensional embeddings). The tree topology is not predetermined and none of the label assignments are present during inference. Other methods that produce a distribution over node label assignment in trees (or more generally in graphs) either assume conditional independence of the label assignment, operate on a fixed graph topology, or require part of the node labels to be observed. Our method defines a Markov Network with the corresponding topology of the random tree and an associated Gibbs distribution. We parameterize the Gibbs distribution with a Graph Neural Network that operates on the random tree and the node embeddings. This allows us to estimate the likelihood of node assignments for a given random tree and use MCMC to sample from the distribution of node assignments.

We evaluate our method on the tasks of node classification in trees on the Stanford Sentiment Treebank dataset. Our method outperforms the baselines on this dataset, demonstrating its effectiveness for modeling joint distributions of node labels in random trees.

Keywords: Markov Networks · Node Classification · Graph Neural Networks

1 Introduction

The node classification task is concerned with assigning a label to the nodes of a graph. For example, let us consider a graph where the nodes represent publications and the edges coming out of a node represent the citations of that publication. On this data structure, we can define the task of assigning the topic (from a fixed set of topics) of each publication as a node classification task.

Typically, publications that cite each other more commonly belong to the same topic. Therefore, a topic assignment or a particular node is generally not conditionally independent from the topic assignment of its neighboring nodes.

To achieve this property a model performing this task needs to produce a joint probability distribution over all nodes in the graph. The representation of this distribution grows exponentially in the number of nodes, and modeling this distribution from data quickly faces the curse of dimensionality.

© The Author(s), under exclusive license to Springer Nature Switzerland AG 2024
I. Miliou et al. (Eds.): IDA 2024, LNCS 14641, pp. 105–116, 2024.
https://doi.org/10.1007/978-3-031-58547-0_9

We utilize the topology of the graph to factorize the distribution into conditionally independent groups of variables, studied under the framework of Probabilistic Graphical Models (PGM) [10]. PGMs represent individual random variables as nodes in a graph and the dependency between the two corresponding random variables as an edge. As our data structure is also a graph we can form a direct correspondence between our data and the PGM model, by assigning random variables to each node and utilizing the edges to represent the conditional dependence between the random variables.

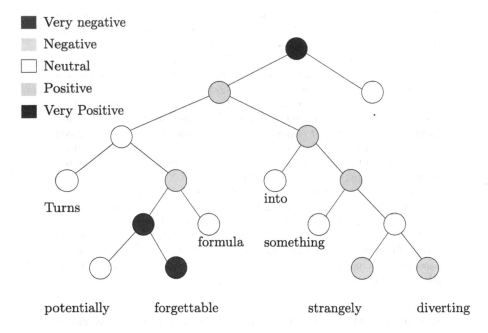

Fig. 1. Example of the node classification task for sentiment analysis. The node labels and their relation encode the structural composition of sentiment.

The task of assigning a topic to the set of publications in a broader sense is the classification of relational data, in which the entities have a number of properties and we know the structure of the relationship between the entities. Therefore, our goal is to develop models that express the relationship between the properties of each entity (or node in the graph representation), the relationships between the entities (or the edges), and the joint distribution over the class assignment to all the nodes in the graph.

For this, a Markov Network with the corresponding graph topology is usually employed to represent the joint distribution over the class assignments. In a Markov Network, every node corresponds to a random variable, and every edge corresponds to a bidirectional dependence.

Typically, the node classification task is solved by finding the marginal distribution over class assignments using a Graph Neural Network (GNN) [6,19], a model that operates on the node properties and their neighborhood in the graph. GNNs have seen considerable success in the node classification task in recent years [9,22].

When the graph structure is predetermined (i.e. all the data points share the same topology) and some of the nodes have assigned labels, determining the joint distribution over the remaining labels has been successfully implemented with a mean-field approximation parameterized by a graph neural network in [17].

In this paper, we consider the case of node classification of a set of graphs with varying topology sampled from an underlying distribution, where none of the labels are known. The mean-field approximation reduces to assuming full independence of the node assignment. This is because in a general mean-field assumption-based method we incorporate the knowledge of known labels into the distribution over unknown labels, while assuming the posterior distribution over unknown labels is factorized into independent components. If there is no knowledge about node labels available, the mean-field variational approximation has no label information to incorporate into the posterior distribution over other node labels and will therefore assume complete independence between all node labels.

To address this, our method, Neural Factor Trees, is able to combine and aggregate local information, as well as utilize the graph topology to factorize a joint probability distribution over node labels in a graph. This factorized distribution can then be sampled to obtain node labels. These samples take the dependencies between node labels into account and give more accurate solutions to the node classification task. Specifically, we produce a Gibbs Distribution that parameterizes a joint distribution over a Markov Network to find the node labels. In this paper, we implement this method to operate on trees. Markov Networks with a tree as structure, as an acyclic graph, have the benefit that inference is tractable and hence provide a way to prove the validity of our method. The method presented in this paper is hypothesized to generalize to graphs of arbitrary topology through approximate inference, however, this is out of scope for this paper.

The main contributions of this paper are as follows:

- Specification of the Node Classification task when a dataset of graphs is given and instances of the problem consist of completely unlabelled graphs (Sect. 3.1).
- Usage of a Graph Neural Network to parameterize a factor set that defines a Gibbs Distribution (Sect. 3).
- Implementation of this method on a sentiment analysis dataset and empirical evidence that this method outperforms methods adopted from semi-supervised Graph Node Classification (Section 4).

2   Related Work

2.1   Learning Probabilistic Graphical Models

Markov Networks are trained through gradient ascent of the model parameters [10] after running approximate inference using some form of belief propagation [13,14,24]. Utilizing neural networks to learn the parameters of Conditional Random Fields (CRF) has also been investigated in [15], and although the authors regard linear chain CRFs as their main field of application, this can be generalized to arbitrary graphs. However, these methods all assume multiple draws from the same Markov Network. Since we are interested in learning a probability distribution where we witness every graph in the training set only once, the aforementioned methods can not be applied in our problem setting. However, the gradient ascent-based learning of these models inspires the idea of combining learning PGMs with other forms of gradient-based learning.

2.2   Node Classification Using Graph Neural Networks

In recent years, Graph Neural Networks have achieved significant success in semi-supervised graph node classification [2,9,22]. Graph Neural Networks are capable of learning high-dimensional representations of nodes [7]. These high-dimensional node embeddings are obtained by aggregating local information between nodes to obtain and incorporate neighborhood information in node embeddings. The process of emitting and aggregating local information is called message passing [5]. Because they can be trained end-to-end with backpropagation [18], Graph Neural Networks can achieve state-of-the-art performance on many problems defined on graphs, such as semi-supervised node classification on the CORA [20] and Site-Seer [4] data sets. However, in their original implementation, message-passing algorithms performing graph node classification ignore dependencies between node labels because these methods assume independence between node labels. Nevertheless, the ability of Graph Neural Networks to learn high-dimensional node embeddings is a powerful property that we will leverage in order to model structural dependencies in the joint distribution.

There are methods that attempt to leverage the dependencies between labels in the graph [23]. However, these methods either involve a variant of the Label Propagation algorithm [25] or make assumptions about the correlation between label classes [12]. In our case, the Label Propagation algorithm is not applicable since in our problem setting there are no known labels to propagate. A promising method that does not assume prior knowledge about the correlations between the labels of neighboring vertices is the Graph Markov Neural Network method [17]. Here, a Graph Neural Network is used to find a Gibbs distribution that factorizes the distribution over a Conditional Random Field [17]. The authors propagate known node labels along with node features through the graph and use this to infer the unknown node labels, assuming a mean-field factorization over the unknown labels. The usage of Conditional Random Fields in [17] was the inspiration to model Graph Node Classification as a Conditional Random

Field in this paper, although the mean-field assumption made in this paper is less relevant in our problem setting. Nevertheless, the methods presented in [17] compute a local conditional distribution of node labels, and with that, it is the method from the state-of-the-art that is most closely related to our problem formulation.

All things considered, there are multiple methods in the current literature that solve the (semi-supervised) node classification task. In our problem setting of having a training dataset of completely labeled graphs and having to classify previously unseen graphs, however, none of these methods is designed for this problem formulation. In the method section, we present a method that uses a Graph Neural Network to parameterize a Gibbs Distribution to solve this problem.

3 Method

3.1 Problem Formulation

In this paper, we study the node classification task in trees. This entails that we are given a dataset $\mathcal{D}$ of trees $\mathcal{G}_i = (V_{\mathcal{G}_i}, E_{\mathcal{G}_i})$ with node set $V_{\mathcal{G}_i}$ and edge set $E_{\mathcal{G}_i}$. Each of the nodes $v \in V_{\mathcal{G}_i}$ in a graph has node attributes x_v and a node label y_v. The attribute set for all nodes in a tree is denoted by $x_{V_{\mathcal{G}_i}}$, and all node labels in a graph are written as $y_{V_{\mathcal{G}_i}}$. To reduce notational clutter, we will define the problem for a single graph and hence refer to x_v, x_V, y_v, and y_V. The problem is to infer y_V given x_V and the tree topology. In other words, our goal is to model the probability distribution $p(y_V \mid x_V, \mathcal{G}_i)$.

In general, we can model the joint distribution over the node labels of a tree with a Gibbs Distribution that factorizes over a Markov Network. A Gibbs Distribution is defined as a distribution P_Φ over a multivariate random variable $\mathbf{y} = \{y_1, y_2, \cdots, y_n\}$ parameterized by a set of factors $\Phi = \{\phi_1(\mathbf{D}_1), \phi_2(\mathbf{D}_2), \cdots, \phi_m(\mathbf{D}_m)\}$ if it is defined as follows:

$$P_\Phi(y_1, y_2, \cdots, y_n) = \frac{1}{Z} \tilde{P}_\Phi(y_1, y_2, \cdots, y_n)$$

We call $\tilde{P}_\Phi(y_1, y_2, \cdots, y_n)$ the unnormalized Gibbs Measure, defined as:

$$\tilde{P}_\Phi(y_1, y_2, \cdots, y_n) = \phi_1(\mathbf{D}_1) \times \phi_2(\mathbf{D}_2) \times \cdots \times \phi_m(\mathbf{D}_m)$$

and

$$Z = \sum_{y_1, y_2, \cdots, y_n} \tilde{P}_\Phi(y_1, y_2, \cdots, y_n)$$

the normalization constant or partition function. The sets $\mathbf{D}$ in the case of a Markov Network are cliques in the graph, where $\bigcup_i \mathbf{D}_i = V$. In trees, this means that the sets $\mathbf{D}$ are the endpoints of the edges in the trees.

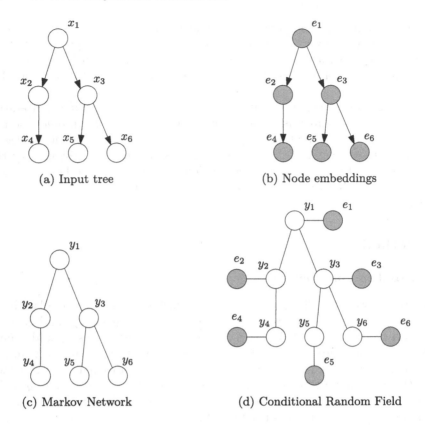

(a) Input tree   (b) Node embeddings

(c) Markov Network   (d) Conditional Random Field

Fig. 2. Different graphs utilized in the construction of our method. In 2a we see an instance of the node classification problem, and in 2c, we see the associated Markov Network with this instance. A Graph Neural Network takes an instance of the problem and produces high dimensional node embeddings (2b). These node embeddings are combined with the Markov Network to form a Conditional Random Field (2d).

3.2 Approach

The challenge in learning how to parameterize the Gibbs Distribution is that the topology of the input tree in our setting is not fixed, so we cannot directly learn the factor functions. In our problem setting, the data-generating model generates node attributes along with a graph topology. This is a significant challenge in learning a Gibbs distribution in this problem setting, as the distribution varies in functional form between data points. To deal with this we develop the following method. This method is based on the idea that the Gibbs distribution over node labels is fully specified by the factor functions, so if we can estimate the factor set that best describes the underlying data generating probability distribution, we are able to solve instances of the node classification problem and obtain a more expressive distribution than if we conduct independent classification for all nodes. Therefore, we want to learn a function that takes the topology of the graph and node attributes and outputs a Gibbs Distribution that defines a joint probability distribution over node labels. Because this function has to learn the generative model that generates node embeddings and the graph topology, as well as perform Bayesian inference in this generative model, we use a Graph Neural Network (GNN) as a universal function approximator that can learn this function. To illustrate our method we also provide a visual representation of an example tree with 6 nodes in Fig. 2.

In our method, which is called Neural Factor Trees, we construct a Markov Network with the same topology as the input tree (Fig. 2a and 2c), where the individual nodes correspond to the random variables of the Markov Network. We then use a GNN model to develop a node representation based on the node properties of the tree and its neighborhood (Fig. 2a and 2b) These high dimensional node embeddings are referred to as e_V. We construct a CRF where each random variable is associated with an observed variable that carries the embedding produced by the GNN (Figure 2d). The GNN is a key component, as it operates on graphs with various topologies and it can learn to produce a node representation that takes into account the node properties and the node's neighborhood, including the properties of the neighbors and the topology of the neighborhood. To specify the joint distribution over the random variables we represent the CRF as a factor graph (Fig. 3).

We use a linear Neural Network layer to compute the node factors from the node embeddings. The edge factors are produced by concatenating the embeddings of the endpoint nodes and passing this vector through a linear Neural Network layer. The embedding of the parent node here comes first, in order to keep a consistent permutation scheme.

3.3 GNN Design

For the GNN model, we use a Message Passing Neural Network (MPNN) and specifically the Gated Graph Neural Networks model [11]. In this algorithm, we use a Multilayer Perceptron to calculate the messages emitted by all vertices. Note that the official algorithm in [11] restricts this Multilayer Perceptron to a single layer, whereas we do not impose this constraint. We found that using a transfer function with 2 layers in the Multilayer Perceptron provided us with a slight improvement in our model evidence. This design choice was originally made to make our GNN unit more expressive. In every node, we collect and sum incoming messages and update node representations using a Gated Recurrent Unit [1]. The update rules are as follows:

$$h_V^0 = [\mathbf{X}_V \parallel \mathbf{0}]$$
$$a_i^t = \sum_{j \in \mathcal{N}(i)} f_\theta(h_j^t) \tag{1}$$
$$h_i^{t+1} = \text{GRU}(a_i^t, h_i^t)$$

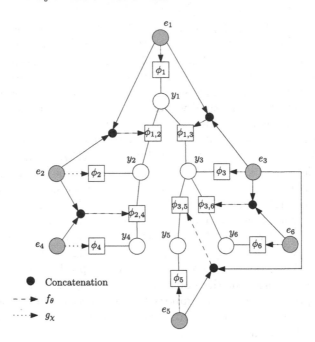

Fig. 3. Factor graph construction. The node embeddings e determine the node and edge factors ϕ through linear Neural Networks. This fully specifies a Gibbs Distribution over the resulting Markov Network. With d being the number of classes a node can have and $|e|$ the size of the node embedding, $f_\theta : \mathbb{R}^{|e|+|e|} \to \mathbb{R}^{d \times d}$ is a function with trainable parameters θ that determines the edge factors from the embeddings. $g_\chi : \mathbb{R}^{|e|} \to \mathbb{R}^d$ is a function with trainable parameters χ that determines the node factors from the embeddings.

Here, $||$ is the concatenation operator. Because we use a Gated Recurrent Unit (GRU) [1] as an update function, every vertex is able to distill relevant information from incoming messages and update its representation accordingly.

As this method only utilized Gated Graph Neural Network and linear layers, the model is fully differentiable with respect to its parameters. As is common in training Probabilistic Graphical Models, we optimize the log-likelihood of the produced distribution given a set of samples from the true distribution $\mathcal{D}$. The log-likelihood is a function of the produced factor set Φ, which in our case is determined by the model $p_\psi (\Phi \mid \mathbf{G}_{\mathcal{D}_i}, \mathbf{X}_{\mathcal{D}_i})$. Here, to avoid visual clutter, we let Φ_ψ be shorthand for $p_\psi (\Phi \mid \mathbf{G}_{\mathcal{D}_i}, \mathbf{X}_{\mathcal{D}_i})$, and Z_{Φ_ψ} for the normalization constant of the Gibbs Distribution produced by Φ_ψ.

$$\mathcal{L}(\psi : \mathcal{D}) = \sum_{i=1}^{|\mathcal{D}|} \sum_{\phi \in \Phi_\psi} (\log \phi(\mathbf{D}_\phi)) - \log Z_{\Phi_\psi} \tag{2}$$

Here, ψ is the set of all trainable parameters of the model, including θ and χ in Fig. 3. As we have trees in our problem statement, the normalization constant is tractable to compute [10]. With this setup, we have defined a fully differentiable method and an adequate quality metric with which to train our method.

3.4 Classifying Nodes

With the produced Gibbs Distribution we can construct solutions to the node classification problem. In order to do this we utilize a Markov Chain Monte Carlo (MCMC) [8] algorithm to produce samples from the produced probability distribution. We utilize the Gibbs Sampling [3] algorithm. The advantage of the Gibbs Sampling algorithm is that it produces unbiased samples from the distribution on which it operates. The disadvantage of Gibbs Sampling, as with most MCMC algorithms, is that it is computationally expensive to run. Alternatively, because we only consider trees in this paper, we can use the Max-Product algorithm to obtain the mode of the Gibbs Distribution. However, since this research is inspired by a desire to model the full distribution over node labels and is therefore also interested in other probable assignments of node labels, we use the Gibbs Sampling algorithm to obtain a histogram of probable node label assignments.

4 Evaluation

4.1 Dataset

The Stanford Sentiment Treebank (SST) [21] dataset is a dataset of fully labeled parse trees of sentences from movie reviews. Each parse tree consists of a single sentence and every node of the parse tree is labeled by its sentiment (Fig. 1). The combination of nodes in the tree allows for the analysis of compositional relations of sentiment in language. The dataset consists of 11855 fully labeled sentences, where every node gets a sentiment score out of 5 categories: very negative, negative, neutral, positive, and very positive. The task is to classify all nodes in each tree. Due to the compositional relations of sentiment in these sentences, we cannot assume conditional independence of the label assignment.

We use GloVe [16] word embeddings as node attributes x_v. The unlabeled version of this graph corresponds to Fig. 2a of our method. We use GloVe embeddings to isolate the performance of our method and not rely on the word embeddings we learn as a by-product of our method. The GloVe embeddings are not to be confused with the node embeddings our method produces after a pass through its GNN.

4.2 Experiments

The performance of our method can be evaluated in multiple ways; a theoretical and an empirical performance metric. Since we can compute the log-likelihood of a label configuration for the joint probability distribution, we have access to a theoretical performance metric. Finding the maximum likelihood estimate

Table 1. Results on the SST Dataset

	GNN	GMNN	Neural Factor Trees
Log-likelihood	−0.522	−0.515	**−0.469**
Accuracy	73.9%	74.0%	**78.3%**

of a Gibbs Distribution in trees can be done with the Max-Product algorithm. Furthermore, we can sample the produced Gibbs Distribution and provide the empirical mode of the sample distribution. This provides us with a maximum likelihood estimate of the label assignments of the tree under our produced Gibbs Distribution and we are therefore able to compute the accuracy of our method.

We test the performance of our method[1] against a base-line Graph Neural Network model that assumes full independence of the class label of the nodes, as well as a Graph Markov Neural Network (GMNN) [17], which uses a mean-field approximation of the distribution over node labels. In the related work GMNN has been identified as a state-of-the-art method for semi-supervised graph node classification that is closely related to our problem formulation. However, since there is no information about node labels available during inference, we expect this mean-field approximation to reduce to independent classification. Nevertheless, because of its effectiveness in computing local conditional distributions, we use GMNN as a baseline method to test our method against. A disadvantage of our method compared to GMNN is that the maximum a posteriori assignment of labels under the posterior distribution in GMNN is easily found, as it is the node-wise maximum a posteriori estimate of the categorical distributions in every node. For our method, we have to employ the Max-Product algorithm to find the maximum a posteriori estimate or report the mode of the empirical dataset sampled from the Gibbs Distribution, which is significantly more computationally expensive.

4.3   Results

In Table 1 we display the results of our method on the SST dataset. The log-likelihoods presented in this table are aggregated over the number of nodes in a tree, as a large tree would otherwise have a smaller log-likelihood due to the individual contributions by the nodes. We see that the data has a higher log-likelihood under the distributions we produce than results from both independent classification using Graph Neural Networks and GMNN methods, which means that the learned distributions better represent the dataset. We also see this reflected in the accuracy scores. An interesting detail is that GMNN does not outperform independent classification by a significant margin, supporting our

---

[1] The code for these experiments is available on Github (https://github.com/wouterwln/NeuralFactorTrees).

hypothesis that the mean-field assumption reduces to independent classification when none of the labels are known.

5 Conclusion

In this paper, we described a novel way of addressing local dependencies in graph labels when conducting random tree node classification. We have developed the Neural Factor Trees method; a method that is able to take the topological properties of the graph into consideration and can produce a joint probability distribution over node labels. Our method can be trained end-to-end using backpropagation. We show that our method is able to outperform existing methods on the SST Dataset.

The method described in this paper operates on trees, and this property is leveraged in the computation of the log-likelihood. The fact that the graphs in the dataset are trees ensures the computation of the normalization constant is tractable in linear time. If this tree property is dropped, we are no longer guaranteed to be able to compute the normalization constant in polynomial time. This is a known problem in learning Probabilistic Graphical Models and might also be a hurdle when generalizing our method to non-tree graphs. We, therefore, identify this as the main limitation of our method, as this tree property is not always satisfied. Furthermore, we only model the dependencies between node labels of nodes that are connected in the input tree. However, these (conditional) dependencies might exist for more than 1 step in the graph. With our current method, we are not able to model these interactions, instead depending on the dependency between labels of two distant nodes flowing over the path between these nodes in the graph.

An extension of this work might investigate an alleviation of these drawbacks and might investigate the performance of our method under approximate inference (e.g. belief propagation) for the computation of the partition function. This would extend our method to operate on any graph instead of trees, and the interesting research avenue is how our method performs under different forms of approximate inference.

References

1. Cho, K., Van Merriënboer, B., Bahdanau, D., Bengio, Y.: On the properties of neural machine translation: encoder-decoder approaches. arXiv preprint arXiv:1409.1259 (2014)
2. Defferrard, M., Bresson, X., Vandergheynst, P.: Convolutional neural networks on graphs with fast localized spectral filtering. Adv. Neural Inf. Process. Syst. **29** (2016)
3. Geman, S., Geman, D.: Stochastic relaxation, gibbs distributions, and the bayesian restoration of images. IEEE Trans. Pattern Anal. Mach. Intell. **6**, 721–741 (1984)
4. Getoor, L.: Link-based classification. In: Advanced Methods for Knowledge Discovery from Complex Data, pp. 189–207. Springer, New York (2005). https://doi.org/10.1007/1-84628-284-5_7

5. Gilmer, J., Schoenholz, S.S., Riley, P.F., Vinyals, O., Dahl, G.E.: Neural message passing for quantum chemistry. In: International Conference on Machine Learning, pp. 1263–1272. PMLR (2017)
6. Gori, M., Monfardini, G., Scarselli, F.: A new model for learning in graph domains. In: Proceedings of the 2005 IEEE International Joint Conference on Neural Networks, 2005, vol. 2, pp. 729–734. IEEE (2005)
7. Hamilton, W.L., Ying, R., Leskovec, J.: Inductive representation learning on large graphs. In: Proceedings of the 31st International Conference on Neural Information Processing Systems, pp. 1025–1035 (2017)
8. Hastings, W.K.: Monte Carlo sampling methods using Markov chains and their applications. Biometrika **57**(1), 97 (1970)
9. Kipf, T.N., Welling, M.: Semi-supervised classification with graph convolutional networks (2017)
10. Koller, D., Friedman, N.: Probabilistic Graphical Models: Principles and Techniques. MIT Press (2009)
11. Li, Y., Tarlow, D., Brockschmidt, M., Zemel, R.: Gated graph sequence neural networks. arXiv preprint arXiv:1511.05493 (2015)
12. Luan, S., et al.: Is heterophily a real nightmare for graph neural networks to do node classification? arXiv preprint arXiv:2109.05641 (2021)
13. Murphy, K.P., Weiss, Y., Jordan, M.I.: Loopy belief propagation for approximate inference: an empirical study. arXiv preprint arXiv:1301.6725 (2013)
14. Pearl, J.: Reverend Bayes on inference engines: a distributed hierarchical approach. School of Engineering and Applied Science, Cognitive Systems Laboratory (1982)
15. Peng, J., Bo, L., Xu, J.: Conditional neural fields. Adv. Neural. Inf. Process. Syst. **22**, 1419–1427 (2009)
16. Pennington, J., Socher, R., Manning, C.D.: Glove: global vectors for word representation. In: Proceedings of the 2014 Conference on Empirical Methods in Natural Language Processing (EMNLP), pp. 1532–1543 (2014)
17. Qu, M., Bengio, Y., Tang, J.: Gmnn: graph Markov neural networks (2020)
18. Rumelhart, D.E., Hinton, G.E., Williams, R.J.: Learning representations by back-propagating errors. Nature **323**(6088), 533–536 (1986)
19. Scarselli, F., Gori, M., Tsoi, A.C., Hagenbuchner, M., Monfardini, G.: The graph neural network model. IEEE Trans. Neural Netw. **20**(1), 61–80 (2008)
20. Sen, P., Namata, G., Bilgic, M., Getoor, L., Galligher, B., Eliassi-Rad, T.: Collective classification in network data. AI Mag. **29**(3), 93–93 (2008)
21. Socher, R., et al.: Recursive deep models for semantic compositionality over a sentiment treebank. In: Proceedings of the 2013 Conference on Empirical Methods in Natural Language Processing, pp. 1631–1642 (2013)
22. Velickovic, P., Cucurull, G., Casanova, A., Romero, A., Liò, P., Bengio, Y.: Graph attention networks (2018)
23. Wang, H., Leskovec, J.: Unifying graph convolutional neural networks and label propagation. arXiv preprint arXiv:2002.06755 (2020)
24. Zhang, Z., Wu, F., Lee, W.S.: Factor graph neural networks. Adv. Neural. Inf. Process. Syst. **33**, 8577–8587 (2020)
25. Zhu, X.: Semi-supervised Learning with Graphs. Carnegie Mellon University (2005)

Self-supervised Siamese Autoencoders

Friederike Baier[1], Sebastian Mair[2], and Samuel G. Fadel[3]([figure])

[1] Leuphana University of Lüneburg, Lüneburg, Germany
[2] Uppsala University, Uppsala, Sweden
sebastian.mair@it.uu.se
[3] Linköping University, Linköping, Sweden
samuel@nihil.ws

Abstract. In contrast to fully-supervised models, self-supervised representation learning only needs a fraction of data to be labeled and often achieves the same or even higher downstream performance. The goal is to pre-train deep neural networks on a self-supervised task, making them able to extract meaningful features from raw input data afterwards. Previously, autoencoders and Siamese networks have been successfully employed as feature extractors for tasks such as image classification. However, both have their individual shortcomings and benefits. In this paper, we combine their complementary strengths by proposing a new method called SidAE (Siamese denoising autoencoder). Using an image classification downstream task, we show that our model outperforms two self-supervised baselines across multiple data sets and scenarios. Crucially, this includes conditions in which only a small amount of labeled data is available. Empirically, the Siamese component has more impact, but the denoising autoencoder is nevertheless necessary to improve performance.

Keywords: Self-supervised learning · representation learning · Siamese networks · denoising autoencoder · pre-training · image classification

1 Introduction

Fully-supervised machine learning models usually require large amounts of labeled training data to achieve state-of-the-art performance. For many domains, however, labeled training data is often costly and more challenging to acquire than unlabeled data. Existing unsupervised or self-supervised pre-training approaches successfully reduce the amount of labeled training data needed for achieving the same or even higher performance [13]. Generally speaking, the aim is to pre-train deep neural networks on a self-supervised task such that after pre-training, they can extract meaningful features from raw data, which can then be used in a so-called *downstream task* like classification or object detection.

In the context of image recognition, earlier work focuses on designing specific pretext tasks. These include, e.g., generation-based methods such as inpainting

I. Miliou et al. (Eds.): IDA 2024, LNCS 14641, pp. 117–128, 2024.
https://doi.org/10.1007/978-3-031-58547-0_10

[23], colorization [29], and image generation (e.g., with generative adversarial networks [8]), context-based methods involving Jigsaw puzzles [20], predicting the rotation angle of an image [7], or relative position prediction on patch level [6]. Yet, resulting representations are rather specific to these pretext tasks, suggesting more general semantically meaningful representations should be invariant to certain transformations instead of covariant [3,19].

Following this argument, many recent state-of-the-art models have been built on Siamese networks [3,4,9,10]. The idea is to make the model learn that two different versions of one entity belong to the same entity and that the factors making the versions differ from each other do not play a role in its identification. The simple Siamese (SimSiam) model [4] is designed to solve the common representation collapse issue arising from that strategy.

Another family of models used for pre-training neural networks, acting as feature extractors, are autoencoders. They build on the principle of maximizing the mutual information between the input and the latent representation [26]. However, good features should not contain as much information as possible about the input, but only the most relevant parts [24]. This is commonly done by keeping the dimensionality of the latent space lower than that of the input and by adding noise to the image. This is the idea of a denoising autoencoder [25].

By themselves, both approaches to learning representations lead to simple and effective solutions. This simplicity allows us to leverage both approaches to design an arguably even more powerful feature extractor, as both strategies result in non-conflicting ways of learning representations.

In this paper, we propose the combination of a Siamese network and a denoising autoencoder to create a new model for self-supervised representation learning that comprises advantages of both and could therefore compensate for the shortcomings of the individual components. We introduce a new model named SidAE (Siamese denoising autoencoder), which aims to adopt the powerful learning principles of both Siamese networks with multiple views of the sample input and denoising autoencoders with noise tolerance. Our experiments show that SidAE outperforms its composing parts in downstream classification in a variety of scenarios, either where all labeled data is available for the downstream task or only a fraction of it, both with or without fine-tuning of the feature extractor.

2   Self-supervised Representation Learning

Self-supervised learning leverages vast amounts of unlabeled data. The key is to define a loss function using information that is extracted from the input itself. Generally, self-supervision can be used in various domains, including text, speech, and video. In this paper, we focus on the domain of image recognition.

We consider two stages: (i) self-supervised pre-training and (ii) a supervised downstream task. The goal of the pre-training stage is to train a so-called *encoder* neural network which learns to extract meaningful information from raw input signals. In the second stage, we utilize the encoder as a feature extractor. Then, a classifier, e.g., a simple fully-connected layer, is trained on top of the representations extracted by the encoder using a supervised loss and labeled data.

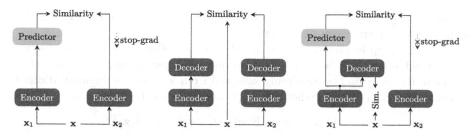

Fig. 1. Siamese networks (left) and (denoising) autoencoders (middle), compared to our proposed model SidAE (right). The left and right illustrations should be symmetric, but we show only one side of the symmetry for brevity. The middle illustration shows two views due to our experimental setting (for a fair comparison), although a vanilla denoising autoencoder usually has only one.

The weights of the encoder can be fine-tuned (optimized) or kept frozen (not optimized) for the downstream task.

The main challenges are determining which features are meaningful and how to tweak models to produce representations that satisfy certain properties. It has been suggested that representations should contain as much information about the input as possible (InfoMax principle; [17]). Yet, others argue that task-specific representations should only contain as much signal as needed for solving a specific task, as task-unrelated information could be considered noise and might even harm performance (InfoMin principle; [24]). In general, representations should be invariant under certain transformations, i.e., so-called *nuisance factors*, such as lighting, color grading, and orientation, as long as those factors do not change the identity of an object. One line of reasoning is that a model that can predict masked parts or infer properties of the data should have some kind of *higher-level understanding* of the content [1,12].

Siamese Networks. Siamese (also known as joint-embedding [1]) architectures are designed in a way that a so-called *backbone network* should learn that two distorted versions of an image still represent the same object. Intuitively, the reasoning is that the features extracted from the same entity should be similar to each other, even under those distortions.

Our proposed model builds on SimSiam [4]. Due to its simplicity compared to other Siamese architectures, it is well-suited as a basis for more complex architectures. In general, Siamese networks consist of two branches encoding two different *views*, x_1 and x_2, of the same input x, which we detail below. Each branch in SimSiam contains both an encoder and a so-called *predictor network*, where the encoders for the two branches share their weights. In other approaches such as in MoCo [10] and BYOL [9], the weights of one network are a lagging moving average of the other. Figure 1 (left) illustrates the architecture of SimSiam, showing the stop-gradient operation only for one side for brevity.

In SimSiam and other state-of-the-art models, the encoder comprises a neural network, often referred to as *backbone*. In some cases, there is also an additional so-called *projector network*. A projector is also used in SimCLR [3], where it improves the downstream performance.

The first step in pre-training is to extract two different *views* of an (vectorized) input image $\mathbf{x} \in \mathbb{R}^{d_{in}}$. These views are obtained by applying two sets of augmentations, t_1 and t_2, randomly sampled from a family of augmentations T. The encoder $\mathrm{Enc}(\cdot)$ and predictor $\mathrm{Pred}(\cdot)$ networks are then used as

$$\mathbf{x}_j = t_j(\mathbf{x}), \quad \mathbf{z}_j = \mathrm{Enc}(\mathbf{x}_j), \quad \mathbf{p}_j = \mathrm{Pred}(\mathbf{z}_j),$$

where $\mathbf{x} \in \mathbb{R}^{d_{in}}, \mathbf{x}_j \in \mathbb{R}^{d_{in}}, \mathbf{z}_j \in \mathbb{R}^{d_{hid}}, \mathbf{p}_j \in \mathbb{R}^{d_{hid}}$. Here, d_{in} is the input dimensionality, d_{hid} the hidden or latent dimensionality, and $j \in \{1, 2\}$. Using $d_{hid} = \frac{1}{4}d_{in}$ was found to lead to better results when compared to $d_{hid} = d_{in}$ [4].

With this, the model can be trained by minimizing a distance $D(\mathbf{p}_i, \mathbf{z}_j)$ with $i, j \in \{1, 2\}$ and $i \neq j$. Hence, the loss is minimized by forcing the representations to be close to each other. Both negative cosine similarity and cross-entropy can be used as $D(\cdot, \cdot)$, but earlier reports show that the former leads to superior results [4]. Crucially, a stop-gradient operation is applied on $\mathbf{z}_j$. Thus, while backpropagating through one of the branches, the other branch is treated as a fixed transformation. More specifically, the loss for SimSiam is calculated as

$$L_{\mathrm{si}} = \frac{1}{2}D(\mathbf{p}_1, \mathrm{stopgrad}(\mathbf{z}_2)) + \frac{1}{2}D(\mathbf{p}_2, \mathrm{stopgrad}(\mathbf{z}_1)), \tag{1}$$

where $\mathrm{stopgrad}(\cdot)$ denotes the stop-gradient operation. This is essential to avoid collapsing solutions [4] and works as a replacement for earlier attempts, such as using one branch with weights as a lagging moving average of the other [9,10].

Denoising Autoencoders. Autoencoders are another family of self-supervised models and build on the concept of maximizing the mutual information between some input $\mathbf{x}$ and corresponding latent code $\mathbf{z}$ [26]. Yet, they need to be restricted in some way such that the model does not simply learn an identity mapping where $\mathbf{z}$ is equal to $\mathbf{x}$, usually achieved by using a smaller dimensionality for $\mathbf{z}$. For a denoising autoencoder, the input $\mathbf{x}$ is distorted, e.g., by applying Gaussian noise, and this corrupted version $\tilde{\mathbf{x}}$ is used as input to the model $\mathbf{x}' = \mathrm{Dec}(\mathrm{Enc}(\tilde{\mathbf{x}}))$. The loss $L_{\mathrm{dae}} = D(\mathbf{x}', \mathbf{x})$ is minimized to make $\mathbf{x}$ and $\mathbf{x}'$ as similar as possible. Note that the denoising autoencoder does not see $\mathbf{x}$, only its corrupted version $\tilde{\mathbf{x}}$, but still aims to reconstruct $\mathbf{x}$. Hence, the model should learn to focus more on the crucial information to reconstruct the original inputs, and optimally, the representations encode only relevant signals without noise.

3   A Siamese Denoising Autoencoder

The newly proposed model contains parts of SimSiam [4] and a denoising autoencoder. The design of the denoising autoencoder was inspired by works of [16]. To the best of our knowledge, the exact architecture and application as shown here have not yet been presented. Note that components that appear in multiple models (i.e., the encoder, decoder, and predictor networks) are the same in SimSiam, SidAE, and the denoising autoencoder.

3.1 Motivation

Usually, a denoising autoencoder is trained on a single corrupted view of the data, while Siamese networks use two views. Additionally, Siamese networks do not aim to reconstruct the original inputs from the information its encoder extracted, while the denoising autoencoder does. We hypothesize combining both approaches results in a more powerful method since they compensate for the shortcomings of each other. This difference in behavior can be used to the benefit of self-supervision, i.e., allowing the denoising autoencoder to access two different views of the same sample. Additionally, the encoder used by the Siamese part is encouraged to keep enough information for the decoder to reconstruct the inputs. This symbiosis is the fundamental design principle in SidAE.

Intuitively, in contrast to pre-training a network with SimSiam, using SidAE should make the network learn that two views come from the same object. In SimSiam, the original input is not directly used to optimize the model, as only augmented versions are fed into the model, and the loss is computed using representations in the latent space. In SidAE, however, disrupted local information is mapped to the original, undisrupted global information by the denoising autoencoder part. Hence, the model is explicitly encouraged to encode information about the original input $\mathbf{x}$ into the latent representations $\mathbf{z}_1$ and $\mathbf{z}_2$.

With this, compared to a denoising autoencoder, SidAE should not just be optimized to encode as much information as possible from the original input into the latent codes (while being robust to noise). It should also ignore specific nuisance factors, which is achieved by the Siamese part of its network. This should prevent the features from containing irrelevant details. In what follows, we will describe the architecture and specify the loss function of SidAE. A visual overview of SidAE is given in Fig. 1 (right).

3.2 Architecture

Input. SidAE uses two views, $\mathbf{x}_1$ and $\mathbf{x}_2$, which are obtained by applying two augmentations, t_1 and t_2, which we sample from a set of augmentations T. We employ the same augmentation pipeline as in SimSiam [4] for CIFAR-10 which consists of the following transformations stated in PyTorch [22] objects: (i) `ColorJitter` with brightness $= 0.4$, contrast $= 0.4$, saturation $= 0.4$, and hue $= 0.1$ applied with probability $p = 0.8$, (ii) `RandomResizedCrop` with scale in $[0.2, 1]$, (iii) `RandomGrayscale` applied with probability $p = 0.2$, and (iv) `RandomHorizontalFlip` applied with probability $p = 0.5$. For Fashion-MNIST and MNIST, we additionally apply a `GaussianBlur` with a standard deviation in $[0.1, 2.0]$ with probability $p = 0.5$.

Encoder. The backbone network in the encoder is a ResNet-18 [11]. As in Sim-Siam, we replace the last fully-connected layer with a 3-layer projection MLP.

Decoder. After the inputs $\mathbf{x}_1$ and $\mathbf{x}_2$ are encoded into the latent representations $\mathbf{z}_1$ and $\mathbf{z}_2$, they are fed into the decoder, producing the reconstructions $\mathbf{x}_1'$ and $\mathbf{x}_2'$.

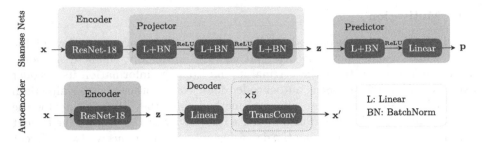

Fig. 2. The components of the models used in our experiments.

The decoder mirrors the encoder in that it up-samples the data step-by-step to reconstruct an output of the original input size. For this purpose, a fully-connected layer followed by five transposed convolution layers are stacked on each other. Their kernels have a size of 3×3, [output] padding of 1, and stride of 2.

Predictor. The predictor is a two-layer MLP, the same as in SimSiam [4]. It maps the latent representations z_1 and z_2 to another latent space to produce two embeddings p_1 and p_2. Both z_i and p_j ($i, j \in \{1, 2\}$ and $i \neq j$) have the same dimensionality since they are compared to each other later. We depict the full structure with the encoder, projector, decoder, and predictor in Fig. 2.

Loss. The loss for SidAE comprises two terms, one for the Siamese and one for the autoencoder part of the network. The former is applied for comparing z_1 to p_2 and z_2 to p_1, and the latter compares the reconstructions x'_1 and x'_2 to the raw inputs, x_1 and x_2, respectively. In SimSiam, the subcomponents of the loss are scaled by $\frac{1}{2}$, which amounts to scaling the whole loss by $\frac{1}{2}$. We do the same for the autoencoder part of the loss. Additionally, we introduce a weighting term on the Siamese (si) and autoencoder (dae) parts of the loss by setting a parameter $w \in [0, 1]$. This parameter controls the magnitude of the relative importance of each mechanism in the model. The full loss L_{sidae} of SidAE is given by

$$L_{\text{dae}} = \frac{1}{2} D_{\text{mse}}(x, x'_1) + \frac{1}{2} D_{\text{mse}}(x, x'_2),$$

$$L_{\text{si}} = \frac{1}{2} D_{\text{ncs}}(p_1, \text{stopgrad}(z_2)) + \frac{1}{2} D_{\text{ncs}}(p_2, \text{stopgrad}(z_1)),$$

$$L_{\text{sidae}} = w L_{\text{dae}} + (1 - w) L_{\text{si}},$$

where D_{mse} is a mean squared error and D_{ncs} is the negative cosine similarity.

4   Experiments

We now perform several experiments regarding our proposed SidAE model, compare it against relevant baselines, evaluate the influence of the parameter w, and assess how the amount of pre-training impacts a downstream classification task on several real-world data sets. For clarity and to avoid repetitiveness, we refer below to a denoising autoencoder whenever an autoencoder is mentioned.

4.1   Experimental Setup

Data. We evaluate on four real-world data sets: *CIFAR-10* [14], *MNIST* [15], *Fashion-MNIST* [28], and *STL-10* [5]. All contain images from ten classes and come with pre-defined train/test splits. STL-10 is designated for self-supervised or unsupervised pre-training. Thus, we use the unlabeled data for pre-training our models and the train and test sets for training and evaluating the model on the classification task. For the other data sets, we use the same training protocol as [4]. We utilize the training set for self-supervised pre-training and for training the classifier as the downstream task. In addition, we simulate harder tasks in which we only use a small fraction (i.e., 1%) of the labeled training data for supervised training on the downstream task. The subsets for each data set are drawn uniformly at random but are kept the same throughout all runs to enable a fair comparison. We always use the full training set (or the unlabeled set in the case of STL-10) for pre-training. To use the same architectures for all data sets, we resize all images to a resolution of 32×32.

Baselines. We compare SidAE against *SimSiam* [4] and a *denoising autoencoder* [25]. The encoder of the autoencoder uses a ResNet-18 with a three-layer projection MLP as used in SimSiam and SidAE and the decoder is like in SidAE. The latent dimensionality of SidAE is set to $d_{\text{hid}} = 2048$ [4]. Since SidAE sees two noisy views at a time, we also provide the autoencoder with two noisy views that are constructed in the same way. The autoencoder uses the MSE loss. We also compare against a fully-supervised ResNet-18 for the classification task.

Setup. We use Python and implement all models in PyTorch [22]. All experiments run on a machine with two 32-core AMD Epyc CPUs, 512 GB RAM, and an NVIDIA A100 GPU with 40GB memory. We re-implement SimSiam based on [4] and use the same settings for pre-training. As an optimizer, we use stochastic gradient descent using a cosine decay schedule with an initial learning rate of 0.03, momentum of 0.9, weight decay of 0.0005, and batch size of 512.

We pre-train the backbone using SidAE, SimSiam, and a denoising autoencoder for 200 epochs. The frozen pre-trained backbones are used as feature extractors for our supervised downstream task of image classification. Each pre-trained model is evaluated at different stages of pre-training, i.e., in a range from 25 to 200 epochs, to assess how the duration of pre-training affects the quality of learned representations. We replace the projector MLP within the backbone for the classifier with a fully-connected layer with an input dimension equal to $\mathbf{z}$. The classifier is trained for 50 epochs using a stochastic gradient descent with a learning rate of 0.05, a momentum of 0.9, and a batch size of 256. Since the backbone weights are frozen, only the weights of the last fully-connected layer are trained. We later evaluate the scenario of fine-tuning the frozen weights.

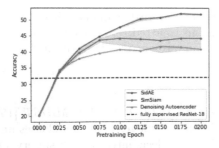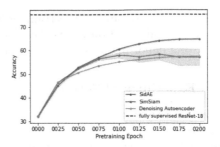

Fig. 3. Classification accuracy on CIFAR-10 (averaged over 5 runs incl. std. err.) after different pre-training stages using a frozen pre-trained backbone. For downstream training, 1% (left) and 100% (right) of training data are used.

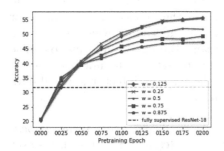

Fig. 4. SidAE: The influence of the weight w on CIFAR-10.

Table 1. Classification accuracies for SidAE ($w = 0.5$), SimSiam, and a denoising autoencoder averaged over five runs (std. err.).

Data	Pre-Train	SidAE	SimSiam	Autoencoder
CIFAR-10	100%	**64.75** (0.27)	57.08 (3.16)	57.44 (0.31)
	1%	**51.67** (0.28)	44.06 (3.74)	40.67 (0.25)
Fashion-MNIST	100%	**88.78** (0.12)	85.10 (0.09)	87.67 (0.05)
	1%	**81.62** (0.22)	77.84 (0.19)	80.20 (0.31)
MNIST	100%	**97.80** (0.13)	96.59 (0.60)	97.23 (0.11)
	1%	**94.65** (0.49)	79.42 (5.15)	88.41 (0.46)
STL-10	STL-10	**61.18** (0.22)	56.74 (0.74)	48.39 (0.22)
	CIFAR-10	**66.20** (0.31)	56.64 (0.25)	57.08 (3.16)

4.2 Results

Figure 4 shows the effect of pre-training in downstream classification on CIFAR-10. The more pre-training epochs we use, the better the downstream accuracy. In the case where we only have 1% of labeled training data for the classification task, only 25 pre-training epochs of the self-supervised methods are enough to outperform a fully-supervised classifier. When 100% of the labeled training data is available, it does not necessarily outperform a supervised baseline without fine-tuning the encoder. Notably, the performance gap for the supervised baseline between 100% and 1% is way higher than for the self-supervised models. These observations support that supervised models need a large amount of training data, performing rather poorly if only little labeled data is available. Note that SimSiam yields the largest standard errors, whereas SidAE is more stable. Overall, SidAE (with $w = 0.5$) outperforms both the autoencoder and SimSiam.

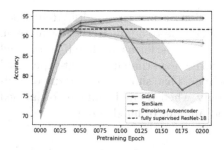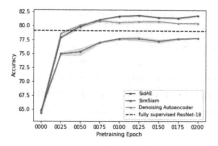

Fig. 5. Classification accuracies on MNIST (left) and Fashion-MNIST (right) using 1% of the training data for downstream training.

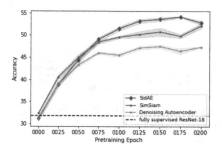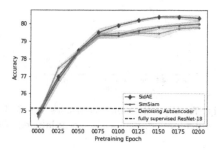

Fig. 6. Classification accuracies on the fine-tuned downstream task (CIFAR-10) at different pre-training stages using encoders pre-trained with different models using 1% (left) or 100% (right) of the training data for the downstream task.

The Influence of w in SidAE. We now investigate the influence of the weight w within the loss. Figure 4 depicts the influence for the 1% case on CIFAR-10. The performance of the classifier when using SidAE with $w \in \{0.125, 0.25\}$ is indeed better than the previously used choice of $w = 0.5$. The results are similar for the 100% case and hence not shown.

Other Data Sets. Table 1 shows the results after 200 pre-training epochs for the other data sets. As before, SidAE outperforms its baselines. Figure 5 depicts the results for MNIST and Fashion-MNIST using 1% of the training data for downstream training. Here, SimSiam is especially unstable on MNIST after 100 epochs. On both data sets, the autoencoder performs better than Sim-Siam. Besides, our proposed SidAE model yields better results than the fully-supervised baseline. The results for using 100% of the training data for downstream training yield the same outcome and are hence not shown.

Fine-Tuning the Downstream Task. So far, we kept the pre-trained weights of the encoder frozen when training the downstream classifier. This is useful to evaluate the capabilities of the feature extractors from the pre-training procedure. We now fine-tune the weights of the encoder during supervised downstream training. We set $w = 0.125$ within SidAE due to the previous experiment (Fig. 4).

Figure 6 shows the downstream classification accuracies on CIFAR-10 when we allow for fine-tuning, i.e., allowing the backbone to adapt to the downstream task. As before, our proposed SidAE model outperforms both self-supervised baselines in almost all cases. In contrast to earlier experiments, all self-supervised models now surpass the supervised baseline. This shows some form of self-supervised pre-training is beneficial not only when a small amount of labeled data is available, but also if there is a good amount of labeled data.

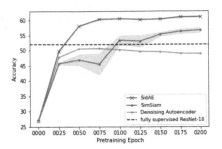 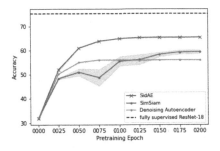

Fig. 7. Classification accuracies on STL-10 (left) and CIFAR-10 (right) using an encoder pre-trained with the unlabeled training data of STL-10.

Results on STL-10. We now pre-train the backbone on the unlabeled training data of STL-10 and use the frozen backbone as a feature extractor for the classification task on STL-10 and also on CIFAR-10. This allows us to evaluate whether the feature extraction can be pre-trained on data sets with similar characteristics. We set $w = 0.25$ for SidAE as it yielded the highest downstream accuracy on both data sets. Similar results are obtained for $w = 0.125$. The classification accuracies on STL-10 and CIFAR-10 are depicted in Fig. 7 on the left- and right-hand side, respectively. For STL-10, we can observe that SidAE and SimSiam both outperform the fully-supervised baseline while the denoising autoencoder falls short. The situation differs for CIFAR-10, where the fully-supervised baseline is still better. However, note that the performance of SidAE is very similar to the case when the backbone was pre-trained on CIFAR-10 itself (Fig. 4). Overall, SidAE performs much better than both self-supervised baselines.

5   Related Work

Many models build on Siamese networks [3,4,9,10,18]. The idea is to make the model learn that two different versions of one object represent the same object. The main challenge is to avoid collapsing representations, i.e., all inputs map to the same feature vector. One widely applied strategy is to use a contrastive loss like InfoNCE [3,19,21], which requires a large amount of negative samples for good performance. Various approaches have been presented in this context, such as using large batch sizes [3], a memory bank to store negatives [19,27], or a *queue* of negatives retaining negatives from previous batches [10]. Another approach avoids pairwise comparison by mapping embeddings of images to prototype vectors found by online clustering [2]. Those alleviate but do not solve the issues of dealing with negative samples. Newer methods follow simpler ways to address the collapsing issue. BYOL [9] circumvents the need for negative samples by applying a momentum encoder, where the weights in one of the branches of the Siamese network are a lagging moving average of the other one. This makes it difficult for the model to converge to a collapsed state but makes the optimization process more complex. SimSiam [4] uses a stop-gradient operation on one of the branches instead, showing that it is sufficient to prevent collapsing solutions. More recently, [18] show that the stop-gradient operation implicitly introduces a constraint that encourages feature decorrelation, explaining why the simple architecture of SimSiam performs well.

6   Conclusion

We proposed combining the benefits of Siamese networks and denoising autoencoders for learning meaningful data representations. To achieve this, we introduced a new model called SidAE (Siamese denoising autoencoder). We empirically evaluated the representations learned by our model in various scenarios on several real-world data sets using classification as a downstream task. There, we first pre-trained our model on unlabeled training data in a self-supervised fashion before using the extracted features for learning the supervised classifier. In our experiments, we compared SidAE to SimSiam and a denoising autoencoder.

SidAE consistently outperformed both self-supervised baseline models in terms of mean downstream classification accuracy in all our experiments. Furthermore, compared to SimSiam, SidAE leads to results that are more stable regarding initialization seeds and the amount of used pre-training epochs.

Acknowledgements. This work was partially supported by the Wallenberg AI, Autonomous Systems and Software Program (WASP) funded by the Knut and Alice Wallenberg Foundation.

References

1. Assran, M., et al.: Self-supervised learning from images with a joint-embedding predictive architecture. In: CVPR (2023)
2. Caron, M., Misra, I., Mairal, J., Goyal, P., Bojanowski, P., Joulin, A.: Unsupervised learning of visual features by contrasting cluster assignments. In: NeurIPS (2020)
3. Chen, T., Kornblith, S., Norouzi, M., Hinton, G.: A simple framework for contrastive learning of visual representations. In: ICML (2020)
4. Chen, X., He, K.: Exploring simple Siamese representation learning. In: CVPR (2021)
5. Coates, A., Ng, A., Lee, H.: An analysis of single-layer networks in unsupervised feature learning. In: AISTATS (2011)
6. Doersch, C., Gupta, A., Efros, A.A.: Unsupervised visual representation learning by context prediction. In: ICCV (2015)

7. Gidaris, S., Singh, P., Komodakis, N.: Unsupervised representation learning by predicting image rotations. In: ICLR (2018)
8. Goodfellow, I.J., et al.: Generative adversarial nets. In: NeurIPS (2014)
9. Grill, J.B., et al.: Bootstrap your own latent-a new approach to self-supervised learning. In: NeurIPS (2020)
10. He, K., Fan, H., Wu, Y., Xie, S., Girshick, R.: Momentum contrast for unsupervised visual representation learning. In: CVPR (2020)
11. He, K., Zhang, X., Ren, S., Sun, J.: Deep residual learning for image recognition. In: CVPR (2016)
12. Henaff, O.: Data-efficient image recognition with contrastive predictive coding. In: ICML (2020)
13. Jing, L., Tian, Y.: Self-supervised visual feature learning with deep neural networks: a survey. IEEE Trans. Pattern Anal. Mach. Intell. $43(11)$, 4037–4058 (2020)
14. Krizhevsky, A., Hinton, G.: Learning multiple layers of features from tiny images. Technical report (2009)
15. LeCun, Y., et al.: Learning algorithms for classification: a comparison on hand-written digit recognition. Neural Netw.: Stat. Mech. Perspect. $261(276)$, 2 (1995)
16. Li, Y., Fang, C., Yang, J., Wang, Z., Lu, X., Yang, M.H.: Universal style transfer via feature transforms. In: NeurIPS (2017)
17. Linsker, R.: How to generate ordered maps by maximizing the mutual information between input and output signals. Neural Comput. $1(3)$, 402–411 (1989)
18. Liu, K.J., Suganuma, M., Okatani, T.: Bridging the gap from asymmetry tricks to decorrelation principles in non-contrastive self-supervised learning. In: NeurIPS (2022)
19. Misra, I., Maaten, L.v.d.: Self-supervised learning of pretext-invariant representations. In: CVPR (2020)
20. Noroozi, M., Favaro, P.: Unsupervised learning of visual representations by solving jigsaw puzzles. In: Leibe, B., Matas, J., Sebe, N., Welling, M. (eds.) ECCV 2016. LNCS, vol. 9910, pp. 69–84. Springer, Cham (2016). https://doi.org/10.1007/978-3-319-46466-4_5
21. Oord, A.v.d., Li, Y., Vinyals, O.: Representation learning with contrastive predictive coding. arXiv preprint arXiv:1807.03748 (2018)
22. Paszke, A., et al.: PyTorch: an imperative style, high-performance deep learning library. In: NeurIPS (2019)
23. Pathak, D., Krahenbuhl, P., Donahue, J., Darrell, T., Efros, A.A.: Context encoders: feature learning by inpainting. In: CVPR (2016)
24. Tian, Y., Sun, C., Poole, B., Krishnan, D., Schmid, C., Isola, P.: What makes for good views for contrastive learning? In: NeurIPS (2020)
25. Vincent, P., Larochelle, H., Bengio, Y., Manzagol, P.A.: Extracting and composing robust features with denoising autoencoders. In: ICML (2008)
26. Vincent, P., Larochelle, H., Lajoie, I., Bengio, Y., Manzagol, P.A., Bottou, L.: Stacked denoising autoencoders: learning useful representations in a deep network with a local denoising criterion. J. Mach. Learn. Res. $11(12)$ (2010)
27. Wu, Z., Xiong, Y., Yu, S.X., Lin, D.: Unsupervised feature learning via non-parametric instance discrimination. In: CVPR (2018)
28. Xiao, H., Rasul, K., Vollgraf, R.: Fashion-MNIST: a novel image dataset for benchmarking machine learning algorithms (2017)
29. Zhang, R., Isola, P., Efros, A.A.: Colorful image colorization. In: Leibe, B., Matas, J., Sebe, N., Welling, M. (eds.) ECCV 2016. LNCS, vol. 9907, pp. 649–666. Springer, Cham (2016). https://doi.org/10.1007/978-3-319-46487-9_40

Equivariant Parameter Sharing
for Porous Crystalline Materials

Marko Petković[1,2]([✉]) [ID], Pablo Romero Marimon[1] [ID], Vlado Menkovski[1,2] [ID],
and Sofía Calero[1,2] [ID]

[1] Eindhoven University of Technology, Eindhoven, The Netherlands
p.romero.marimon@student.tue.nl,
{m.petkovic1,v.menkovski,s.calero}@tue.nl
[2] Eindhoven Artificial Intelligence Systems Institute, Het Kranenveld 12, Eindhoven,
The Netherlands

Abstract. Efficiently predicting properties of porous crystalline materials has great potential to accelerate the high throughput screening process for developing new materials, as simulations carried out using first principles models are often computationally expensive. To effectively make use of Deep Learning methods to model these materials, we need to utilize the symmetries present in crystals, which are defined by their space group. Existing methods for crystal property prediction either have symmetry constraints that are too restrictive or only incorporate symmetries between unit cells. In addition, these models do not explicitly model the porous structure of the crystal. In this paper, we develop a model which incorporates the symmetries of the unit cell of a crystal in its architecture and explicitly models the porous structure. We evaluate our model by predicting the heat of adsorption of CO_2 for different configurations of the mordenite and ZSM-5 zeolites. Our results confirm that our method performs better than existing methods for crystal property prediction and that the inclusion of pores results in a more efficient model.

Keywords: Graph Neural Networks · Porous Materials · Symmetries

1 Introduction

Deep Learning has shown to be of great use in materials science, in tasks like property prediction and high-throughput screening of potential materials [5]. In these workflows, many materials are first simulated using first principles methods, such as Density Functional Theory (DFT) and classical simulation, such as Molecular Dynamics (MD), to find candidate materials to synthesize. However, these simulations are often computationally expensive and can take days or weeks to simulate a single new material. With Deep Learning, it is possible to accelerate the process of finding suitable materials, by developing data-driven surrogate models. These models scale significantly better than first principle simulators and allow for efficient search of the space of potential candidates [27]. Graph Neural Network (GNN) architectures are commonly used for modeling molecules and materials [22] as these objects can effectively be represented as a graph. However, general-purpose GNNs are too restrictive as they incorporate only a part of the symmetries and periodicity present in crystal structures.

To overcome these limitations, for crystalline materials, multiple GNN architectures have recently been proposed [2,4,16,24,31,32] that accurately predict the properties of materials. These methods are specific extensions of general-purpose GNN that preserve the geometric structure of the crystal in their data representation. Despite preserving the geometric structure, none of the proposed models explicitly encode any information regarding pores, as the empty space does not lie on the data domain, and is thus not taken into account. Furthermore, they do not make use of the crystal symmetries in the material representation, since they are typically equivariant to a symmetry group larger than the space group of the crystal. We hypothesize that model architectures that do not explicitly model the porous structure of porous materials will struggle to infer the relevance of atom arrangements around pores for different properties.

Zeolites are a type of porous, crystalline materials of particular interest, as they are easily synthesizable [17]. They are used in applications such as gas separation and are a potential method for carbon capture [26]. The crystal structure of zeolites consists of TO_4 tetrahedra. In these tetrahedra, the T-atoms can either be aluminium or silicon, and both have different influences on the properties of the material. All four corners of the tetrahedra are shared, which results in a porous material. In Fig. 1(a, b), the porous structure of the Mordenite (MOR) and ZSM-5 (MFI) zeolites can be seen. The ability to capture CO_2 of a zeolite can be measured by its heat of adsorption in kJ/mol and is calculated as follows: $-\Delta H = \Delta U - RT$. Here, ΔU is the difference in internal energy before and after adsorption, R is the universal gas constant, and T is the temperature. The heat of adsorption can be influenced by the structure of the different types of zeolites and the amount and distribution of aluminium atoms in the framework [3,19,33].

Due to the difference in charge between aluminium and silicon atoms, it is necessary to balance the charge when aluminium atoms are present in a framework. To achieve this, cations such as sodium are inserted in the crystal structure. Since the cations are positively charged, they additionally attract CO_2 through Coulombic forces, thus increasing the ability of the material to adsorb CO_2. However, while the cations increase the adsorption strength, they also occupy physical space in the pores of the material. When multiple sodium cations are inside a pore, they can restrict CO_2 from entering it. As a result, the adsorption capacity of the material decreases. It is unclear which distributions of aluminium and silicon in different zeolites are optimal to maximize the heat of adsorption.

In this paper, we propose a novel GNN architecture that exploits information regarding the porous structure, as well as the symmetry of these materials, using parameter-sharing based on their space groups. This model allows us to effectively model the properties of porous crystalline materials.

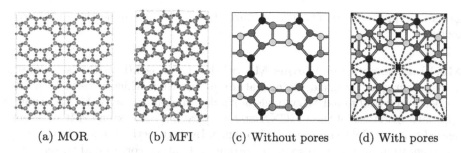

| (a) MOR | (b) MFI | (c) Without pores | (d) With pores |

Fig. 1. (a)-(b): Four unit cells of all silica MOR and MFI viewed along the z-axis and y-axis respectively. Images were generated using iRASPA [7]. (c)-(d): Weight sharing scheme for MOR (z-axis). Nodes/edges of the same color share parameters in their node/edge update functions. Dashed edges are between atoms (circle) and pores (squares). Solid edges are between atoms.

We empirically validate our approach by modeling the heat of adsorption of CO_2 for the MOR and MFI zeolite. Our contributions are threefold: 1) We adapt the Equivariant Crystal Networks (ECN) architecture from [16] to be equivariant with respect to the symmetry group of the unit cell. 2) We extend this architecture to explicitly model pores and show how this modification improves property prediction performance. 3) We introduce a new dataset containing different configurations of aluminium and silicon for the MOR and MFI zeolites, along with the CO_2 heat of adsorption values for the different configurations.

2 Related Work

Machine Learning Methods for Crystals. Due to the success of different GNN architectures in modeling molecules, similar GNN architectures have been proposed for predicting material properties. Crystal Graph Convolutional Neural Networks (CGCNN) [31] are one of the first architectures for crystals, which include periodicity in the data representation and are invariant with respect to permutations of atomic indices. In the MEGNet architecture [2], a global state is used to improve the generalization of the model. Continuous filter convolutions have been introduced in the SchNet architecture [24], which as a result can model the precise relative locations of atoms better when calculating local correlations. Another approach has been proposed in DimeNet [10,11], where the network also takes directional information between atoms into account. In ALIGNN [4], the GNN processes simultaneously the graph and the line graph representation of the crystal, which takes the angles between edges into account. In addition, a transformer based architecture [32] has been proposed, which additionally encodes the periodic nature of the crystal. More recently a new approach has been proposed [16], which proposes a parameter-sharing scheme for message-passing. In this method, multiple unit cells are modelled, where parameters are shared based on the symmetry group of the crystal lattice. As such, the model gains in expressivity by encoding a part of the crystal symmetries in its architecture.

Machine Learning in Porous Materials. Existing ML methods for porous materials frequently make use of feature engineering, which is used to predict properties by traditional ML models or shallow neural networks [15,34]. Another approach made use of the CGCNN architecture [28] and extended it with engineered features [29] to improve performance. In their method, nodes in the graph representation do not correspond to atoms but rather correspond to secondary building units (SBUs), which consist of multiple atoms.

3   Crystal Symmetries

Unit Cell. Zeolites are crystalline materials, meaning that they contain an infinitely repeating pattern in all directions. This pattern can be described by the set of integral combinations of linearly independent lattice basis vectors $\mathbf{a}_i$:

$$\Lambda = \left\{ \sum_i^3 m_i \mathbf{a}_i \mid m_i \in \mathbb{Z} \right\} \tag{1}$$

The crystal lattice has an associated translation group T_Λ, which captures translational symmetry. A unit cell is a subset of the lattice, which tiles the space of the crystal when translated by lattice vectors and is the minimum repeating pattern of the crystal. The unit cell is defined by the basis vectors as follows:

$$U = \left\{ \sum_i^3 x_i \mathbf{a}_i \mid 0 \le x_i < 1 \right\} \tag{2}$$

The unit cell of a crystalline material contains a set of atomic positions, which is defined as $S = \{\mathbf{x}_i \mid \mathbf{x}_i \in U\}$, where $\mathbf{x}_i$ is the position of the atom in the unit cell. In addition to the set of atomic positions, we also define the set of pores contained in the unit cell. We define each pore using the atoms directly surrounding the pore. We represent each pore by the location of its center, as well as its surface area along which diffusion happens in terms of $\mathring{A}^2$. This results in the following set of pores: $P = \{(\mathbf{x}_{p_i}, area(p_i)) \mid p_i \in U\}$, where p_i is the pore and $\mathbf{x}_{p_i}$ is the centre of the pore.

Space Groups. In crystalline materials, there are often multiple symmetries present inside the unit cell, defined by a space group G. The space group G is the set of isometries that maps the crystal structure onto itself. Each element of the space group can be expressed as a linear transformation $\mathbf{W}$ and a translation $\mathbf{t}$, represented by a tuple $(\mathbf{W}, \mathbf{t})$. When mapping a vector $\mathbf{x}$ using an element of the space group, it is mapped to $\mathbf{W}\mathbf{x} + \mathbf{t}$.

Inside a unit cell, every element of the space group G maps the atomic/pore positions in the unit cell onto itself. While the type of atom at a certain position in the unit cell might change as a result of a transformation, the material remains the same. Therefore, each group action g of the space group can be considered a permutation of the atoms and pores in the unit cell.

Group Orbits. The orbit of an element is created by applying all of the different elements of a space group G to it. If the element is a vector $\mathbf{x}$, its orbit is the set of vectors to which the element can be moved by the group action. The orbit of $\mathbf{x}$ is defined as follows:

$$G \cdot \mathbf{x} = \{g \cdot \mathbf{x} \mid g \in G\} \tag{3}$$

4 Methods

Crystal Representation. In our crystal representation, we only consider the set of atoms inside of the unit cell, as the content of a unit cell fully defines the porous structures and symmetry of the material. Each atom in the unit cell is represented by a feature vector $\mathbf{t}_i$ that is a one-hot encoding of the atom type. Next to this, we represent each pore inside the unit cell with a feature vector $\mathbf{p}_i$, which contains its surface area, as well as the number of atoms surrounding it.

To represent the topology of the atoms and pores we construct a graph, where each atom is represented with a node. When the crystal contains clearly defined covalent bonds these can be used as edges in the graph, like in the case of zeolites. Pores are included in the graph representation by adding additional edges between the pore nodes and each atom on the boundary of the pore. By including these nodes, all atoms around the same pore are reachable from each other at most in two steps. Without pore nodes, this number could have been significantly larger, particularly for crystals with larger pores.

The notion of a pore has a certain analogy to the global feature vector introduced in MEGNet [2]. However, our approach is distributed in the geometry of the crystal which in turn allows the GNN-based model to learn locally distributed features, which is a more parameter-efficient solution.

Based on [24], we make use of radial basis functions to encode the distance between two neighboring nodes in the graph. We calculate the edge embedding $\mathbf{e}_{ij}$ as in Eq. 4, where γ and μ are hyperparameters. When calculating the distance between two atoms, we respect the periodic boundary conditions set by the unit cell by using the minimum image convention. Thus, we treat the opposite boundaries as a single boundary and consider the atoms and pores as neighbors and therefore sharing an edge.

$$\mathbf{e}_{ij} = \exp\left(-\gamma(\|\mathbf{x}_i - \mathbf{x}_j\| - \mu)^2\right) \tag{4}$$

Since we are developing a network architecture to predict properties based on the silicon and aluminium configuration in zeolites, we do not explicitly encode the oxygen atoms as nodes, as only the atoms placed in the T-sites of each TO_4 tetrahedron can change, while oxygen atoms always remain in the same position.

Equivariant Message Passing. Since the space group acts as a permutation on the atoms and pores in the unit cell, we can describe the action of a group element using Eq. 5. Here, π_g^t and π_g^p are the permutations of the atoms and pores as a result of group action g.

$$\left(g\mathbf{t}_i = \mathbf{t}_{\pi_g^t(i)} \wedge g\mathbf{p}_j = \mathbf{p}_{\pi_g^p(j)}\right) \forall g \in G \tag{5}$$

As we model different configurations of the same crystal structure using our architecture, the model needs to be equivariant to G. The defined model is based on the message passing framework [12], which we extend by defining parameter-sharing patterns [21] for the message and node update functions.

First, we define how a parameter-sharing pattern for a graph is calculated. Following [16], we define the parameter-sharing pattern as the colored bipartite graph $\Omega \equiv (\mathbb{N}, \alpha, \beta)$. Here, $\mathbb{N}$ is the set of input atoms and pores, α is the edge color function ($\alpha : \mathbb{N} \times \mathbb{N} \rightarrow \{1, ..., C_e\}$) and β is the node color function ($\beta : \mathbb{N} \rightarrow \{1, ..., C_h\}$). C_e and C_h are the amounts of unique edge and node colors respectively. As shown in Eqs. 6 and 7, the color functions take the same value if two edges $((i, j), (k, l))$ or atoms/pores (i, j) lie on each other's orbit.

$$\alpha(i, j) = \alpha(k, l) \iff (k, l) \in G \cdot (i, j) \tag{6}$$
$$\beta(i) = \beta(j) \iff j \in G \cdot i \tag{7}$$

When introducing a parameter-sharing pattern based on the edge and node coloring function, we are effectively introducing an additional message (node) update function for each unique message (node) in the graph representation of the crystal. Following the proof of Claim 6.1 from [16], the model architecture remains equivariant to the space group of the crystal. In Fig. 1 (c,d), the parameter-sharing pattern for MOR can be found, where nodes and edges are colored according to Eqs. 6 and 7. The message passing operation equivariant to the space group is defined in Eqs. 8–10, where $\mathbf{t}_i$ is the atom embedding and $\mathbf{p}_i$ is the pore embedding. Here, superscript h indicates messages between atoms, k messages from pores to atoms and l messages from atoms to pores.

Since different types of crystals have different amounts of atoms and/or different space groups, we cannot share the parameters of message-passing operations between crystals. As a result, each crystal topology requires its own model.

$$\mathbf{m}_{ij}^h = \phi_e^{\alpha(i,j)}(\mathbf{t}_i^t, \mathbf{t}_j^t, \mathbf{e}_{ij}), \quad \mathbf{m}_{ij}^k = \phi_e^{\alpha(i,j)}(\mathbf{t}_i^t, \mathbf{p}_j^t, \mathbf{e}_{ij}), \quad \mathbf{m}_{ij}^l = \phi_e^{\alpha(i,j)}(\mathbf{p}_i^t, \mathbf{t}_j^t, \mathbf{e}_{ij}), \tag{8}$$

$$\mathbf{m}_i^h = \frac{1}{|N_i^h|} \sum_{j \in N_i^h} \mathbf{m}_{ij}^h, \quad \mathbf{m}_i^k = \frac{1}{|N_i^k|} \sum_{j \in N_i^k} \mathbf{m}_{ij}^k, \quad \mathbf{m}_i^l = \frac{1}{|N_i^l|} \sum_{j \in N_i^l} \mathbf{m}_{ij}^l, \tag{9}$$

$$\mathbf{t}_i^{t+1} = \phi_h^{\beta(i)}(\mathbf{t}_i^t, \mathbf{m}_i^h, \mathbf{m}_i^k), \quad \mathbf{p}_i^{t+1} = \phi_h^{\beta(i)}(\mathbf{p}_i^t, \mathbf{m}_i^l). \tag{10}$$

5   Experiments

Network Architecture. In Fig. 2, an overview of the model architecture is presented. First, the edges are embedded using Eq. 4 on their distance (d_{ij}), which

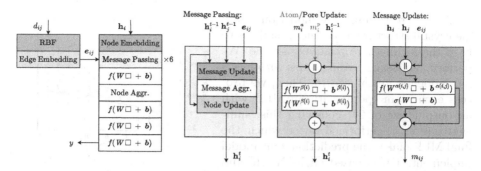

Fig. 2. Overview of the network architecture with pores. The □ and ‖ denote the layer's input and concatenation respectively. The + and * denote elementwise summation and multiplication. $\mathbf{h}_i$ represents the embedding of node i (which can be an atom or a pore), while d_{ij} represents the distance between nodes i and j. m_i^a is the aggregated message received from atoms and m_i^p from pores. f and σ represent the leaky ReLU and sigmoid activation function respectively. $(\boldsymbol{W}/\boldsymbol{b})^{\alpha(i,j)}$ and $(\boldsymbol{W}/\boldsymbol{b})^{\beta(i)}$ denote the set of weights for the value of coloring functions $\alpha(i,j)$ and $\beta(i)$. In the node update function, pore nodes do not receive messages from pores (m_i^p, red). For all aggregations, sum-pooling is used. Finally, the node embedding operation is different for atom and pore nodes.

is followed by a fully connected layer. Simultaneously, both atoms and pores are embedded using two different fully connected layers, as atoms have only one feature while pores have two. In the message update function, the embeddings of the sending and receiving node and the edge embedding are concatenated, which is followed by a fully connected layer that shares weights according to Eq. 6. Each message also receives a weighting factor, calculated by the fully connected layer following the equivariant layer.

The message aggregation step depends on the node type. Because pore nodes are included in the graph, we may distinguish between two types of messages based on whether they are sent from an atom or a pore. To distinguish between the origin of messages, we separately aggregate messages sent from pores (m_i^p) and atoms (m_i^a). Following the concatenation of the different messages and the node embedding, we apply 2 linear layers. The node update block also contains a residual connection [14].

Following multiple message-passing steps, the node features are sum-aggregated, and an MLP is applied to obtain the final prediction. In the model with pores, we only aggregate the pore node features. This way, we implicitly force the model to learn the contribution of each pore to the adsorption capacity.

As we are training the model on a single crystal structure at a time, it suffices to keep the architecture relatively simple. We perform 6 message passing steps, with internal hidden states of size 16. Then, we process each atom (pore) with an MLP with an output size of 24. Following this, sum-aggregation is performed, after which the final MLP makes the prediction. Our model implementation is based on ECN [16] and uses the PyTorch [20] and PyTorch scatter [8] packages as well as the AutoEquiv library [25]. The main difference with ECN is that our architecture only models one unit cell and the symmetries within it, while ECN models multiple unit cell and the symmetries between the them.

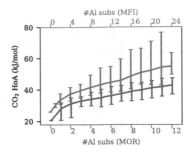

Fig. 3. Heat of Adsorption distributions per amount of Al atoms.

Data Generation. To generate a dataset with porous, crystalline materials, we made use of the Mordenite and ZSM-5 zeolite frameworks. MOR contains 48 T-atoms, and adsorbates diffuse along the z-axis. MFI is a more complex zeolite, containing 96 T-atoms. It has multiple intersecting pores, where adsorbates can diffuse along the x-axis and y-axis. As a result, the adsorption capacity between two configurations with an equal amount of aluminium atoms can vary greatly.

We generated multiple MOR and MFI frameworks with varying aluminium and silicon distributions, and carried out simulations to calculate the corresponding CO_2 heat of adsorption for each configuration. For generating the structures, the Zeoran program [23] was used and atom coordinates where taken from [1]. For MOR, 4992 structures were generated, where each structure contains up to 12 aluminium atoms. For MFI, 3296 structures were generated with up to 24 aluminium atoms. The amount of aluminium atoms was chosen such that they roughly match what is possible in practice. The program makes use of four different algorithms, which places aluminium atoms in either random positions, chains, clusters or homogeneously throughout the zeolite framework. To calculate the CO_2 heat of adsorption Grand-Canonical Monte Carlo simulations were carried out using the Widom Particle Insertion method [30], performed with the RASPA software [6]. Frameworks were considered rigid and the force field parameters for the interactions between the zeolite and the adsorbate were taken from [9]. The force field for carbon dioxide was taken from [13].

In Fig. 3, the 95% confidence interval of the heat of adsorption per amount of aluminium atoms is shown. While there is a strong correlation between the amount of aluminium atoms and the heat of adsorption, there is still significant variance in the heat of adsorption for each amount of aluminium substitutions.

Table 1. Performance of different model architectures on CO_2 heat of adsorption prediction for the MOR and MFI datasets.

	MOR		MFI	
	MAE	MSE	MAE	MSE
CGCNN	1.374 ± 0.033	3.414 ± 0.107	2.814 ± 0.047	18.949 ± 0.561
MEGNet	1.260 ± 0.086	2.785 ± 0.290	2.674 ± 0.040	16.533 ± 0.701
Matformer	1.002 ± 0.074	1.843 ± 0.237	2.577 ± 0.372	12.552 ± 2.545
DimeNet++	0.938 ± 0.028	1.568 ± 0.076	2.862 ± 0.027	19.344 ± 0.472
SchNet	0.895 ± 0.016	1.482 ± 0.055	1.876 ± 0.047	$\mathbf{6.826 \pm 0.270}$
ALIGNN	0.828 ± 0.035	1.293 ± 0.102	$\mathbf{1.819 \pm 0.033}$	6.840 ± 0.187
ECN	1.282 ± 0.028	2.984 ± 0.124	2.484 ± 0.046	12.942 ± 0.610
Ours (w/o pores/syms)	1.184 ± 0.048	2.503 ± 0.163	2.717 ± 0.063	16.775 ± 1.093
Ours (w/o syms)	0.901 ± 0.040	1.505 ± 0.094	2.303 ± 0.110	11.841 ± 1.338
Ours (w/o pores)	0.904 ± 0.023	1.546 ± 0.075	2.029 ± 0.058	8.777 ± 0.375
Ours	$\mathbf{0.813 \pm 0.010}$	$\mathbf{1.286 \pm 0.038}$	1.902 ± 0.024	8.184 ± 0.288

Model Evaluation. To evaluate the predictive performance of our models, we made an uninformed, random assignment of samples to the training (90%) and testing set (10%). We compare the performance of our model to different baselines [2, 4, 10, 16, 24, 31, 32]. For each baseline, we use the hyperparameters from their original papers. In addition, we conducted an ablation study, where we excluded pores and symmetries from our model to asses their contribution.

Since predicting the heat of adsorption is a regression task, we made use of the Huber loss function.

Table 2. Parameter count.

Model	Parameters
CGCNN	0.11M
MEGNet	0.19M
Matformer	2.77M
DimeNet++	1.74M
SchNet	0.44M
ALIGNN	4.01M
ECN	2.81M
Ours (MOR)	0.03M
Ours (MFI)	0.15M

We used the AdamW optimizer [18] with a learning rate of 0.001, and trained each model for 200 epochs. We report the mean-absolute error (MAE) and mean-squared error (MSE). To obtain confidence bounds, we trained each model 10 times with random weight initialization. The code for the model implementations and the zeolite dataset are available on www.github.com/marko-petkovic/porousequivariantnetworks.

As can be seen in Table 1, our model obtains the best results for MOR, and achieves competitive results with SchNet and ALIGNN on MFI, despite using significantly fewer parameters (Table 2). In addition, our model outperformed the ablated versions. For MFI, the other baselines achieve significantly worse results. We speculate that this behaviour is caused by the spatial features of the zeolites not explicitly being encoded in the graph representation.

In Fig. 4, we plot the true against the predicted heat of adsorption values for our best model with pores on the MOR and MFI datasets are shown. We see that most predictions for both models are accurate. For higher heat of adsorption values, the models perform slightly worse. This may be due to an insufficient amount of training examples present with a high heat of adsorption.

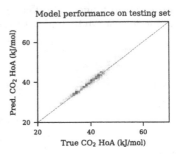

Fig. 4. Predicted against true heat of adsorption.

In addition, we carried out experiments to compare the data efficiency of our model, ALIGNN and SchNet. Here, we trained each model using different fractions ($\frac{1}{8}, \frac{1}{4}, \frac{1}{2}, \frac{3}{4}$ and 1) of the training set, and evaluated them on the same testing set. In Fig. 5, we see that our model achieves a data efficiency comparable to ALIGNN and better than SchNet for MOR. In the case of MFI, we see that our model has a slightly lower data efficiency than ALIGNN, while performing slightly better than SchNet for low amounts of training data.

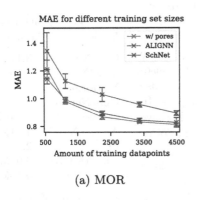

(a) MOR

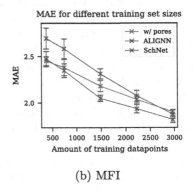

(b) MFI

Fig. 5. Data efficiency (MAE on test set) with different amounts of training data.

6   Discussion

We have proposed a new type of network which can exploit both symmetries inside the unit cell as well as the structure of porous crystalline materials. Our method achieved excellent performance on the CO_2 heat of adsorption prediction task, and has also shown a better parameter efficiency and a competitive data efficiency. This class of models has a significant potential to accelerate the high throughput screening of porous materials, by quickly narrowing the search space for candidate materials.

References

1. Baerlocher, C., McCusker, L.B., Olson, D.H.: Atlas of zeolite framework types. Elsevier (2007)
2. Chen, C., Ye, W., Zuo, Y., Zheng, C., Ong, S.P.: Graph networks as a universal machine learning framework for molecules and crystals. Chem. Mater. **31**(9), 3564–3572 (2019)
3. Choi, H.J., Jo, D., Hong, S.B.: Effect of framework si/al ratio on the adsorption mechanism of co2 on small-pore zeolites: Ii. merlinoite. Chem. Eng. J. **446**, 137100 (2022)
4. Choudhary, K., DeCost, B.: Atomistic line graph neural network for improved materials property predictions. npj Computational Materials **7**(1), 185 (2021)
5. Choudhary, K., et al.: Recent advances and applications of deep learning methods in materials science. npj Comput. Mater. **8**(1), 59 (2022)
6. Dubbeldam, D., Calero, S., Ellis, D.E., Snurr, R.Q.: Raspa: molecular simulation software for adsorption and diffusion in flexible nanoporous materials. Mol. Simul. **42**(2), 81–101 (2016)
7. Dubbeldam, D., Calero, S., Vlugt, T.J.: iraspa: Gpu-accelerated visualization software for materials scientists. Mol. Simul. **44**(8), 653–676 (2018)
8. Fey, M.: PyTorch Scatter (2023)
9. Garcia-Sanchez, A., Ania, C.O., Parra, J.B., Dubbeldam, D., Vlugt, T.J., Krishna, R., Calero, S.: Transferable force field for carbon dioxide adsorption in zeolites. J. Phys. Chem. C **113**(20), 8814–8820 (2009)
10. Gasteiger, J., Giri, S., Margraf, J.T., Günnemann, S.: Fast and uncertainty-aware directional message passing for non-equilibrium molecules. In: Machine Learning for Molecules Workshop, NeurIPS (2020)
11. Gasteiger, J., Groß, J., Günnemann, S.: Directional message passing for molecular graphs. In: International Conference on Learning Representations (2020)
12. Gilmer, J., Schoenholz, S.S., Riley, P.F., Vinyals, O., Dahl, G.E.: Neural message passing for quantum chemistry. In: International Conference on Machine Learning, pp. 1263–1272. PMLR (2017)
13. Harris, J.G., Yung, K.H.: Carbon dioxide's liquid-vapor coexistence curve and critical properties as predicted by a simple molecular model. J. Phys. Chem. **99**(31), 12021–12024 (1995)
14. He, K., Zhang, X., Ren, S., Sun, J.: Deep residual learning for image recognition. In: Proceedings of the IEEE Conference on Computer Vision and Pattern Recognition, pp. 770–778 (2016)
15. Jablonka, K.M., Ongari, D., Moosavi, S.M., Smit, B.: Big-data science in porous materials: materials genomics and machine learning. Chem. Rev. **120**(16), 8066–8129 (2020)
16. Kaba, S.O., Ravanbakhsh, S.: Equivariant networks for crystal structures. In: Advances in Neural Information Processing Systems (2022)
17. Khaleque, A., et al.: Zeolite synthesis from low-cost materials and environmental applications: a review. Environ. Adv. **2**, 100019 (2020)
18. Loshchilov, I., Hutter, F.: Decoupled weight decay regularization. arXiv preprint arXiv:1711.05101 (2017)
19. Moradi, H., Azizpour, H., Bahmanyar, H., Rezamandi, N., Zahedi, P.: Effect of si/al ratio in the faujasite structure on adsorption of methane and nitrogen: a molecular dynamics study. Chem. Eng. Technol. **44**(7), 1221–1226 (2021)

20. Paszke, A., et al.: Pytorch: An imperative style, high-performance deep learning library. Advances in Neural Information Processing Systems **32** (2019)
21. Ravanbakhsh, S., Schneider, J., Poczos, B.: Equivariance through parameter-sharing. In: International Conference on Machine Learning, pp. 2892–2901. PMLR (2017)
22. Reiser, P., Neubert, M., Eberhard, A., Torresi, L., Zhou, C., Shao, C., Metni, H., van Hoesel, C., Schopmans, H., Sommer, T., et al.: Graph neural networks for materials science and chemistry. Commun. Mater. **3**(1), 93 (2022)
23. Romero-Marimon, P., Gutiérrez-Sevillano, J.J., Calero, S.: Adsorption of carbon dioxide in non-löwenstein zeolites. Chemistry of Materials (2023)
24. Schütt, K., Kindermans, P.J., Sauceda Felix, H.E., Chmiela, S., Tkatchenko, A., Müller, K.R.: Schnet: A continuous-filter convolutional neural network for modeling quantum interactions. Advances in Neural Information Processing Systems **30** (2017)
25. Shakerinava, M.: AutoEquiv (2 2021), www.github.com/mshakerinava/AutoEquiv
26. Sneddon, G., Greenaway, A., Yiu, H.H.: The potential applications of nanoporous materials for the adsorption, separation, and catalytic conversion of carbon dioxide. Adv. Energy Mater. **4**(10), 1301873 (2014)
27. Stein, H.S., Gregoire, J.M.: Progress and prospects for accelerating materials science with automated and autonomous workflows. Chem. Sci. **10**(42), 9640–9649 (2019)
28. Wang, R., Zhong, Y., Bi, L., Yang, M., Xu, D.: Accelerating discovery of metal-organic frameworks for methane adsorption with hierarchical screening and deep learning. ACS Applied Materials & Interfaces **12**(47), 52797–52807 (2020)
29. Wang, R., Zou, Y., Zhang, C., Wang, X., Yang, M., Xu, D.: Combining crystal graphs and domain knowledge in machine learning to predict metal-organic frameworks performance in methane adsorption. Microporous Mesoporous Mater. **331**, 111666 (2022)
30. Widom, B.: Some topics in the theory of fluids. J. Chem. Phys. **39**(11), 2808–2812 (1963)
31. Xie, T., Grossman, J.C.: Crystal graph convolutional neural networks for an accurate and interpretable prediction of material properties. Phys. Rev. Lett. **120**(14), 145301 (2018)
32. Yan, K., Liu, Y., Lin, Y., Ji, S.: Periodic graph transformers for crystal material property prediction. Adv. Neural. Inf. Process. Syst. **35**, 15066–15080 (2022)
33. Yang, C.T., Janda, A., Bell, A.T., Lin, L.C.: Atomistic investigations of the effects of si/al ratio and al distribution on the adsorption selectivity of n-alkanes in brønsted-acid zeolites. The Journal of Physical Chemistry C **122**(17), 9397–9410 (2018)
34. Zhang, C., Xie, Y., Xie, C., Dong, H., Zhang, L., Lin, J.: Accelerated discovery of porous materials for carbon capture by machine learning: A review. MRS Bull. **47**(4), 432–439 (2022)

Subgraph Mining for Graph Neural Networks

Adem Kikaj[1,2]([✉]) [iD], Giuseppe Marra[1,2] [iD], and Luc De Raedt[1,2] [iD]

[1] Department of Computer Science, KU Leuven, Leuven, Belgium
{giuseppe.marra,luc.deraedt,adem.kikaj}@kuleuven.be
[2] Leuven.AI - KU Leuven Institute for AI, Leuven, Belgium

Abstract. While Graph Neural Networks (GNNs) are state-of-the-art models for graph learning, they are only as expressive as the first-order Weisfeiler-Leman graph isomorphism test algorithm. To enhance their expressiveness one can incorporate complex structural information as attributes of the nodes in input graphs. However, this typically demands significant human effort and specialised domain knowledge. We demonstrate the feasibility of automatically extracting such information through subgraph mining and feature selection. Our experimental evaluation, conducted across graph classification tasks, reveals that GNNs extended with automatically selected features obtained using subgraph mining can achieve comparable or even superior performance to GNNs relying on manually crafted features.

Keywords: Graph neural networks · Expressivity · Subgraph mining

1 Introduction

Graphs as abstract mathematical structures can represent relational data such as biological compounds and social networks. Numerous graph-based learning methodologies have been proposed to address tasks such as comparative analysis, classification, regression, etc. Broadly, graph learning methodologies can be categorised into two main domains: sub(graph) mining [6] and deep learning approaches [10,25]. Unfortunately, the existing approaches to graph learning have limitations in terms of scalability, interpretability, and expressivity.

Graph mining, also known as traditional graph learning, relies mainly on hand-crafted features as a signal to distinguish and analyse graphs between each other. There exist multiple graph mining techniques [6,23], such as GSPAN [28] and GASTON [22]. These techniques are interpretable, however, they typically have high computational complexity due to involving the subgraph isomorphism test as a sub-routine [23].

Deep learning techniques such as Graph Neural Networks (GNNs) are state-of-the-art for graph learning. With these methods features of graphs can be learned automatically in an end-to-end fashion. However, there are two problems with GNNs. First, they are of a black-box nature, lacking interpretability.

I. Miliou et al. (Eds.): IDA 2024, LNCS 14641, pp. 141–152, 2024.
https://doi.org/10.1007/978-3-031-58547-0_12

Second, there is a trade-off between complexity and expressivity, as standard GNNs are as expressive as the 1-WL isomorphism test [9,14] and higher-order GNNs with increased expressivity come at a computational cost [10,20].

To address the problems of scalability, interpretability, and expressivity for graph learning techniques, we propose AUTOGSN an automated framework bridging the gap between traditional graph and deep learning.

Message Passing Neural Networks (MPNNs) [8] is the most popular GNN framework and is very practical for large graphs since the memory complexity increases linearly with the graph size. However, they fall into the problem of limited expressivity as shown by Xu *et al.* [10,27]. As a consequence, GNNs based on MPNNs, fail to detect certain graph structures such as *cycles* or *cliques*. To address the problem of limited expressiveness while keeping the computational cost feasible, researchers have introduced MPNN-based approaches capable of capturing higher-order graph structures. Relevant to this paper are Graph Substructure Networks (GSN) [3], where subgraph occurrences are used to enhance the feature representation graphs.

AUTOGSN leverages subgraph mining techniques to generate substructures that enhance the performance of substructure-based MPNNs. Our approach selects various off-the-shelf subgraph mining techniques to generate a set of candidate substructures for performance improvement. Furthermore, we employ different feature selection strategies to identify the most useful substructures. The selected substructures are then encoded at the node or edge level for the given graphs. To summarise, this work makes the following contributions: 1) We propose an automated framework to select substructures for use with substructure-based GNNs; 2) We show that our framework can be used to automatically select 'useful' substructures to use with subgraph-based MPNNs; 3) We perform an extensive evaluation on graph classification tasks.

The paper is structured as follows: In Sect. 2 we introduce notions and definitions used throughout the paper. Next, in Sect. 3 our proposed framework is described. Section 4 experimental evaluation. Section 5 discusses the related work. Section 6 concludes the work and describes potential future directions.

2   Preliminaries

In this section, we introduce concepts and terminology.

Graph Theory. A *graph* $G = (V(G), E(G))$ consists of a set $V(G)$ of vertices or nodes, and a set $E(G)$ of edges. Edges are represented as tuples in the form (u, v), where $u, v \in V(G)$. Additionally, the nodes and edges can be associated with labels using a labelling function l, i.e. $l(u)$ or $l(u, v)$. Given two graphs G and H, G is *isomorphic* to H if they are structurally equivalent. More formally, if there exists a bijective function $f : V(G) \rightarrow V(H)$ such that for any edge $(u, v) \in E(G)$ there is a corresponding edge $(f(u), f(v)) \in E(H)$ and vice versa. Two isomorphic graphs are denoted by $H \simeq G$. A *subgraph* $S = (V(S), E(S))$

of a graph $G = (V(G), E(G))$ is a graph that is obtained by selecting a subset of nodes and edges from G such that $V(S) \subseteq V(G)$ and $E(S) \subseteq E(G)$. *Subgraph isomorphism* is the problem of deciding, given two graphs G and S, whether G contains a subgraph G_k that is isomorphic to S, i.e. $G_k \simeq S$. The *Weisfeiler-Leman (WL)* test [14] is a fast heuristic to check whether two graphs are isomorphic or not. The WL test iteratively aggregates the labels of nodes and their neighbourhoods, and it hashes the aggregated labels into unique new labels. For a given graph $G = (V(G), E(G), X_v)$, where X_v is the set of node features, a hash for every node is provided:

$$h_v^k = \text{HASH}(h_v^{k-1}, \{\{h_u^{k-1} : u \in \mathcal{N}(v)\}\}) \; \forall v \in V(G), \tag{1}$$

where h_v^k represents the feature vector of node v at the k-th iteration, and $\mathcal{N}(v)$ denotes the neighbours of a node v. The algorithm decides that the two graphs are non-isomorphic if the hashed labels of the nodes between the two graphs differ at some iteration.

Graph Neural Networks. Message Passing Neural Networks (MPNNs) [27] are a general framework for GNNs. Specifically, node-to-node information is propagated by iteratively aggregating neighbouring node information to a central node. After k iterations of aggregation, a node's representation captures the structural information within its k-hop neighbourhood. Formally, the k-th layer of an MPNN is:

$$a_v^{(k)} = AGGREGATE^{(k)}(\{h_u^{(k-1)} : u \in \mathcal{N}(v)\}), \tag{2}$$

$$h_v^{(k)} = COMBINE^{(k)}(h_v^{k-1}, a_v^k) \tag{3}$$

where $h_v^{(k)}$ represents the feature vector of node v at the k-th iteration and $\mathcal{N}(v)$ represents all the neighbouring nodes of node v. For node classification, $h_v^{(K)}$ of the final iteration is used for prediction, while for graph classification, the READOUT function is used to obtain the graph's representation h_G:

$$h_G = READOUT(\{h_v^{(K)} \mid v \in V(G)\}). \tag{4}$$

The 1-dimensional form of the WL test (1-WL) is analogous to neighbour aggregation (message-passing) in Graph Neural Networks [3,9,27]. Most of the work done on improving GNN expressivity is based on mimicking the higher-order generalisation of WL, known as k-WL [20]. *Graph Substructure Network* (GSN) are GNNs that use more informative topological features [3]. GSNs make minimal changes to the basic MPNN architecture (Eq. 2) by enriching the message passing with more complex predefined structural features, such as the number of cliques or cycles to which nodes or edges belong. The message passing of GSNs is defined as follows:

$$h_v^{(k+1)} = COMBINE^{(k+1)}(h_v^k, a_v^{k+1}), \tag{5}$$

$$a_v^{(k+1)} = \begin{cases} A^{(k+1)}(\{h_v^{(k)}, h_u^{(k)}, \mathbf{x}_v^V, \mathbf{x}_u^V, \mathbf{e}_{u,v} : u \in \mathcal{N}(v)\}) \; \text{(GSN-v)} \\ A^{(k+1)}(\{h_v^{(k)}, h_u^{(k)}, \mathbf{x}_{u,v}^E, \mathbf{e}_{u,v} : u \in \mathcal{N}(v)\}) \; \text{(GSN-e)} \end{cases} \tag{6}$$

Algorithm 1: AUTOGSN

Input : D_t, D_v ▷ The training and validation sets of graphs
Output: *score* ▷ A score for the current subgraph mining and selection strategy
1: $S = MINING(D_t)$
2: $S' = SELECTION(S)$
3: $D_t^{aug} = COUNTING(D_t, S')$
4: $D_v^{aug} = COUNTING(D_v, S')$
5: $GNN.train(D_t^{aug})$
6: $score = GNN.evaluate(D_v^{aug})$

---

where $\mathbf{x}_v^V$ and $\mathbf{x}_{u,v}^E$ are the additional features obtained from the provided topological features counts (e.g. number of cliques or cycles).

3   AutoGSN

We propose AUTOGSN, a framework for increasing the expressivity of GNNs by augmenting the feature representation of the input graphs with automatically discovered subgraph patterns. Similar to GSN [3], AUTOGSN encodes substructural information in the message passing of graph neural networks, by enriching node or edge representations. However, AUTOGSN automates the substructure selection, alleviating the need for human knowledge and allowing the discovery of unknown informative structural features. AUTOGSN operates as a hyperparameter search that optimises a set of mining and feature selection tools, instead of directly searching over the set of possible structural features, as in [3].

3.1   Subgraph Mining

The initial step in AUTOGSN involves subgraph generation, i.e. extracting a set of informative subgraphs $S = \{H_1, H_2, ..., H_m\}$ from an input graph dataset $D = \{G_1, G_2, ..., G_n\}$. AUTOGSN employs off-the-shelf subgraph mining techniques. Clearly, there are different criteria to determine which subgraphs are informative, which leads us to investigate three classes of mining techniques: *(i)* frequent subgraph mining, *(ii)* discriminative subgraph mining and *(iii)* rule mining.

Frequent subgraph mining (FSM) systems, such as GSPAN [28] and GASTON [22], compute all connected frequent subgraphs within a graph database. Specifically, when provided with a set of graphs D and a minimum support threshold *minSup*, FSM identifies all graphs S_D that are subgraphs in at least *minSup* of the graphs in D. They are often used as features to perform classification or regression. One downside of frequent subgraph mining systems is that they generate an prohibitive amount of patterns in unlabelled graphs, i.e. graphs with only one edge type and one node type.

Discriminative subgraph mining systems, like GAIA [11], work on a set of graphs D that are classified and that can be used to discriminate between two or more classes. This is usually formulated in terms of scoring higher than a

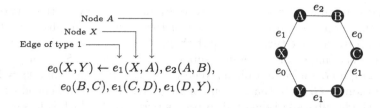

Fig. 1. AUTOGSN - Conversion of *if-then* rules to graphs.

threshold w.r.t. a criterion such as entropy or χ^2, or alternatively, finding the top-k best scoring patterns. Discriminative frequent subgraph mining being an extension of FSMs systems only operate on fully labelled graphs for the same reasons of rising complexity with unlabelled subgraph isomorphism.

Rule mining systems, including ANYBURL [18] and RDFRULES [29], have been proposed in the knowledge graph community. They focus on learning to complete a partial knowledge graph by learning *if-then* rules, like: $h \leftarrow b_1, b_2, ..., b_n$ which states that the edge h must be in the knowledge graph if all the edges $b_1, b_2, ..., b_n$ are in the graph. In the context of knowledge graphs, both h and b_i are binary predicates of the form $r(X, Y)$, where r is the relation type and X and Y are variables, i.e. placeholders for entities. An example of an if-then rule is shown in Fig. 1 (left). In order to adapt rule miners for knowledge graphs to our subgraph generation task, we go through the following steps. First, we build a unique big knowledge graph by the union of single graphs contained in the original dataset. Second, after mining the rules, we translate them into a set of subgraphs. Given a rule ρ, let $\mathcal{X}$ be the set of all variables in the rule and let $\mathcal{R} = \{h\} \cup \{b_1, b_2, ..., b_n\}$. We translate each rule ρ into a subgraph $S = (\mathcal{X}, \mathcal{R})$, where the variables are turned into the nodes of the subgraph and the (binary) relations into its edges. An illustration of this process is shown in Fig. 1. While subgraph mining techniques generate non-isomoprhic subraphs per construction, rule mining techniques may generate rules that correspond to isomorphic subgraphs. Therefore, we *filter* the generated rules such that only unique graphs are kept. We compare all the generated graphs through the VF2 algorithm [7].

Notice that, in principle, one could merge the subgraphs mined by different algorithms. We leave such investigation as future work.

3.2 Selection

Controlling the number of generated subgraphs is fundamental because our final goal is to count how many occurrences of the mined subgraphs are present in the input graphs and use this count to augment the graph representation (see Sect. 3.3). However, the output of the mining process is in general a large set of subgraphs, leading to two issues. First, counting may become computationally unfeasible, due to the need to repeatedly call subgraph isomorphism routines for each of the mined subgraphs. Second, encoding the counts of a large number of generated subgraphs introduces a lot of redundancy. To avoid these two problems, in this step, we select only a subset of the generated graphs using feature selection strategies. We turn each input graph $G \in D$ into a table by employing a boolean encoding of the generated subgraphs $S = \{S_1, ..., S_m\}$. In particular, for each graph G, we build a binary vector $x \in \{0,1\}^m$, such that $x_i = 1$ if H_i is a subgraph of G. Note that, when constructing the boolean encoding of the graph, a single subgraph isomorphism test is performed, which is less expensive compared to the subgraph isomorphism counting required in the following step. Once a flattened table is constructed, we perform the selection of subgraphs using existing feature selection methods. Specifically, we use $\mathcal{X}^2$ [15], XGBoost [5], Mutual Information [13], and Random Forests [4]. While subgraph miners often already include a selection process, we find out that executing an additional selection step improved the performances (see Sect. 4).

3.3   Counting

As a result of the selection process, we are left with a set of subgraphs $S' = \{S'_1, S'_2, ..., S'_k\}$, with $k < m$. In this step, we finally exploit the generated substructures to augment the representation of the input graphs.

Following the approach of GSN [3], the goal is to count, in each input graph, how many times a node (resp. an edge, see Fig. 2) appears in edge-induced subgraphs [21] that are isomorphic to the mined substructures. Then, such counts can be exploited in the representation of the input graphs or in the message passing of a graph neural network model. Specifically, given an input graph $G \in D$ and a mined subgraph S_i, we find the set $\mathcal{H}_i$ of all the subgraphs H_i of G that are isomorphic to S_i, i.e. $\mathcal{H}_i = \{H_i : H_i \subseteq G, H_i \simeq S_i\}$.

We define the vertex structural feature $x_i(v)$ as follows:

$$x_i(v) = |\{H_i : H_i \subseteq G, H_i \simeq S_i, v \in N(H_i)\}| \qquad (7)$$

i.e., how many subgraphs H_i of G and isomorphic to S_i contain v. Similarly, we define the edge structural feature $x_i(v, u)$ as follows:

$$x_i(v, u) = |\{H_i : H_i \subseteq G, H_i \simeq S_i, (v, u) \in E(H_i)\}| \qquad (8)$$

i.e., how many subgraphs H_i of G and isomorphic to S_i contain (v, u). The corresponding structural feature vectors are defined as $x(v) = [x_1(v), ..., x_k(v)]$ and $x(v, u) = [x_1(v, u), ..., x_k(v, u)]$

Discussion. AUTOGSN involves computationally intensive operations, such as isomorphism testing and subgraph counting. Nonetheless, when compared with methodologies that incorporate analogous principles within the neural network architecture [16, 20], two distinct advantages emerge. Firstly, by being employed as a preprocessing phase prior to the training procedure, these computations are applied only once, as opposed to being repeated during each forward pass of the network. Secondly, by performing such operations outside the computational graph of the network, we circumvent the constraints imposed by the tensor

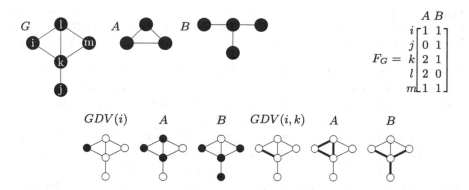

Fig. 2. F_G represents the constructed Graphlet Degree Vector (GDV) of all the nodes of graph G for the subgraphs A and B. Matches of the subgraphs that touch node i, and matches of the subgraphs that touch edge (i, k) are shown at the bottom.

operations of deep learning frameworks like PYTORCH or TENSORFLOW and instead gain the flexibility to leverage specialised software tailored to each of these computational steps.

4 Experiments

In this section, we evaluate empirically our proposed framework on different graph classification datasets. AUTOGSN[1] is used to generate and select subgraphs, that are then encoded in *GSNs* [3]. For a fair comparison, we have reproduced the results using the TUDataset benchmark framework [19]. We show that:

A1: AutoGSN obtains state-of-the-art performance in the majority of the datasets (Table 1). We evaluate AUTOGSN on datasets from the TUD benchmarks [19], which is a standard benchmarking platform for GNNs. We select six datasets from domains of bioinformatics and social networks. We compare AUTOGSN with GIN [27] and GSNs [3]. We build upon the same evaluation protocol of [27]. Specifically, we create 10-fold cross-validation sets and we average the corresponding accuracy curves. Since results are very dependent on weights initialisation, we perform 5 runs with different seeds for each fold, as suggested in [19]. The corresponding 50 curves (10 folds × 5 seeds) are then averaged, obtaining an averaged validation curve. We select the single epoch that achieved the maximum averaged validation accuracy, and we report also the standard deviation over the 50 runs at the selected epoch. Table 1 lists datasets, best-performing subgraphs generated by AUTOGSN used in our benchmark, and their respective accuracy on the test set.

A2: Some of substructures extracted by AutoGSN match human-designed features. Human-designed features used in GSNs [3] reach state-of-the-art results. Since we know these features, we can see if any of those features

---

[1] https://github.com/ademkikaj/AutoGSN.

Table 1. Graph classification accuracy on TUD Dataset. Feature selection method are denoted with *RF* - Random Forest, *MI* - Mutual Information, $\mathcal{X}^2$, *XGB* - XGBoost. The best performing AUTOGSN setting is reported for AUTOGSN-v and AUTOGSN-e.

Dataset	MUTAG	PTC_MR	PROTEINS	NCI1	IMDB-B	IMDB-M
GIN	89.4 ± 5.1	64.5 ± 2.5	75.8 ± 1.2	83.4 ± 2.7	$\mathbf{75.0 \pm 0.9}$	51.6 ± 0.8
GSN-v	90.9 ± 1.0	65.2 ± 1.2	74.4 ± 0.6	82.5 ± 0.3	71.9 ± 0.4	46.5 ± 0.9
GSN-e	90.4 ± 1.0	64.9 ± 1.6	73.9 ± 0.3	83.3 ± 0.5	72.9 ± 0.3	51.6 ± 0.4
AUTOGSN-v	$\mathbf{94.4 \pm 0.6}$	$\mathbf{66.6 \pm 0.8}$	76.1 ± 0.5	$\mathbf{83.5 \pm 0.2}$	72.7 ± 0.4	51.0 ± 0.3
Subgraphs	GSPAN	ANYBURL	GSPAN	ANYBURL	RDFRULES	RDFRULES
	RF	$\mathcal{X}^2$	MI	RF	RF	RF
AUTOGSN-e	93.5 ± 1.0	66.3 ± 0.3	75.4 ± 0.4	83.3 ± 0.4	72.7 ± 0.2	$\mathbf{51.8 \pm 0.3}$
Subgraphs	GSPAN	GASTON	GSPAN	ANYBURL	RDFRULES	RDFRULES
	MI	XGB	$\mathcal{X}^2$	MI	RF	RF

exist in generated features by AUTOGSN. Table 2 shows the generated set of AUTOGSN features that exist in the set of human-designed features for each dataset used in our experiments. In analysing the MUTAG dataset, AUTOGSN extracts non-closed substructures-specifically labelled paths with size greater than 10. We explore whether deeper GNNs can learn similar path-based substructures, finding that deeper GNNs on MUTAG fail to capture additional information (Fig. 4). On the other hand, our assessment indicates that injecting substructures into GNNs enhances performance, enabling effective training with shallower networks. Therefore, our conclusion suggests that by incorporating informative substructures, we can effectively train shallower GNNs.

Table 2. Set of AUTOGSN features similar to human-designed ones. Feature selection method are denoted with *RF* - Random Forest, *MI* - Mutual Information, $\mathcal{X}^2$, *XGB* - XGBoost.

Dataset	Human-Designed	AUTOGSN	AUTOGSN Setting
MUTAG	Cycles $k = [3, 12]$	Cycles $k = \{5, 6, 9, 10, 11\}$	ANYBURL & MI
PTC_MR	Cycle $k = [6]$	Cycle $k = \{6\}$	ANYBURL & XGB
PROTEINS	Cliques $k = [3, 4]$	Clique $k = \{3\}$	ANYBURL & $\mathcal{X}^2$
NCI1	Cycle $k = [3]$	Cycle $k = \{3\}$	RDFRULES & RF
IMDB-BINARY	Cliques $k = [3, 5]$	Clique $k = \{3\}$	RDFRULES & RF
IMDB-MULTI	Cliques $k = [3, 5]$	Clique $k = \{3\}$	RDFRULES & RF

A3: GNNs extended with AutoGSN features are more expressive than those using human-crafted features (Fig. 3). We perform an experiment on a synthetic dataset composed of 4152 strongly regular graphs [17]. No pair of different graphs in the dataset are isomorphic. In this experiment, we generate subgraph features with AUTOGSN and, then, we use them to perform a single forward pass of a GSN architecture with random weights. By comparing the output embeddings, we are able to understand whether the network is able to discriminate the non-isomorphic graphs, which is a signal of their expressivity [3]. In Fig. 3, we show the failure percentage with different classes of features, either human-designed [3] or automatically extracted by AUTOGSN. GSNs extended with automatically mined features and encoded in nodes (AUTOGSN-v) manage to distinguish almost as many graphs as GSNs with human-crafted features. This support our claim that, with no human intervention, AUTOGSN is still able to highly increase the expressivity of GNNs, with no computational increase in the forward pass, but only as a preprocessing counting step.

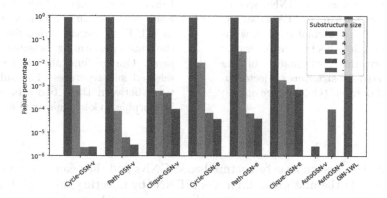

Fig. 3. Strongly regular graphs isomorphism test (smaller is better).

A4: Rule miners tend to extract better substructures than subgraph mining approaches. While there is no overall best performing miner, as showed in Table 1, rule miners tend to provide the better features on average. This is particularly evident in large datasets, where only rule miners are able to extract relevant features.

A5: The selection procedure of substructures significantly changes the performance. The subgraph miners together with feature selection strategies can drastically change the performance of subgraph-enhanced GNNs. In Fig. 5, we show the best and worst performing selected substructures. In this experiment we increase the number of injected substructures, the first single substructure is the most important one according to the feature selection strategy. In this case is worth mentioning that the set of tested substructures are not isomorphic, meaning that specialised miners like GSPAN are much better at finding good substructures compared to rule-miners.

5   Related Work

Graph Neural Networks and Substructures. Since the limitations of MPNNs have been uncovered, there has been a substantial amount of work investigating GNNs that are capable of capturing more topological information. These

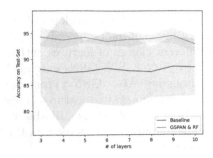

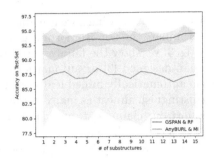

Fig. 4. Performance of GNNs according to their depth on MUTAG dataset. Blue line is the original dataset without any encoded substructures. Orange line shows the performance of the best selected substructures injected to MUTAG dataset. (Color figure online)

Fig. 5. Performance of the best and worst selected substructures injected to MUTAG dataset. Blue line shows the best performing selected structures. Orange line shows the worst selected substructures. The substructures between these two sets are not isomorphic. (Color figure online)

GNNs are also known as substructure-based GNNs and they can be divided into two groups: 1) the first group improves MPNNs by injecting subgraph information into the aggregation process [30,31] (similar to GDV, see Fig. 2) and 2) the second group decomposes the graph into a few subgraphs and then merges their embeddings to produce an embedding of the entire graph [1,2,26].

Subgraph Mining Techniques. The goal of subgraph mining algorithms is to identify an optimal set of subgraphs. These algorithms can be divided into two categories based on the type of supervision that they use. The first category is referred to as Frequent Subgraph Mining (FSM), where no external supervision is used. The second category, known as Discriminative Frequent Subgraph Mining (DFSM), uses supervised or semi-supervised methods for subgraph mining. Frequent Subgraph Mining techniques [12,22,23,28] are focused on efficiently finding subgraphs from a set of graphs or a knowledge graph with a frequency no less than a given support threshold. The limitation of these methods is that the graph pattern search space grows exponentially with the pattern size [6]. On the other hand, Discriminative Frequent Subgraph Mining [11,24] techniques search for patterns (subgraphs) from a set of all possible subgraphs based on the label (target) information.

Rule Mining on Knowledge Graphs. Rule mining on knowledge graphs is a data-driven approach used to discover patterns or subgraphs within large graphs. This method involves predicting missing nodes or edges in the knowledge graph. One key advantage of knowledge graph rule mining techniques is their scalability, as they primarily focus on closed and connected rules or subgraphs. Some of the well-known knowledge graph completion methods are AnyBURL [18] and RDFRules [29].

6 Conclusion

We introduce AutoGSN, a framework for enhancing the expressiveness of GNNs by incorporating complex structural information automatically obtained through subgraph mining techniques.

The results of our experiments demonstrate that AutoGSN can achieve performance comparable to, or even superior to, GNNs relying on manually crafted features. This finding not only opens up new possibilities for improving the effectiveness of GNNs but also alleviates the burden of requiring specialised domain knowledge and significant human effort to engineer features.

In the future, we plan to extend our framework with more advanced hyperparameter tuning techniques, moving towards a fully automated machine learning pipeline. Moreover, we will investigate how to more closely integrate the mining component and neural components following a neurosymbolic approach.

Acknowledgments. This research received funding from the Flemish Government (AI Research Program) and the KU Leuven Research Fund (iBOF/21/075). GM has also received funding from KU Leuven Research Fund (STG/22/021).

References

1. Alsentzer, E., Finlayson, S., Li, M., Zitnik, M.: Subgraph neural networks. In: NeurIPS, vol. 33, pp. 8017–8029 (2020)
2. Bodnar, C., et al.: Weisfeiler and lehman go topological: message passing simplicial networks. In: ICML, pp. 1026–1037. PMLR (2021)
3. Bouritsas, G., Frasca, F., Zafeiriou, S., Bronstein, M.M.: Improving graph neural network expressivity via subgraph isomorphism counting. IEEE Trans. Pattern Anal. Mach. Intell. **45**(1), 657–668 (2022)
4. Breiman, L.: Random forests. Mach. Learn. **45**, 5–32 (2001)
5. Chen, T., Guestrin, C.: XGBoost: a scalable tree boosting system. In: Proceedings of the 22nd ACM SIGKDD International Conference on Knowledge Discovery and Data Mining, pp. 785–794 (2016)
6. Cheng, H., Yan, X., Han, J.: Mining graph patterns. In: Frequent Pattern Mining, pp. 307–338 (2014)
7. Cordella, L.P., Foggia, P., Sansone, C., Vento, M.: A (sub) graph isomorphism algorithm for matching large graphs. IEEE Trans. Pattern Anal. Mach. Intell. **26**(10), 1367–1372 (2004)
8. Gilmer, J., Schoenholz, S.S., Riley, P.F., Vinyals, O., Dahl, G.E.: Neural message passing for quantum chemistry. In: ICML, pp. 1263–1272. PMLR (2017)
9. Huang, N.T., Villar, S.: A short tutorial on the Weisfeiler-Lehman test and its variants. In: ICASSP 2021-2021 IEEE International Conference on Acoustics, Speech and Signal Processing (ICASSP), pp. 8533–8537. IEEE (2021)
10. Jegelka, S.: Theory of graph neural networks: representation and learning. arXiv preprint arXiv:2204.07697 (2022)

11. Jin, N., Young, C., Wang, W.: GAIA: graph classification using evolutionary computation. In: Proceedings of the 2010 ACM SIGMOD International Conference on Management of Data, pp. 879–890 (2010)
12. Ketkar, N.S., Holder, L.B., Cook, D.J.: Subdue: compression-based frequent pattern discovery in graph data. In: Proceedings of the 1st International Workshop on Open Source Data Mining: Frequent Pattern Mining Implementations, pp. 71–76 (2005)
13. Kraskov, A., Stögbauer, H., Grassberger, P.: Estimating mutual information. Phys. Rev. E **69**(6), 066138 (2004)
14. Leman, A., Weisfeiler, B.: A reduction of a graph to a canonical form and an algebra arising during this reduction. Nauchno-Technicheskaya Informatsiya **2**(9), 12–16 (1968)
15. Liu, H., Setiono, R.: Chi2: feature selection and discretization of numeric attributes. In: Proceedings of 7th IEEE ICTAI, pp. 388–391. IEEE (1995)
16. Maron, H., Ben-Hamu, H., Serviansky, H., Lipman, Y.: Provably powerful graph networks. In: NeurIPS, vol. 32 (2019)
17. McKay, B.: http://users.cecs.anu.edu.au/~bdm/data/graphs.html
18. Meilicke, C., Chekol, M.W., Ruffinelli, D., Stuckenschmidt, H.: Anytime bottom-up rule learning for knowledge graph completion. In: IJCAI, pp. 3137–3143 (2019)
19. Morris, C., Kriege, N.M., Bause, F., Kersting, K., Mutzel, P., Neumann, M.: TUDataset: a collection of benchmark datasets for learning with graphs. arXiv preprint arXiv:2007.08663 (2020)
20. Morris, C., et al.: Weisfeiler and Leman go neural: higher-order graph neural networks. In: Proceedings of the AAAI Conference on Artificial Intelligence, vol. 33, pp. 4602–4609 (2019)
21. NetworkX. https://networkx.org/documentation/stable/reference/algorithms/isomorphism.vf2.html#subgraph-isomorphism
22. Nijssen, S., Kok, J.N.: The Gaston tool for frequent subgraph mining. Electron. Notes Theor. Comput. Sci. **127**(1), 77–87 (2005)
23. Ribeiro, P., Paredes, P., Silva, M.E., Aparicio, D., Silva, F.: A survey on subgraph counting: concepts, algorithms, and applications to network motifs and graphlets. ACM Comput. Surv. (CSUR) **54**(2), 1–36 (2021)
24. Saigo, H., Nowozin, S., Kadowaki, T., Kudo, T., Tsuda, K.: gBoost: a mathematical programming approach to graph classification and regression. Mach. Learn. **75**, 69–89 (2009)
25. Scarselli, F., Gori, M., Tsoi, A.C., Hagenbuchner, M., Monfardini, G.: The graph neural network model. IEEE Trans. Neural Networks **20**(1), 61–80 (2008)
26. Sun, Q., et al.: SUGAR: subgraph neural network with reinforcement pooling and self-supervised mutual information mechanism. In: Proceedings of the Web Conference 2021, pp. 2081–2091 (2021)
27. Xu, K., Hu, W., Leskovec, J., Jegelka, S.: How powerful are graph neural networks? arXiv preprint arXiv:1810.00826 (2018)
28. Yan, X., Han, J.: gSpan: graph-based substructure pattern mining. In: 2002 IEEE International Conference on Data Mining, Proceedings, pp. 721–724. IEEE (2002)
29. Zeman, V., Kliegr, T., Svátek, V.: RDFRules: making RDF rule mining easier and even more efficient. Semant. Web **12**(4), 569–602 (2021)
30. Zhang, M., Li, P.: Nested graph neural networks. In: NeurIPS, vol. 34, pp. 15734–15747 (2021)
31. Zhao, L., Jin, W., Akoglu, L., Shah, N.: From stars to subgraphs: uplifting any GNN with local structure awareness. arXiv preprint arXiv:2110.03753 (2021)

Applications

Super-Resolution Analysis for Landfill Waste Classification

Matías Molina[1]([✉]) [iD], Rita P. Ribeiro[1,2] [iD], Bruno Veloso[1,3] [iD],
and João Gama[1,3] [iD]

[1] INESC TEC, Porto, Portugal
matias.d.molina@inesctec.pt
[2] Faculty of Sciences, University of Porto, 4169-007 Porto, Portugal
rpribeiro@fc.up.pt
[3] Faculty of Economics, University of Porto, 4200-464 Porto, Portugal
{bveloso,jgama}@fep.up.pt

Abstract. Illegal landfills are a critical issue due to their environmental, economic, and public health impacts. This study leverages aerial imagery for environmental crime monitoring. While advances in artificial intelligence and computer vision hold promise, the challenge lies in training models with high-resolution literature datasets and adapting them to open-access low-resolution images. Considering the substantial quality differences and limited annotation, this research explores the adaptability of models across these domains. Motivated by the necessity for a comprehensive evaluation of waste detection algorithms, it advocates cross-domain classification and super-resolution enhancement to analyze the impact of different image resolutions on waste classification as an evaluation to combat the proliferation of illegal landfills. We observed performance improvements by enhancing image quality but noted an influence on model sensitivity, necessitating careful threshold fine-tuning.

Keywords: Waste Detection · Image Classification · Super-resolution

1 Introduction

Illegal landfills are places where waste material is dumped, violating management laws. Illegal dumping has a tremendous impact on our ecosystem, affecting our economy and health. As a result, various governments and institutions worldwide have invested additional resources to prevent the proliferation of waste dumping [3,13,15].

Aerial images from satellites, planes and, more recently, drones are an important source of information for earth observation that helps in fighting against environmental crime [1,2]. In recent years, with the advance of artificial intelligence and particularly computer vision approaches due to the increasing computing capabilities and the exponential growth of available annotated data, it has become possible to build different and accurate models for image classification, as exemplified in numerous image recognition challenges, including the

I. Miliou et al. (Eds.): IDA 2024, LNCS 14641, pp. 155–166, 2024.
https://doi.org/10.1007/978-3-031-58547-0_13

Large Scale Visual Recognition Challenge (LSVRC) [5], especially following the introduction of Convolutional Neural Networks (CNNs) [8]. However, the need for an initial amount of images to train such models makes it difficult to apply in waste detection. To build robust machine learning methods for landfill classification, obtaining a significant amount of landfill locations and access to their aerial images is crucial. The number of aerial landfill datasets is limited, and they are commonly high-resolution images. However, the location of these images is not typically provided due to confidential information agreements. On the other hand, although it is possible to obtain free access to different European satellite images (e.g. Sentinel-2 via Copernicus[1]), these images are typically of very low resolution. In this context, we face two challenges: i) The classification models are based on high-resolution literature datasets with no geolocation provided, and they only contain the standard Red, Green, and Blue (RGB) bands. ii) The open-access satellite image banks provide low-resolution information, but we need information about illegal landfill locations. These facts encourage us to investigate the adaptability of the classification models built on the literature dataset to different resolution query domains, as in the case of the freely available image banks. However, the tremendous quality difference between the two types of images and the lack of annotations present a hard limitation. For a comprehensive evaluative process on the road of building machine learning and computer vision waste detection algorithms, we motivate the evaluation of cross-domain classification together with super-resolution enhancement, which is the main goal of our work.

The main contributions of this manuscript lie in leveraging a dataset with exceptionally high resolution to construct a binary classification model and systematically evaluate the model's performance across various resolutions. Furthermore, we suggest an assessment of the model's capabilities when trained with higher-resolution images and subsequently applied to downscaled samples. Following these experiments, we advocate utilising a super-resolution model to enhance the quality before the classification process.

2   Related Work

This section explores two essential aspects: literature datasets used for landfill classification tasks and methodologies employed to enhance image resolution. Each subsection provides crucial insights into these key components of our study.

2.1   Image Classification for Landfills Discovery

The high performance of deep learning (DL) in image classification, particularly in identifying landfills in aerial photographs, relies on the availability of high-quality datasets to build the models. Unfortunately, such datasets have been lacking in the field, posing a significant challenge in developing scalable and

---

[1] Copernicus hub: https://scihub.copernicus.eu/.

accurate methods. To address this critical gap, Torres and Fraternali [17] introduced the AerialWaste dataset for the specific task of landfill detection. What sets this dataset apart is that it has been meticulously annotated by professional photo interpreters, making it valuable for building well-accurate image classifiers. AerialWaste contains 10434 images of very-high-resolution images generated from different sources: AGEA Orthophotos (1000×1000 pixels and 0.2 m Ground Sampling Distance, GSD), WORLDView-3 (700×700 pixels and 0.3 m GSD) and GoogleEarth (1000×1000 pixels and 0.5 m GSD). They also provide 3478 positive and 6956 negative samples, indicating the presence or absence of waste, respectively. The authors also provide the results of training a deep residual model for binary classification using the ResNet-50 [6] architecture, initialized with transfer learning from ImageNet [5] and augmented with a Feature Pyramid Network (FPN) [11]. Another deep learning model applied to landfill monitoring is RetinaNet [12] with DenseNet [7] as the backbone to identify landfills as an object detection task in satellite images of the Shanghai district [4,16]. One important conclusion of this study is the positive impact of data augmentation.

Despite the different literature models to address classification, detection or segmentation for landfills, in most cases, the primary challenge involves the lack of annotations and the different sizes of the contained waste. The recent public Aerial Waste dataset provides an important amount of annotated images and the evaluation of the binary classification through ResNet-50+FPN. We use this dataset for our experiments, and the ResNet model is the base for our classification benchmark.

2.2   Image Quality Improvement

Super-resolution (SR) is one of the most popular techniques that aim to enhance the resolution and quality of images. It is particularly useful when dealing with low-resolution images. The general idea behind SR is to learn to generate missing details in low-resolution images by correlating them with their original high-resolution pair. Various models have been explored in the literature to address super-resolution for different tasks [9,14]. The standard optimization approach minimizes the mean squared error (MSE) between the original high-resolution image and the high-resolution version constructed from the low-resolution input. In addition, it is common to consider the convergence to the peak signal-to-noise ratio (PSNR) as a complementary measure. One of the most popular models uses a generative adversarial network (GAN) approach (SRGAN) [14] and employs a deep residual network (ResNet) architecture. Based on SRGAN, Lim et al. propose an enhanced model (Enhanced Deep residual network for Super-Resolution, EDRS) [10] as a simplified version of the SRGAN architecture by removing the unnecessary modules. The authors also suggest minimizing the L1 norm instead of the MSE or L2 norm. Such modification increases the original performance, improving the computation time. In the following sections, we explain our proposal based on using ResNet for classification and the EDSR baseline for our experiments.

Fig. 1. Super-Resolution (SR) example: low-resolution input (left), SR output (center) and high-resolution ground truth (right).

In our work, we extend and incorporate the SR framework into the classification pipeline to address the specific challenges of landfill classification at different resolution settings. We perform an exhaustive evaluation to understand how super-resolution can serve as a pre-process, mitigating the resolution gap and enhancing the performance of our classification model.

In Fig. 1 it is possible to visualize an example of applying our implementation of the EDSR model on a downscaled sample of the Aerial Waste dataset.

3   Methodology

Since our study focuses on evaluating waste detection through image classification at different resolution domains, we leverage the advances in computer vision and deep learning for this task. We use high-resolution aerial images annotated with the presence or absence of waste.

We propose reproducing the ResNet-50 model with transfer learning from ImageNet and trained with aerial waste, as proposed in [17]. To analyze the impact of the different scales, we are not augmenting the architecture with FPN. The first experiment is to train the model by downscaling the entire dataset. It provides us with a general behaviour of the tasks throughout different resolutions. After this, we use the model trained with the highest resolution to query the different downscaled test samples to understand how much information is missing at each resolution domain. Finally, we propose using a super-resolution to improve each downscaled sample to the original image to analyze its impact when classifying.

In summary, the three experimental scenarios are:

I. Waste classification at different resolutions: ResNet-50 for binary classification at different resolutions given by the different image size dimensions {256, 128, 64, 32, 16, 8}. The output of this experiment aims to answer the following question: *How good is the model performance while trained with data of inferior quality?*

II. High-resolution training for waste classification: ResNet-50 for binary classification, train with high-resolution images and querying with different lower resolutions. The output of this experiment aims to answer the following question: *How good is the model performance when it is queried with data of inferior quality?*

III. Waste classification and super-resolution enhancement: ResNet-50 for binary classification and Enhanced Deep Super-Resolution network (EDSR) for image quality improvement. The EDSR is trained to be applied to images of different resolutions to match the resolution of the images on which the ResNet model has been trained. The output of this experiment aims to answer the following question: *How good is the classification model when it is combined with a super-resolution enhancement as a pre-processing step?*

In all the cases, we use the provided train and test split and also split the training part to use the 20% for validation. We found the best learning rate by a grid search into {1e-2, 1e-3, 1e-4, 1e-5}, with 1e-4 being the best value. Moreover, as the ResNet is pre-trained with ImageNet, we did an initial validation, freezing a different number of layers to find the best possible option to avoid overfitting. Freezing the first two layers best works for us (as also worked in [17]).

3.1 Experimental Setup

Metrics. We evaluate the classification model with the standard Accuracy, Precision, Recall and F1 Score metrics for a comprehensive analysis. In addition, we also consider the True Positive Rate (TPR) and False Positive Rate (FPR) for a deeper analysis, as is discussed in Sect. 3.2.

For super-resolution, we use the two widely used metrics, Peak Signal-to-Noise Ratio (PSNR) and Structural Similarity Index (SSIM) [10].

Experiment I. The initial experiment involves training ResNet classifiers on different dataset resolutions. Initially, the dataset is resized to 512×512, and subsequent experiments with a size of 256×256 yielded similar performance. Therefore, 256×256 is chosen as the initial size for further experiments, representing the highest resolution.

The experiment outputs ResNet-50 models trained on images of sizes 256×256, 128×128, 64×64, 32×32, 16×16, and 8×8. This corresponds to scaling

Table 1. Different downscales and respective resolution (meters) for each source.

Downscale	Size (px)	AGEA (m)	WV3 (m)	Google (m)
0 (initial)	256	0.78	0.82	1.95
1 (1/2)	128	1.56	1.64	3.9
2 (1/4)	64	3.10	3.28	7.8
3 (1/8)	32	6.24	6.56	15.6
4 (1/16)	16	12.48	26.24	31.2
5 (1/32)	8	24.96	52.48	62.4

the images down by factors of $1/2^5$ (1/2, 1/4, 1/8, 1/16, and 1/32, respectively). Considering the original dataset size and resolutions, as described in Sect. 2.1, three different resolutions are obtained for each image source, and the resulting resolutions are calculated with respect to each resizing. These details are summarized in Table 1. Despite the original dataset having various sizes and resolutions, this table serves as a reference, emphasizing that beyond the size of 32 × 32, we start approaching the resolution of open and accessible satellite imagery banks. We denote different models by indicating the input size in a sub-index: $ResNet_{256}$, $ResNet_{128}$, ..., $ResNet_8$.

Experiment II. The goal of the second experiment is to analyze the behaviour of the model trained with our higher size ($ResNet_{256}$) and apply it across all different resolutions (sizes of 128, 64, 32, 16 and 8). In this experiment, the model is trained only once using 256-size images. After that, it queries using the same testing split with different image sizes.

Experiment III. The final experiment aims to enhance the quality of the downscaled images to improve the final performance of the classification task. We use super-resolution to enhance the different downscaled images to the original input size of 256 and then classify them with the same ResNet model trained with images of size 256. We implement other EDSR models to learn how to improve each downscaled image to the original size of 256 × 256.

In this case, we trained only one model for classification $ResNet_{256}$, and we built different super-resolution models $SR_{128\rightarrow256}$, $SR_{64\rightarrow256}$, .., $SR_{8\rightarrow256}$.

Training and Evaluation. The ResNet model was trained using the original split provided by the dataset authors. The validation was performed using the 20% of the training set. We stop the training at the epoch in which the loss training function converges, and the validation loss function deviates from it. We use the ResNet-50 architecture and the ImageNet pre-trained options. Depending on the input size setting, we needed between 17 and 20 epochs to train the model. Despite that, the process behaviour is similar in all the cases.

The Super Resolution model (EDSR) was trained using the dataset resize to 256 × 256 as high-resolution and each of one of its downscaled versions as low-resolution. Continuing the high- and low-resolution input pairs. In this case, we follow three different metrics: the loss function (L1 loss), Pick Signal to Noise Radio (PSNR) and the structural similarity index measure (SSMI). We focused on the loss and PSNR curves more than the SSMI since the different downscaled images make it more challenging to achieve a good SSMI even though it also converges.

3.2   Results

The results of experiments I and II, involving training and testing the model at different resolutions, are summarized in Table 2. For a more comprehensive comparison, refer to Fig. 2, where solid lines depict the different metrics of training and testing the classification model at each resolution. While the model shows reasonable performance across different resolutions, a noticeable decline occurs beyond a certain point, especially when the resolution is downscaled to size 32. Dashed lines represent the performance of training the model with high-resolution images (size 256) and testing with each lower resolution. An important observation is that despite is possible to obtain a suitable performance at lower resolutions, from a certain point in advance, the performance declines with a stronger tendency (e.g., training with a model for image size 256 is significantly less effective for images of size 32, 16, or 8). This suggests that (taking our reference in Table 1) training a model for a resolution of 1 m does not ensure good performance at 10 m. We demonstrate quantitatively that these observations align with the literature concerning the lack of annotation and the resolution gap. These challenges hinder the classification model adaptation.

Table 2. Evaluation of models trained on different resolutions (Experiment I) and maximum resolution (Experiment II) by image size.

Size (px)	Experiment I				Experiment II			
	F1-score	Recall	Precision	Accuracy	F1-score	Recall	Precision	Accuracy
256	0.8600	0.8793	0.8416	0.9044	0.8366	0.8092	0.8659	0.8944
128	0.8381	0.8598	0.8175	0.8891	0.7819	0.7563	0.8093	0.8591
64	0.8139	0.8598	0.7727	0.8687	0.6767	0.5954	0.7837	0.8100
32	0.7931	0.8241	0.7644	0.8564	0.2036	0.1161	0.8279	0.6967
16	0.6684	0.6034	0.7489	0.8000	0.0000	0.0000	0.0000	0.6660
8	0.4280	0.3057	0.7173	0.7271	0.0000	0.0000	0.0000	0.6660

Experiment III is summarized in Table 3. It shows the results of the super-resolution enhancement by setting a standard classification threshold at 0.5 compared to the results after choosing the best possible threshold.

In Fig. 3, we observe the impact of using super-resolution (SR) to improve the quality when querying the model trained with higher-resolution images. The final result shows intermediate performance between Experiment I and Experiment II, which means the SR approach improves when querying with lower resolutions than the training. Nevertheless, it is evident how this methodology impacts the model sensitivity (high recall) that biases the F1 Score up. To better understand the behaviour of the model sensitivity when SR is applied, we analyzed the true positive rate and the false positive rate throughout different threshold values during testing. The impact of changing the default classification threshold during testing could be examined in Fig. 4. The plot shows a model with lesser sensitivity than the default threshold while maintaining the F1 score with no significant decrease.

Regarding the threshold selection process, we could refer to the different Receiver Operating Characteristic curves (ROC) shown in Fig. 5. The ROC

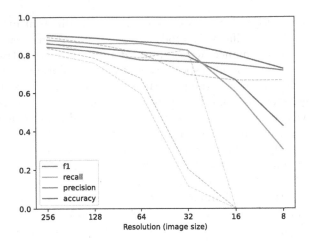

Fig. 2. Evaluation of models trained on different resolutions (Experiment I) - solid lines - and on maximum resolution (Experiment II) - dashed lines - by image size.

Table 3. Evaluation of the model with Super-Resolution enhancement on downscaled images (Experiment III): default vs best threshold by image size.

Size (px)	Default threshold				Best threshold			
	F1-score	Recall	Precision	Accuracy	F1-score	Recall	Precision	Accuracy
256	0.8600	0.8793	0.8416	0.9044	0.8600	0.8793	0.8416	0.9044
128	0.5674	0.9943	0.3970	0.4937	0.6402	0.8908▼	0.4997▲	0.6656
64	0.5636	0.9954	0.3931	0.4852	0.6295	0.8897▼	0.4871▲	0.6503
32	0.5812	0.9897	0.4114	0.5236	0.6185	0.8414▼	0.4890▲	0.6534
16	0.6061	0.9471	0.4456	0.5889	0.5988	0.7299▼	0.5076▲	0.6733
8	0.5931	0.8402	0.4583	0.6150	0.5918	0.7425▼	0.4920▲	0.6580

Table 4. Best thresholds for SR enhancement (Experiment III) based on the ROC curves by image size.

Size	128	64	32	16	8
Threshold	0.97	0.98	0.97	0.92	0.76

curves indicate high values for both TPR and FPR rates when the threshold is set to 0.5 (shown as a red circle). The best threshold, thus, that makes a good compromise between both rates is plotted as a black star. It was revealed that higher thresholds can lead to a better compromise between both rates (Table 4). We consider a 95% as enough confidence level, and consequently, we set to 0.95 all the thresholds greater than that. Another interesting observation is that the threshold maximizing both rates decreases as the original image resolution is reduced. These thresholds converge to the default one as the resolution decreases. We can conclude that the model's sensitivity can be adjusted, and its utility

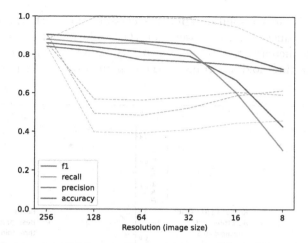

Fig. 3. Evaluation of models trained on different resolutions (Experiment I) - solid lines - and SR enhancement with default threshold (Experiment III) - dashed lines - by image size.

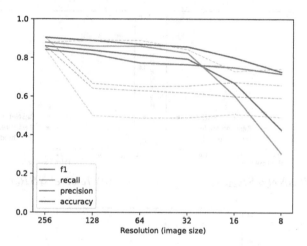

Fig. 4. Evaluation of models trained on different resolutions (Experiment I) - solid lines - and SR enhancement with best threshold (Experiment III) - dashed lines - by image size.

depends on the specific problem domain. A more sensitive model may be preferred in cases like environmental crimes, particularly in the challenging task of detecting potential illegal landfills. This allows environmental agencies and governments to verify potential issues on-site. However, the model might incur a higher false negative rate in other scenarios, potentially missing a possible illegal landfill. This study motivates further exploration of various domain applications to provide a more comprehensive and cross-functional evaluation.

In summary, super-resolution enhances classification performance when the resolution of the query image is lower than the resolutions used during training. However, it also increases the model's sensitivity. As a result, it becomes essential to fine-tune the threshold according to the domain's specific requirements.

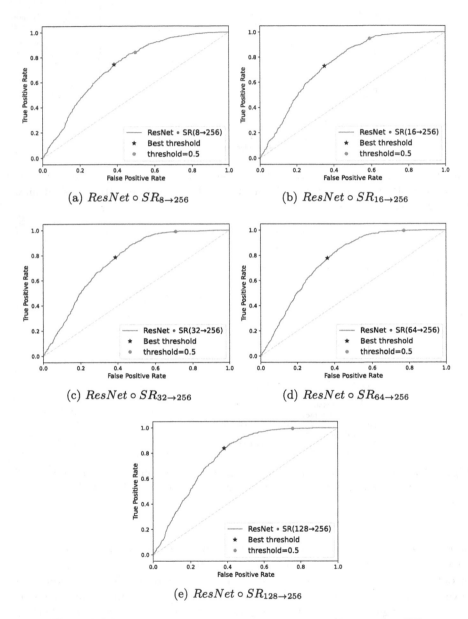

(a) $ResNet \circ SR_{8 \rightarrow 256}$ 　　　　 (b) $ResNet \circ SR_{16 \rightarrow 256}$

(c) $ResNet \circ SR_{32 \rightarrow 256}$ 　　　　 (d) $ResNet \circ SR_{64 \rightarrow 256}$

(e) $ResNet \circ SR_{128 \rightarrow 256}$

Fig. 5. ROC curves obtained with SR enhancement (Experiment III).

4   Conclusions

This study makes contributions by introducing a dual-model approach, combining a classification model and a super-resolution model, to address challenges in waste detection related to varying image resolutions. The methodology focused on combining both models to analyze the performance of waste classification and the pertinence of the resolution enhancement motivated by the lack of information in conjunction with the different resolutions as a crucial challenge in literature. The research design involved a series of experiments to fairly compare image quality's influence on building classification models and explore potential improvements. The experiments reveal important insights into the performance of waste detection models at various image resolutions. It evidences that the classifiers can be effectively trained at different resolutions. Still, their effectiveness declines as the resolution decreases, especially after the third downscaling, which resolutions correlate with the low resolution provided by the various open-access satellite imagery services. While training with high-resolution images initially yields favourable results, performance deteriorates as the resolution decreases. This indicates that a model designed for higher resolutions performs poorly at lower resolutions, especially after the first downscaling. These observations align with the challenges highlighted in existing literature, demonstrating the impact of the scarcity of annotations and the resolution gap.

Furthermore, we observe improvement in terms of performance when applying super-resolution to enhance image quality. This approach notably impacts the model's sensitivity, leading to higher recall and, consequently, increased F1 Score. The study also explores the model's sensitivity at varying threshold values, showcasing the importance of adjusting thresholds to match specific problem domains. In summary, super-resolution improves the waste classification performance for lower-resolution images. Still, it intensifies the model's sensitivity, underscoring the need for threshold fine-tuning tailored to specific domain requirements.

For future research, we intend to assess different super-resolution models and verify the variability using various public and collected datasets.

Acknowledgments. This work was supported by the EMERITUS project, funding from the European Union's Horizon Europe research and innovation programme under Grant Agreement 101073874.

References

1. ESA: Earth observation finds a role in environmental treaties. https://www.esa.int/Applications/Observing_the_Earth/Earth_observation_finds_a_role_in_environmental_treaties. Accessed Nov 2023
2. ESA: Satellite technology to help fight crime. https://www.esa.int/Enabling_Support/Preparing_for_the_Future/Space_for_Earth/Satellite_technology_to_help_fight_crime. Accessed Nov 2023

3. Europol: Environmental crime in the age of climate change, threat assessment 2022 (2022). https://www.europol.europa.eu/publications-events/publications/ environmental-crime-in-age-of-climate-change-2022-threat-assessment. Accessed Nov 2023
4. Abdukhamet, S.: Landfill detection in satellite images using deep learning. Shanghai Jiao Tong University Shanghai, Shanghai, China (2019)
5. Deng, J., Dong, W., Socher, R., Li, L.-J., Li, K., Fei-Fei, L.: Imagenet: a large-scale hierarchical image database. In: 2009 IEEE Conference on Computer Vision and Pattern Recognition, pp. 248–255. IEEE (2009)
6. He, K., Zhang, X., Ren, S., Sun, J.: Deep residual learning for image recognition. In: IEEE Conference on Computer Vision and Pattern Recognition, pp. 770–778 (2016)
7. Huang, G., Liu, Z., Van Der Maaten, L., Weinberger, K.Q.: Densely connected convolutional networks. In: IEEE Conference on Computer Vision and Pattern Recognition, pp. 4700–4708 (2017)
8. Krizhevsky, A., Sutskever, I., Hinton, G.E.: Imagenet classification with deep convolutional neural networks. In: Pereira, F., Burges, C.J., Bottou, L., Weinberger, K.Q. (eds.) Advances in Neural Information Processing Systems, vol. 25. Curran Associates, Inc. (2012)
9. Ledig, C., et al.: Photo-realistic single image super-resolution using a generative adversarial network. In: IEEE Conference on Computer Vision and Pattern Recognition, pp. 4681–4690 (2017)
10. Lim, B., Son, S., Kim, H., Nah, S., Lee, K.M.: Enhanced deep residual networks for single image super-resolution. In: IEEE Conference on Computer Vision and Pattern Recognition Workshops, pp. 136–144 (2017)
11. Lin, T.-Y., Dollár, P., Girshick, R., He, K., Hariharan, B., Belongie, S.: Feature pyramid networks for object detection. In: IEEE Conference on Computer Vision and Pattern Recognition, pp. 2117–2125 (2017)
12. Lin, T.-Y., Goyal, P., Girshick, R., He, K., Dollár, P.: Focal loss for dense object detection. In: IEEE International Conference on Computer Vision, pp. 2980–2988 (2017)
13. Mager, A., Blass, V.: From illegal waste dumps to beneficial resources using drone technology and advanced data analysis tools: a feasibility study. Remote Sens. **14**(16), 3923 (2022)
14. Nasrollahi, K., Moeslund, T.B.: Super-resolution: a comprehensive survey. Mach. Vision Appl. **25**, 1423–1468 (2014)
15. Rodríguez-Robles, D., García-González, J., Juan-Valdés, A., Morán-del Pozo, J.M., Guerra-Romero, M.I.: Overview regarding construction and demolition waste in Spain. Environm. Technol. **36**(23), 3060–3070 (2015)
16. Torres, R.N., Fraternali, P.: Learning to identify illegal landfills through scene classification in aerial images. Remote Sens. **13**(22) (2021)
17. Rocio Nahime Torres and Piero Fraternali: Aerialwaste dataset for landfill discovery in aerial and satellite images. Sci. Data **10**(1), 63 (2023)

Predicting Performance Drift in AI Models of Healthcare Without Ground Truth Labels

Ylenia Rotalinti[1,2(✉)], Puja Myles[2], and Allan Tucker[1]

[1] Department of Computer Science, Brunel University London, London, UK
ylenia.rotalinti@mhra.gov.uk
[2] Medicines and Healthcare Products Regulatory Agency, London, UK

Abstract. Many real-world applications of machine learning involve handling data that is collected over an extended period of time. The longer this time-period, the more likely that the underlying characteristics of this data are to change, potentially leading to a degradation in prediction accuracy and impacting decision-making. This phenomenon, commonly referred to as data drift, poses a risk in the context of medical AI regulation and monitoring. Regulatory bodies must regularly assess previously approved models for their performance on new data, realistically even in scenarios where prediction labels are not yet available, making the tracking of model performance unfeasible. In this paper, our contribution involves introducing a comprehensive framework to estimate the performance drift of a model when evaluated on new unlabelled target data. We introduce a method that assesses both i) the uncertainty in model predictions and ii) the discrimination error between training batches and subsequent test batches, serving as key indicators for identifying drift in AI model performance. We test our framework on simulated drift data where we can control the nature of change, and high-fidelity synthetic primary care data focused on the UK Covid-19 pandemic. Promising results emerge from our experiments, suggesting that the proposed metrics can effectively monitor potential changes in the performance of AI health products post-deployment even in the absence of labelled data.

Keywords: performance drift · healthcare data · machine learning

1 Introduction

AI models, like other medical devices, must satisfy regulators that they are safe and fit for their intended purpose with appropriate evidence. However, the underlying characteristics of health data are likely to change over time, introducing a challenge for regulatory bodies. To ensure patient safety, it is necessary to understand if the model has significantly changed since it was approved either due to the change in the actual algorithm logic or because of data drift [1,3], which could mean that the model is out of date in light of new data. If it has, both manufacturers and regulators need to know whether it remains safe and fit for purpose and if instructions on use have to be modified.

I. Miliou et al. (Eds.): IDA 2024, LNCS 14641, pp. 167–178, 2024.
https://doi.org/10.1007/978-3-031-58547-0_14

The work in this paper investigates methodologies to address drift detection in the context of medical AI regulation and monitoring. Predicting data drift has an important impact on the decision processes of regulators for new approved medical devices. For instance, such insights can drive decisions to update the model accordingly or even withdraw the product from the market if necessary. As the monitoring process is likely to involve assessing models on new data when labels are not readily available, we focus on introducing a framework in unsupervised settings. We exploit a regression-based method that incorporates two different domain-drift detection metrics. The method aims to predict the performance drift of a model on a target domain in the absence of ground truth labels. The code to reproduce the experiments conducted is available on GitHub[1].

2   Related Work

As many real-world applications of machine learning involve handling data that is collected over an extended period of time, learning in non-stationary environments has been extensively studied [4]. In these scenarios, the likelihood of a change in the underlying characteristics of the data increases and thereby resulting in sub-optimal predictions and decisions. To address this issue, several error-based approaches have been developed for drift detection in supervised settings [6,12].

Recent attention has been also directed towards methodologies capable of detecting drift in the absence of true labels. This is because subsequent to the initial model fitting, no additional labels are available during the deployment phase in a test set [19]. Zliobaite [13] presents an analytical view of the conditions that must be met to allow drift detection in a delayed labelled setting. Ackerman et al. [14] proposed a method that utilises feature space rules, known as data slices.

The purpose of this research differs from the methods mentioned above since the aim is to explore drift in the context of regulatory requirements to realistically detect changes in approved AI models that have been deployed in a population, based on availability of new incoming data. Here, we extend our previous work [2] by focusing on predicting performance change in a more realistic scenario where ground truth labels are not readily available, contrasting the previous emphasis on general concept drift detection.

The project involves case studies on synthetic data to simulate multiple batch releases of primary care data from the UK's Clinical Practice Research Datalink (CPRD) [15]. In contrast to the stream learning paradigm, where a sequence of data elements with associated timestamps is continuously collected, the experimental methodologies detailed in this paper are tailored to address a batch learning scenario. Given that primary care data is frequently delivered in discrete batches, often varying in size, the study's design accommodates this characteristic. As an illustration, the CPRD releases updates to their anonymised primary care data in monthly batches [15].

---

[1] https://github.com/yleniarotalinti/predicting-performance-drop.

The next section presents a formalized framework for drift detection, providing consolidated definitions from a statistical perspective.

3 Methods

From a medical device regulation standpoint, disentangling distinct sources of drift and investigating the nature of change is crucial to inform decisions about potential intervention. Moreover, medical software may have additional complexities to consider due to different levels of transparency and interpretability that can complicate the assessment of significant changes. In this context, a mathematical definition of how data changes is necessary to rigorously identify different sources of drift. The next paragraph provides formal definitions of drift and potential sources of change.

3.1 Probabilistic Sources of Drift

Supervised learning tasks involve labelled data, where each observation is a pair consisting of a feature vector X_i and a target variable y_i. Additionally, detecting changes in data samples requires a time-ordered sequence of data with associated timestamps. According to [1], an evolving concept can be defined as the joint distribution of X and y at time t, denoted as $P_t(X, y)$. In this work, we assume a batch learning scenario, where blocks of data b (potentially different in size) are released and processed all at once at a given time. In this context, data drift is formally defined as a change in the joint distribution between two batches of data b_1 and b_2. Therefore, drift occurs if $P_{b1}(X, y) \neq P_{b2}(X, y)$. According to [8], we can express the joint probability distribution as:

$$P_b(X, y) = P_b(X) \times P_b(y|X) = P_b(y) \times P_b(X|y) \qquad (1)$$

where:

- $P_b(X)$ denotes the covariate distribution;
- $P_b(y|X)$ indicates the target y posterior probability given the covariates X;
- $P_b(y)$ is the prior probability distribution of the target variable;
- $P_b(X|y)$ indicates the posterior probability of the covariates X given the target y.

As evident from the above equation, change can be triggered by different sources depending on which factor has shifted. Kelly et al. [9] suggest that in the context of classification learning, $P(Y|X)$ is the most crucial of these drift subjects as such a change will necessitate an update of a model to maintain accuracy and prevent performance degradation. However, Tsymbal [10] argues that tracking covariate drift and prior probability drift is also essential as they could affect model performances too.

An additional challenge is posed by the fact that ground truth labels are often not readily available at the time of analysis in many real-world scenarios. Some annotations might arrive with a delay or not arrive at all due to labelling costs. The unavailability of labels precludes the application of any learning algorithms that rely on traditional performance drift detection approaches. The next section presents the regression-based methodology we designed to predict performance drift in unsupervised settings.

3.2   Drift Detection Framework

As highlighted in the introduction, we assume a batch-learning scenario. Contrary to scenarios dealing with continuous data updates as in online streaming services, healthcare data are often released in blocks of data at regular time intervals (e.g., monthly, annually) for secondary purposes. Therefore, in the first step of our analysis, the data is split into batches representing different blocks of time to simulate such a scenario. Then, we identify two distinct domains:

1. a **source domain** D_s that represents the training dataset used to fit the parameters of a machine learning algorithm. We assume that data samples from this domain are labelled;
2. a **target domain** D_t that embodies the new data encountered when an approved model is deployed on the market. This data is used to evaluate the accuracy of a product during the monitoring phase after the deployment.

Regression of the Performance Drift. Formally, the aim is to estimate the performance drift $\triangle P$ of a classifier C trained on data (X_s, y_s) drawn from a source domain D_s and tested on data (X_t, y_t) from a target domain D_t. If ground truth labels y_t are available immediately after prediction, we can empirically measure the performance drift of the classifier by computing the difference in test errors between the source and the target domain as:

$$\triangle P = |Cx_s \neq y_s - Cx_t \neq y_t| \tag{2}$$

However, feedback on the predictions is not always immediately available, especially in real-world clinical contexts, as collecting labels is often time-consuming since it requires expert input. For example, annotations in the form of disease labels might come at a later stage than the model evaluation, leading to the infeasibility of any drift detection algorithm relying on ground truth labels. To estimate the performance drift of a model trained on a source domain D_s and tested on batches of data from a target domain D_t, we present a regression-based methodology schematically shown in Fig. 1.

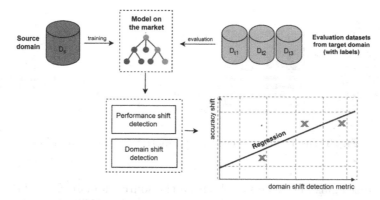

Fig. 1. A regression model is trained on a set of evaluation data from the target domain. The model's performance shifts are subsequently analysed in relation to the computed domain drift detection metrics.

The overall framework was designed to tackle the recurring problem of a lack of ground truth labels in new upcoming batches of data. However, this methodology assumes the availability of a small portion of labelled data from the target domain. Through these evaluation batches of data, we fit a logistic regression model between the shift in the model accuracy and a domain-shift detection metric (see the next paragraph). When a new unlabelled batch of data D_t becomes available, the shift in model performance on D_t can be estimated by evaluating the function learned through regression (Fig. 2).

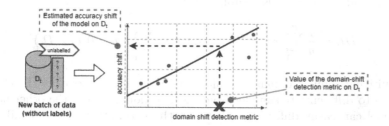

Fig. 2. When a new unlabelled batch of data becomes available, the regression model is used to predict the performance shift when updated using the new batch of data.

The next section presents the details of the two metrics designed to predict the shift in performance of a model assuming the unavailability of labelled examples.

Proposed Metrics to Estimate the Drift in Model Performance. Two families of drift detection metrics have been considered: firstly, a discrimination error-based metric (DE) that relies on the capability of another classification

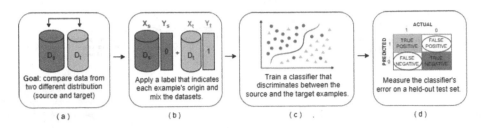

Fig. 3. Discrimination error-based metric (DE) computation.

model to distinguish between samples from the source domain D_s and the target domain D_t; and secondly, a confidence-based metric (AVS) that relies on the uncertainty of the model over its predictions. We now present details on how to compute both drift detection metrics.

Discrimination Error-Based Metric - DE. Inspired by previous work from Ben-David et al. [5], we designed a measure of similarity between distributions from different domains to assess whether data drift has occurred. The algorithm to compute the metric is schematically presented in Fig. 3.

Assume that we aim to compare two data batches D_s and D_t, sampled respectively from the source and the target distributions. As a first step, a new label is added to the batches indicating each example's origin (for example, y_s is set to 0 while y_t is set to 1). Then, a meta-classifier discriminating between the data points x_i from the source domain D_s and the examples x_j from the target domain D_t is learned from the merged data. Finally, the misclassification error DE on a held-out test set is computed as:

$$DE = \frac{1}{n} \sum_{i=1}^{n} p(\hat{y} = 1 | x_i \in D_s) + \frac{1}{m} \sum_{j=1}^{m} p(\hat{y} = 0 | x_j \in D_t) \tag{3}$$

Intuitively, similar domains would generate a bigger error as the classifier struggles to differentiate samples from the two domains. On the other hand, if the model can easily differentiate between the two data batches (small error), we conclude that the examples result from two disparate domains, suggesting a data drift.

The implicit assumption underlying this approach is that any change in the distribution (that results in the meta-classifier correctly identifying the origin of instances) is considered as having an equal impact on the prediction performance. However, this assumption doesn't hold for a drift affecting $P(X)$ without influencing $P(Y|X)$, where a change may not affect prediction performance but could still trigger the meta-classifier.

Confidence-Based Metric - AVS. Although the output of a machine learning algorithm on a binary classification problem is discrete, for many classification algorithms, the weight associated with that decision can be interpreted as the

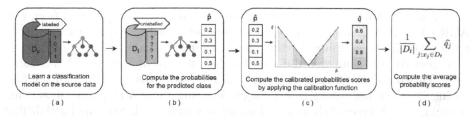

Fig. 4. Confidence-based metric (AVS) computation.

confidence the model has in that outcome. In other words, many machine learning models predict a probability of class membership, which must be interpreted before it can be mapped to a crisp class label. This goal is usually achieved by using a discrimination threshold (such as 0.5): all values equal to or greater than the chosen threshold are mapped to one class while all the others are assigned to the opposite class. In general, this is useful as it provides a measure of the certainty of a prediction. Based on this intuition, we designed a confidence-based metric that aims to estimate the domain drift between the source data and the target data by computing the shift in the average certainty score of the predicted class across the examples. In other words, we measure how the model confidence in regard to its predictions changes over the subsequent batches. In addition, as the associated probabilities to the predicted class label do not reflect its correctness likelihood, a calibration function $f(\hat{p})$ has been proposed to overcome this limitation. The calibrated confidence score $\hat{q}$ is computed as follows:

$$\hat{q} = |2\hat{p} - 1| \tag{4}$$

where $\hat{p}$ represents the probability of the class predicted by the model. Figure 4(c) shows the calibration function. A vector of prediction probabilities $\hat{p}$ is given as input, while a vector of confidence scores $\hat{q}$ is returned as output. For instance, if a model claims 0.2 $(\hat{p})$ as the likelihood associated with the class label predicted, the confidence that it has in that decision would be 0.6 (i.e. $\hat{q} = |2 \cdot 0.2 - 1|$).

The overall framework to compute the confidence-based metric is shown in Fig. 4. Firstly, a classification model is learned from the source domain D_s. The fitted model is then used to predict the class label y_t on a new available batch of data drawn from the target domain D_t. As discussed, the output of this step consists of a probability vector $\hat{p}$ for the predicted class, which provides a measure of the certainty of a prediction toward the 0 or 1 class. The prediction probabilities vector $\hat{p}$ is then given as input to the calibration function $f(\hat{p})$; a confidence score vector $\hat{q}$ is returned as output. Finally, an average over all the probability scores (avr-score) is computed to obtain a singular representative value.

Data. We evaluated our performance drift detection algorithm on two different datasets described in the following sections.

Simulated Data - The framework was tested on an artificial dataset from the AGRAWAL generator [18]. Three drifts were simulated by using various underlying functions to generate different distributions of data and labels in specific time periods. We simulated different batches of data of 5 thousand instances each. The first data batch is considered as the source domain while all the following data batches belong to the target domain. Figure 5 (top left) shows the changing performance of the assessed model over time due to the simulated drifts. The three drifts - located at the 30th, 55th and 80th batches - are identified by vertical dashed lines. In Fig. 5 (top right), we computed the difference between consecutive values to clearly identify any shifts in accuracy of the model and their severity.

UK Primary Care Covid Data - We also explore a case study on prediction of death from Covid-19. This is based on multiple batches of primary care data from the UK's Clinical Practice Research Datalink (CPRD) [15]. A high-fidelity synthetic dataset based on anonymised real primary care data has been generated from the CPRD Aurum database [17]. The dataset includes around 780K observations over 46 features associated with Covid risk factors. Data was collected from 2020 to 2021 in monthly batches.

Evaluation. To evaluate the algorithm, we consider the Mean Absolute Error (MAE), its standard deviation (sd-MAE), and the Max error (MAX) between our predicted values of performance drift and the actual performance drift over the data batches. Furthermore, the predictive power of this regression algorithm on each proposed metric is compared with a baseline: instead of learning a regression model, we compute the mean of the performance drift over all the evaluation datasets. We measure the error when compared with the true drift. As well as exploring the prediction of the magnitude of performance drift, we also explore if the detected drifts coincide with expected periods when actual drifts occur in both the simulated datasets and at key periods of performance drift in the Covid timeline.

4   Results

4.1   Simulated Data

The aim of our experiments involves learning a regression model that predicts the accuracy shift over batches with the domain-drift detection metrics described in the previous paragraph. To achieve this, we sample 70% of data from each batch while the remaining data is used to evaluate the algorithm. Both the discrimination error and the average score metrics are computed by comparing the source domain with all the following batches from the target domain. The regression algorithm is then evaluated on hold-out test sets from the target domain batches to assess its performance drift prediction capabilities. Figure 5 highlights the true performance changes (top left) and the performance shifts (top right) of the assessed model as new data batches become available.

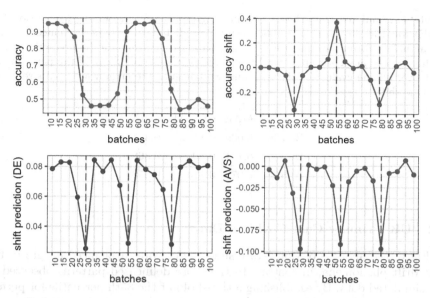

Fig. 5. Simulated Data: Performance (top left), shift in performance (top right), performance shift prediction by DE metric (bottom left) and AVS metric (bottom right)

The results obtained by using the DE and the AVS regression models to predict the performance shift are shown respectively in Fig. 5 (bottom left) and Fig. 5 (bottom right). The red vertical lines denote where a drift is detected based upon applying a threshold set to the 85th percentile for each of these metrics. Notice how both metrics identify the three simulated drifts with sharp changes at the appropriate drift points indicating that a change in the data has occurred and the model needs to be re-assessed.

Table 1 documents how close the regression models come to correctly predicting performance shift. It appears that while both metrics are good at identifying performance drift locations, AVS generally produces better predictions of drift magnitude with a smaller mean absolute error (MAE) though the maximum error (MAX) is higher when compared to DE.

This experimental analysis conducted on simulated data underscores the potential effectiveness of employing metrics such as Discrimination Error (DE) and Average Variance of Scores (AVS) when assessing newly available data batches during the monitoring of approved health apps. This methodology holds promise for early identification of substantial performance shifts, a valuable capability even prior to the availability of labeled data. With this insight obtained from simulated data, our focus shifts to the examination of a synthetic primary care dataset focused on Covid risk factors and outcomes.

Table 1. Simulated Data: Mean Absolute Error (MAE), its standard deviation (sd-MAE), and the Max error (MAX) between the predicted values of performance shift and the actual performance shift over the data batches. To compare the predictive power of this regression algorithm with a baseline, we compute the mean of the performance shift over all the evaluation datasets and we measure the error when compared with the true shift.

	MAE	sd-MAE	MAX
Baseline	0.089	0.111	0.391
DE	0.131	0.106	0.368
AVS	0.079	0.113	0.458

4.2 UK Primary Care Covid-19 Data

In Fig. 6, the analysis of Covid-19 results underscores a nuanced nature in the true drift phenomena, in contrast to the more delineated patterns observed in the simulated data. By establishing a threshold at the 85th percentile for performance change, discernible drift instances become apparent in July 2020, September 2020, and January 2021. Interestingly these shifts respectively coincide with the first lockdown, the appearance of a new Covid variant in the UK, and the point where the roll out of the vaccine reached all vulnerable people [16]. The predicted shift indicated by the DE metric is relatively stable in the series but drops between June and August 2020 (2020-06 and 2020-08) and again between Feb 2021 and Apr 2021 (2021-02 and 2021-04). This matches the actual fluctuations in the performance seen in the true shift. The predicted values by the AVS metric increases between June 2020 and Sept 2020 (2020-06 and 2020-09) and then drops when the fluctuating true shifts occurs from Jan 2021 to April 2021 (2021-01 to 2021-04). It is noteworthy that the primary peaks in the DE metric align with the occurrences of the second and third true shifts (new variant and vaccinated population). It is essential to recognize that DE identifies alterations in the data by assessing the classifier's discriminatory capacity between the two batches. On the other hand, the foremost peaks in the AVS metric correspond to the initial two shifts, encompassing the lockdown and the advent of a new variant, indicating a heightened level of uncertainty in predictive scores during these particular phases of the series.

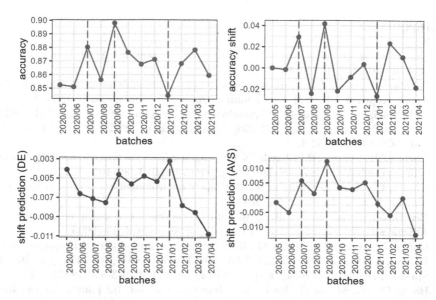

Fig. 6. Covid Data. Performance (top left), shift in performance (top right), performance shift prediction by DE metric (bottom left) and AVS metric (bottom right).

5 Conclusions

This work explores how algorithms can assist in facilitating the post-deployment monitoring of AI healthcare software, particularly in scenarios where ground truth labels are unavailable. The method presented assesses both i) the uncertainty in model performance and ii) the discrimination error between training batches and subsequent test batches for identifying drift in the performance of AI models. The obtained results are encouraging, suggesting the viability of employing such metrics to monitor post-deployment changes in the performance of AI health products to meet regulatory requirements.

It is imperative to acknowledge the multifaceted nature of performance drift, which may be attributed to various factors such as alterations in population demographics, modifications in clinical tests, and updates in data collection methodologies. These factors can significantly impact the predictive performance of an AI model. While our study has provided promising insights, it represents an initial exploration into the utility of metrics in detecting certain facets of performance drift. Future work will involve expanding this research to explore a broader range of metrics that are designed to specifically identify different expected changes, including but not limited to distributional changes, correlation modifications, and alterations in labeling.

Acknowledgements. This project has been made possible by a grant from the £ 3.7 million Regulators' Pioneer Fund launched by the Department for Business, Energy and Industrial Strategy (BEIS).

References

1. Gama, J., et al.: A survey on concept drift adaptation. ACM Comput. Surv. (CSUR) **46**(4), 1–37 (2014)
2. Rotalinti, Y., et al.: Detecting drift in healthcare AI models based on data availability. In: Koprinska, I., et al. (eds.) Joint European Conference on Machine Learning and Principles and Practice of Knowledge Discovery in Databases, ECML PKDD 2022. CCIS, vol. 1753, pp. 243–258. Springer, Cham (2022). https://doi.org/10.1007/978-3-031-23633-4_17
3. Hoens, T.R., Polikar, R., Chawla, N.V.: Learning from streaming data with concept drift and imbalance: an overview. Prog. Artif. Intell. **1**, 89–101 (2012)
4. Ditzler, G., et al.: Learning in nonstationary environments: a survey. IEEE Comput. Intell. Mag. **10**(4), 12–25 (2015)
5. Ben-David, S., et al.: Analysis of representations for domain adaptation. In: Advances in Neural Information Processing Systems, vol. 19 (2006)
6. Gama, J., Medas, P., Castillo, G., Rodrigues, P.: Learning with drift detection. In: Bazzan, A.L.C., Labidi, S. (eds.) SBIA 2004. LNCS (LNAI), vol. 3171, pp. 286–295. Springer, Heidelberg (2004). https://doi.org/10.1007/978-3-540-28645-5_29
7. Baena-García, M., et al.: Early drift detection method. In: Fourth International Workshop on Knowledge Discovery from Data Streams, vol. 6 (2006)
8. Berger, J.O.: Statistical Decision Theory and Bayesian Analysis. Springer, New York (2013). https://doi.org/10.1007/978-1-4757-4286-2
9. Kelly, M.G., Hand, D.J., Adams, N.M.: The impact of changing populations on classifier performance. In: Proceedings of the Fifth ACM SIGKDD International Conference on Knowledge Discovery and Data Mining (1999)
10. Tsymbal, A.: The problem of concept drift: definitions and related work. Comput. Sci. Dept. Trinity Coll. Dublin **106**(2), 58 (2004)
11. Wares, S., Isaacs, J., Elyan, E.: Data stream mining: methods and challenges for handling concept drift. SN Appl. Sci. **1**, 1–19 (2019)
12. Bifet, A., Gavalda, R.: Learning from time-changing data with adaptive windowing. In: Proceedings of the 2007 SIAM International Conference on Data Mining. Society for Industrial and Applied Mathematics (2007)
13. Žliobaite, I.: Change with delayed labeling: when is it detectable?. In: 2010 IEEE International Conference on Data Mining Workshops. IEEE (2010)
14. Ackerman, S., Raz, O., Zalmanovici, M.: FreaAI: automated extraction of data slices to test machine learning models. In: Shehory, O., Farchi, E., Barash, G. (eds.) EDSMLS 2020. CCIS, vol. 1272, pp. 67–83. Springer, Cham (2020). https://doi.org/10.1007/978-3-030-62144-5_6
15. Clinical Practice Research Datalink, 1 June 2022. https://www.cprd.com
16. Covid-19 in the UK, 18 November 2023. https://coronavirus.data.gov.uk/
17. Wolf, A., et al.: Data resource profile: clinical practice research Datalink (CPRD) aurum.". Int. J. Epidemiol. **48**(6), 1740–1740g (2019)
18. Agrawal, R., Imielinski, T., Swami, A.: Database mining: a performance perspective. IEEE Trans. Knowl. Data Eng. **5**(6), 914–925 (1993)
19. Marrs, G.R., Hickey, R.J., Black, M.M.: The impact of latency on online classification learning with concept drift. In: Bi, Y., Williams, M.A. (eds.) Knowledge Science, Engineering and Management: 4th International Conference, KSEM 2010, Belfast, Northern Ireland, UK, 1–3 September 2010, Proceedings, vol. 6291, pp. 459–469. Springer, Heidelberg (2010). https://doi.org/10.1007/978-3-642-15280-1_42

An Interpretable Human-in-the-Loop Process to Improve Medical Image Classification

Joana Cristo Santos[✉][iD], Miriam Seoane Santos[iD],
and Pedro Henriques Abreu[iD]

University of Coimbra, CISUC, Department of Informatics Engineering,
Coimbra 3030-290, Portugal
{jcds,miriams,pha}@dei.uc.pt

Abstract. Medical imaging classification improves patient prognoses by providing information on disease assessment, staging, and treatment response. The high demand for medical imaging acquisition requires the development of effective classification methodologies, occupying deep learning technologies, the pool position for this task. However, the major drawback of such techniques relies on their black-box nature which has delayed their use in real-world scenarios. Interpretability methodologies have emerged as a solution for this problem due to their capacity to translate black-box models into clinical understandable information. The most promising interpretability methodologies are concept-based techniques that can understand the predictions of a deep neural network through user-specified concepts. Concept activation regions and concept activation vectors are concept-based implementations that provide global explanations for the prediction of neural networks. The explanations provided allow the identification of the relationships that the network learned and can be used to identify possible errors during training. In this work, concept activation vectors and concept activation regions are used to identify flaws in neural network training and how this weakness can be mitigated in a human-in-the-loop process automatically improving the performance and trustworthiness of the classifier. To reach such a goal, three phases have been defined: training baseline classifiers, applying the concept-based interpretability, and implementing a human-in-the-loop approach to improve classifier performance. Four medical imaging datasets of different modalities are included in this study to prove the generality of the proposed method. The results identified concepts in each dataset that presented flaws in the classifier training and consequently, the human-in-the-loop approach validated by a team of 2 clinicians team achieved a statistically significant improvement.

Keywords: Interpretability · Classification · Medical Imaging

This work was funded by the FCT - Foundation for Science and Technology, I.P./MCTES through national funds (PIDDAC), within the scope of CISUC R&D Unit - UIDB/00326/2020 or project code UIDP/00326/2020. The work of Joana Cristo Santos was financially supported by the Portuguese Funding Institution Foundation for Science and Technology (FCT) under Grant 2020.05488.BD.

I. Miliou et al. (Eds.): IDA 2024, LNCS 14641, pp. 179–190, 2024.
https://doi.org/10.1007/978-3-031-58547-0_15

1   Introduction

Medical imaging is an essential tool for diagnosing, managing, and treating patients. However, the high demand of medical imaging acquisition and posterior time-consuming clinical assessment requires the development of efficient techniques to aid in this process. Medical image classification is a computational science problem that improves patient prognoses by providing information on disease assessment, staging and treatment response.

Most of the classification methodologies proposed in the literature are based on image enhancement [1,2] and the development of refined neural network architectures [3,4] that outperform state-of-the-art and clinician assessments, occupying deep learning technologies, the pool position for this task. However, the major issue of such techniques relies on their black-box nature which has delayed their use in real-world scenarios.

Interpretability methodologies have become widespread for medical applications due to their capacity to translate black-box models into understandable information. These methodologies are divided mainly into intrinsic interpretability which refers to machine learning models considered interpretable due to their simple structure and post hoc interpretability which refers to the application of interpretation methods after model training [5]. Interpretability strategies can be differentiated into categories such as feature importance which gives the importance of an input feature on the prediction, saliency maps that visually illustrate the importance of different features by coloring a pixel in a given prediction, and model visualization techniques that help visualize patterns detected in an image or the feature distribution in the dataset [6]. These methods are also described according to global interpretability which aims to understand the common patterns on a population level, by studying the model's parameters and the learned relationships [7].

In the healthcare scenario, one of the most promising techniques based on global interpretability strategy is concept-based techniques. These methods can understand the predictions of a deep neural network through user-specified concepts, which can be any data characteristic from a visual pattern to clinical information. However, concepts should be meaningful and coherent to increase human understanding of such models [8]. Concept-based explanations were first formalized with the concept activation vector (CAV) formalism [9] to generate post hoc global explanations for neural networks. CAV [9] is a numerical representation that generalizes a concept in the activation space of a neural network layer and measures the extent of that concept's influence on the prediction for a particular class. CAVs are learned by training a linear classifier to distinguish between the activations produced by a concept's examples and random opposing examples in any layer [9].

Extensions of this work, such as concept activation regions (CAR), intend to improve the generation of global explanations. CAR [10] introduces an extension of the CAV formalism that does not assume linear separability by allowing concept examples to be scattered across different clusters in the neural network latent space. This generalization is implemented by replacing the linear classifiers with kernel-based support vector classifiers. The global explanations provided by the CAV and CAR implementations allow the identification of the relationships that the network learned and can be used to identify possible errors during training.

In the case that these errors are identified, it may be possible to use this knowledge to improve the classification model. However, the identified flaws should present significantly different behavior in comparison to other concepts so that data balancing or augmentation may lead to improved classifier performance. Several works have studied the application of these approaches in medical imaging [11–13], however, none of these studies have after concept detection presented an automatic approach capable of improving the performance of the classifier.

In this work, a study of how the CAV and CAR implementations can be used to identify flaws in the training of a neural network and how these weaknesses can be mitigated in a human-in-the-loop process is developed. The main research question we aim to answer is *"Could the identification of user-defined concepts within a human-in-the-loop approach be able to improve the performance and trustworthiness of a classifier automatically?"*

To reach such goal, three phases have been defined: in the initial phase, the baseline classifier is trained using publicly available datasets; in the second phase, the concept-based interpretability methodologies CAR and CAV are applied to the baseline classifier, and the results are analyzed to identify the concepts that affect the performance of the baseline classifier; lastly, a human-in-the-loop approach validated by a team of 2 clinicians that manipulates the dataset according to the concepts identified previously is implemented to improve the classifier performance. Four medical imaging datasets of different imaging modalities (histology, endoscopy, mammography, and smartphone images) are included in this study to prove the generality of the proposed method. The achieved results significantly improve the performance of the baseline classifiers.

The rest of the paper is organized as follows: In Sect. 2, the methodology implemented in this paper is described. In Sect. 3, the results for the CAV and CAR classifiers and the human-in-the-loop approach are presented. Lastly, in Sect. 4, the conclusion of the study is presented.

2   Methodology

This research study is divided into three main parts: the training of the baseline classifier, the implementation of the CAR and CAV methodologies, and lastly, the implementing a human-in-the-loop approach to improve classifier performance. The experimental setup is illustrated for the mammography imaging dataset in Fig. 1.

2.1   Dataset Description

The four datasets included in this study are BreakHis [14], KID2 [15,16], CBIS-DDSM [17,18], and PAD-UFES-20 [19]. The BreakHis dataset [14] is composed

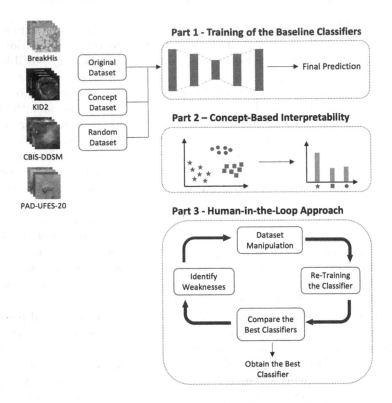

Fig. 1. Experimental Setup of the Research Study.

of 2,480 benign and 5,429 malignant microscopic histological images of breast tumor tissue collected using different magnifying factors. The KID2 [15,16] dataset consists of 1,778 normal and 593 pathology wireless capsule endoscopy images. The CBIS-DDSM [17,18] is composed of 1,872 calcification and 1,696 mass region of interest of mammography images. The PAD-UFES-20 [19] dataset contains 1,089 skin cancer and 1,209 disease smartphone images depicting various skin lesions. These datasets are publicly available and offer a extensive collection of medical images with complementary clinical/patient information. The classification problem is based on binary classification within each dataset, targeting the main classes previously identified.

2.2   Baseline Classifier Description

The Residual Network (ResNet) architecture [20] is used as the baseline classifier to implement the classification problems. ResNet has been implemented as the backbone of several neural networks for classification applications in medical imaging [21,22]. The ResNet50 architecture is chosen for this work due to its simplicity and lower computational requirements and because it presents the best overall performance for the image datasets previously described. This neural network is pre-trained with ImageNet weights and fine-tuned for the classes of each of the classification problems. For the study of interpretability across the neural network, hooks modules are introduced on the first pool layer and layers 1–4 of the ResNet50 architecture.

The ResNet50 network is trained for a maximum of 200 epochs with an early stop of 20 epochs patience, using the Adam Optimizer and learning rate 1e-04. Transformations by horizontal/vertical flips and rotation are implemented during the training to improve the performance of the model. The dataset is divided into 60% training, 30% testing, and 10% validation sets for the ResNet, CAR, and CAV classifiers. To mitigate bias, the simulations of the classifiers are repeated for 30 runs. The implementation code is available in https://github.com/joanacsantos/Mammography_Image_Interpretability.

2.3   Concept-Based Interpretability

The interpretability of the baseline classifiers previously described is obtained using CAV and CAR classifiers. The implementation process of the CAV and CAR classifiers comprises five key steps:

1. **Definition of the concepts** - User-defined concepts are manually selected for each image dataset (the concept selection process is detailed in the following section);
2. **Creation of Training and Testing Sets** - To train classifiers, distinct training and testing sets are created for both concept and random opposing datasets;
3. **Training the Classifiers** - The hidden layers of the baseline classifier are targeted and the CAV/CAR classifier that separates the activations generated by the concept set and the random set is trained. The sensitivity of the prediction is measured by calculating the directional derivative of the prediction in the classifier direction. The resulting coefficient vector of the trained binary classifier is the CAV/CAR of the concept;
4. **Evaluation of Concept Accuracy** - After the training, the accuracy of the classifiers is assessed using the testing dataset;
5. **Application of CAV and CAR to the baseline classifier** - The classifiers are applied to the baseline classifier. To produce a global explanation of a class, the testing score is calculated by the ratio of inputs with positive conceptual sensitivities to the number of inputs for a class.

This process is repeated for each concept and for each baseline classifier. Given that the baseline classifiers are repeated for 30 runs, the CAV and CAR classifiers are similarly trained and tested for each of the 30 classifiers for every dataset.

The analysis of concept-based interpretability in this study is focused on two main components: the analysis of concept accuracy and the interpretation of the global explanations. The concept accuracy analysis assesses the generalization of CAV/CAR classifiers to unseen datasets. The accuracy of the testing dataset is pivotal in determining concept accuracy, with both concept and random datasets used for training and testing the classifiers. The global explanation interpretation compares testing scores for each concept and class to ascertain if CAR concept regions correlate with classification classes.

2.4   Concept Selection

The selected concepts for this work are based on the information available from each dataset. Since the selection of user-defined concepts should be meaningful and coherent to increase human understanding, the datasets included in this work were selected due to the diversity, representation, and clinical relevance of the information available in each dataset. To ensure each defined concept is represented in the creation of the concept and random datasets, 200 image samples for the training set and 30 samples for the testing set are required for each concept [10]. The concepts available for:

- **BreaKHis dataset** are the categories of histological types of tumors. The dataset contains four types of benign breast tumors - adenosis (A), fibroadenoma (F), tubular adenoma (TA), and phyllodes tumor (PT); and four malignant tumors - ductal carcinoma (DC), lobular carcinoma (LC), mucinous carcinoma (MC), and papillary carcinoma (PC);
- **KID2** dataset are the different types of normal and pathology images: normal small bowel, normal stomach, normal esophagus, vascular and inflammatory;
- **CBIS-DDSM** dataset are divided into 4 categories: BI-RADS score (0,2,3,4, 5), breast density score (1,2,3,4), mass and calcification characteristics;
- **PAD-UFES-20** dataset are divided into 4 categories: diagnostic, region of skin lesion, and different indicators for itching, growing and bleeding. The diagnostic category is associated to the types of skin cancers (Basal Cell Carcinoma (BCC)) and diseases (Actinic Keratosis (ACK)).

2.5   Human-in-the-Loop Approach

The information provided by the concept-based interpretability is further explored to improve the performance of the classification problems using a human-in-the-loop approach. This approach with the direct intervention of a team of 2 clinicians is composed of five main steps:

1. **Analysis of the interpretability-based results to identify weaknesses in the training of the classifiers** - After the analysis of the interpretability results, the misclassified classes/concepts that the classifier is not capable of correctly classifying are identified. The misclassification is a consequence of a weakness of the neural networks' learning process which is not just related to low values of accuracy, precision, and recall but also related to the way it learns (the features considered during training - relevant information for a clinician). The identification of the weaknesses in the training allows the understanding of what happened incorrectly during the training process and the best alteration to perform to improve the learning process;

2. **Manipulation of the training dataset to mitigate these weaknesses** - After identifying the classes/concepts that require improvements, an approach to manipulate the training dataset based on elimination, augmentation, and/or balancing techniques is defined;

3. **Re-training of the classifiers** - the ResNet classifier is retrained with the manipulated training dataset;

4. **Comparison between the best classifiers** - The newly trained classifier is compared with the baseline and the other retrained classifiers in terms of performance and stability to understand if it corrected the weakness identified previously. The retrained classifier is only validated if it is considered trustworthy for real-world applicability by the clinicians;

5. **Repetition of the previous steps** - The previous steps are repeated until the clinicians achieve consensus on the performance and trustworthiness of the new classifier or 100 runs of the loop are achieved.

The inclusion of 2 experienced clinicians within the loop allows the validation of the improved classifiers in terms of improvement of performance, stability, and trustworthiness for real-world applicability.

3 Results and Discussion

3.1 Baseline Classifiers

Table 1 presents the results of all classifiers, where the best classification is achieved for the BreaKHis dataset. These results are considered the baseline for this work and it is expected to improve the performance of these classifiers in the next sections.

Table 1. Results for the baseline ResNet classifier.

	BreaKHis	KID2	CBIS-DDSM	PAD-UFES-20
Accuracy	95.92%	85.52%	91.84%	81.98%
Precision	95.14%	80.77%	91.90%	82.10%
Recall	95.44%	81.67%	90.91%	82.05%

3.2 Concept-Based Interpretability Results

Table 2 represents the value of the overall concept and the number of layers for the CAR and CAV classifiers. The comparison between the CAR and CAV classifiers shows that the CAR classifier achieves higher overall concept accuracy than the CAV classifier.

After reviewing the results of the classifiers by layer, the results of the classifiers by concepts is performed to identify the concepts with the best concept accuracy values. These concepts represent concept vectors/regions that the

Table 2. Average Concept Accuracy Values for the CAV and CAR Classifiers according to the ResNet Layer. The best performing classifier for each Dataset and Layer is highlighted in blue.

		Average Concept Accuracy							
		BreaKHis		**KID2**		**CBIS-DDSM**		**PAD-UFES-20**	
		CAV	*CAR*	*CAV*	*CAR*	*CAV*	*CAR*	*CAV*	*CAR*
Layer	*Pool*	0.66	0.72	0.69	0.73	0.59	0.65	0.56	0.61
	Layer 1	0.76	0.78	0.75	0.75	0.64	0.67	0.59	0.61
	Layer 2	0.80	0.82	0.75	0.73	0.67	0.69	0.62	0.65
	Layer 3	0.85	0.86	0.81	0.80	0.72	0.74	0.67	0.68
	Layer 4	0.82	0.84	0.85	0.85	0.72	0.74	0.66	0.68

CAV/CAR classifiers are able to generalize and consequently, generate higher quality global explanations. Due to the high number of concepts considered in the analysis, only the two concept categories with the best concept accuracy values are presented for the datasets CBIS-DDSM and PAD-UFES-20.

Table 3 present the results for the concept accuracy analysis for the CAR/CAV classifiers and the global accuracy analysis for the CAV classifier according to the defined concepts.

In terms of average concept accuracy, the CAR classifier is overall better than the CAV classifier. The vascular and inflammatory concepts in the KID2 dataset and the density 1 concept in the CBIS-DDSM are the only exceptions to these results. The vascular and inflammatory concepts represent concepts within the pathology images that are underrepresented in the datasets (only 25.01%). Due to the unbalancing of the dataset, the linear classifier CAV may be more suitable to classify these two concepts. Similarly, Density 1 is the concept least represented in the density concept category and consequently, the CAV may be better than CAR to classify the concept.

In terms of global explanation analysis, the results for the KID2, BreaKHis, and PAD-UFES-20 show that each concept presents a higher correlation to only one classification class. However, the CBIS-DDSM dataset does not present similar behavior. The CBIS-DDSM dataset presents concepts where there remains a slight increase of global explanation values in the incorrect classification class. Even though the BI-RADS category presents higher concept accuracy values than the density category, this category presents the BI-RADS 4 concept that is not capable of correlating with only one classification class. Consequently, the BI-RADS category is excluded from the human-in-the-loop approach. The PAD-UFES-20 dataset diagnostic and bleeding categories present a high correlation between concepts and classification classes. Although the diagnostic category presents high values of concept accuracy, both categories are included in the human-in-the-loop approach.

Table 3. Average Concept Accuracy and Global Explanation according to the BreasKHis, KID2, CBIS-DDSM and PAD-UFES-20 Dataset Concepts. The best performing classifier for each Concept is highlighted in green. The Classification Class with the highest correlation to each Concept is highlighted in orange and the cases with a significant score in the incorrect Class are highlighted in yellow.

Dataset	Concepts	Concept Accuracy		Global Explanation	
		CAV	CAR	Benign	Malignant
BreaKHis	A	0.86	0.87	0.55	0.01
	F	0.77	0.80	0.86	0.03
	TA	0.87	0.88	0.52	0.02
	PT	0.79	0.81	0.71	0.01
	DC	0.74	0.77	0.02	0.77
	LC	0.77	0.82	0.02	0.43
	MC	0.71	0.73	0.07	0.46
	PC	0.72	0.74	0.06	0.47
		CAV	CAR	Normal	Pathology
KID2	Small Vowel	0.71	0.73	0.46	0.26
	Stomach	0.82	0.82	0.47	0.01
	Esophagus	0.90	0.91	0.25	0.01
	Vascular	0.72	0.71	0.05	0.84
	Inflammatory	0.71	0.69	0.11	0.86
		CAV	CAR	Calcification	Mass
CBIS-DDSM	BI-RADS 0	0.61	0.63	0.28	0.65
	BI-RADS 2	0.77	0.79	0.39	0.03
	BI-RADS 3	0.61	0.65	0.09	0.68
	BI-RADS 4	0.59	0.60	0.56	0.44
	BI-RADS 5	0.70	0.73	0.25	0.43
	Density 1	0.66	0.63	0.26	0.52
	Density 2	0.55	0.58	0.33	0.70
	Density 3	0.56	0.59	0.63	0.39
	Density 4	0.62	0.65	0.53	0.21
		CAV	CAR	Cancer	Disease
PAD-UFES-20	ACK	0.72	0.75	0.14	0.62
	BCC	0.73	0.75	0.83	0.11
	Bleeding	0.68	0.70	0.71	0.10
	No Bleeding	0.68	0.70	0.29	0.90

3.3 Human-in-the-Loop Approach Results

After analyzing the global explanations presented previously, it is identified that there is a high correlation between the concepts and the classification classes. However, in certain concepts, there remains a significant testing CAR score in the incorrect classification class. To improve the performance of the classifiers, it is decided to balance the representation of all the concepts to reduce the testing CAR score in the incorrect classification classes. For this purpose, various manipulation techniques (elimination, augmentation, and balancing) are explored to gradually improve the representation of the concepts within the training set while maintaining a continuous validation of the approach with the team of clinicians. After each application of a manipulation technique, the performance and trustworthiness of the retrained classifier are re-evaluated and the process is repeated until consensus on the performance and trustworthiness of the new classifier between the clinicians or 100 runs of the loop is achieved. The final retrained model for all datasets is based on stratified 10k-fold cross-validation where all the concept categories are balanced between the training and testing sets. The results for these classifiers and respective improvement are presented in Table 4.

Table 4. Results for the Overall Best Strategy of Improvement of Classifier Performance for all Datasets.

	BreasKHis	KID2	CBIS-DDSM	PAD-UFES-20	
				Diagnostic	Bleeding
Accuracy	98.80% (↑2.85%)	89.08% (↑3.56%)	92.57% (↑0.73%)	84.60% (↑2.62%)	85.07% (↑3.09%)
Precision	98.78% (↑3.64%)	85.72% (↑4.95%)	92.62% (↑0.72%)	84.63% (↑2.53%)	85.05% (↑2.95%)
Recall	98.72% (↑3.28%)	85.12% (↑3.45.%)	92.53% (↑1.62%)	84.61% (↑2.56%)	85.13% (↑3.08%)

A Wilcoxon rank-sum statistical test at an alpha value of 0.05 is used to test for a statistically significant difference in means of each metric between the stratified 10k-fold cross-validation based classifiers and the baseline classifiers. All the improvements identified in Table 4 are considered statistically significant, except for the precision value of the CBIS-DDSM dataset.

Even though, most datasets achieve an improvement in classifier performance (3%), the improvements of the CBIS-DDSM dataset are not as significant in comparison to other datasets. This result is a consequence of the weak correlation between concepts and classification classes for the density concepts, that limits the classifier performance improvement. To improve the performance of this dataset, it is consider that there is a need to introduce more images to the original dataset to improve the diversity of the concepts/classes within the training dataset. The augmentation of this dataset should be focused on the cases that present a significant score in the incorrect class identified in Table 3 (BI-RADS 0 and 5, and Density 1, 2, 3 and 4).

pagebreak The unbalanced representation of concepts within the dataset, previously mentioned in the concept-based interpretability results, lead to a reduction of performance of the CAR classifier. However, after applying the 10k-fold cross validation based on the concept, it is possible to verify that the unbalanced datasets (BreaKHis and KID2) present similar improvements to the balanced dataset (PAD-UFES-20).

4 Conclusion

In this work, a study on how concept activation vectors can be used to identify flaws in the training of a neural network and how this weakness can be mitigated in a human-in-the-loop process is presented.

Four medical imaging datasets of different modalities are included in this study to prove the generality of the proposed method. The results identified concepts in each dataset that presented flaws in the classifier training and consequently, the human-in-the-loop approach controlled by a 2 clinicians' team achieved a statistically significant improvement of 3% in accuracy, precision, and recall.

The inclusion of 2 experienced clinicians within the loop allows the validation of the improved classifiers in terms of improvement of performance, stability, and trustworthiness for real-world applicability. These results show that the identification of user-defined concepts can improve the performance of a classifier while providing information on how that concept influences the classification.

References

1. Yu, X., Kang, C., Guttery, D.S., Kadry, S., Chen, Y., Zhang, Y.D.: ResNet-SCDA-50 for breast abnormality classification. IEEE/ACM Trans. Comput. Biol. Bioinf. **18**(1), 94–102 (2020)
2. Shyni, H.M., Chitra, E.: A comparative study of X-ray and CT images in COVID-19 detection using image processing and deep learning techniques. Comput. Methods Programs Biomed. Update **2**, 100054 (2022)
3. Chauhan, T., Palivela, H., Tiwari, S.: Optimization and fine-tuning of DenseNet model for classification of COVID-19 cases in medical imaging. Int. J. Inf. Manage. Data Insights **1**(2), 100020 (2021)
4. Suganyadevi, S., Seethalakshmi, V., Balasamy, K.: A review on deep learning in medical image analysis. Int. J. Multimedia Inf. Retrieval **11**(1), 19–38 (2022)
5. Molnar, C.: Interpretable machine learning (2020). https://christophm.github.io/interpretable-ml-book/
6. Amorim, J.P., Abreu, P.H., Fernández, A., Reyes, M., Santos, J., Abreu, M.H.: Interpreting deep machine learning models: an easy guide for oncologists. IEEE Rev. Biomed. Eng. **16**, 192–207 (2021)
7. Stiglic, G., Kocbek, P., Fijacko, N., Zitnik, M., Verbert, K., Cilar, L.: Interpretability of machine learning-based prediction models in healthcare. Wiley Interdisc. Rev. Data Min. Knowl. Discovery **10**(5), e1379 (2020)
8. Fang, Z., Kuang, K., Lin, Y., Wu, F., Yao, Y.F.: Concept-based explanation for fine-grained images and its application in infectious keratitis classification. In: Proceedings of the 28th ACM International Conference on Multimedia, pp. 700-708. ACM, New York, NY, USA (2020)

9. Kim, B., et al.: Interpretability beyond feature attribution: quantitative testing with concept activation vectors (TCAV). In: International Conference on Machine Learning, vol. 80, pp. 2668–2677. PLMR (2018)

10. Crabbé, J., van der Schaar, M.: Concept activation regions: a generalized framework for concept-based explanations. Adv. Neural. Inf. Process. Syst. **35**, 2590–2607 (2022)

11. Lucieri, A., Bajwa, M.N., Braun, S.A., Malik, M.I., Dengel, A., Ahmed, S.: On interpretability of deep learning based skin lesion classifiers using concept activation vectors. In: International Joint Conference on Neural Networks, pp. 1–10. IEEE (2020)

12. Gamble, P., et al.: Determining breast cancer biomarker status and associated morphological features using deep learning. Commun. Med. **1**(1), 14 (2021)

13. Janik, A., Dodd, J., Ifrim, G., Sankaran, K., Curran, K.: Interpretability of a deep learning model in the application of cardiac MRI segmentation with an ACDC challenge dataset. In: Medical Imaging 2021: Image Processing, vol. 11596, pp. 861–872. SPIE (2021)

14. Spanhol, F.A., Oliveira, L.S., Petitjean, C., Heutte, L.: A dataset for breast cancer histopathological image classification. IEEE Trans. Biomed. Eng. **63**(7), 1455–1462 (2015)

15. Iakovidis, D.K., Georgakopoulos, S.V., Vasilakakis, M., Koulaouzidis, A., Plagianakos, V.P.: Detecting and locating gastrointestinal anomalies using deep learning and iterative cluster unification. IEEE Trans. Med. Imaging **37**(10), 2196–2210 (2018)

16. Koulaouzidis, A., et al.: KID Project: an internet-based digital video atlas of capsule endoscopy for research purposes. Endosc. Int. Open **5**(6), E477–E483 (2017)

17. Lee, R.S., Gimenez, F., Hoogi, A., Miyake, K.K., Gorovoy, M., Rubin, D.L.: A curated mammography data set for use in computer-aided detection and diagnosis research. Sci. Data **4**(1), 1–9 (2017)

18. Heath, M., et al.: Current status of the digital database for screening mammography. In: Karssemeijer, N., Thijssen, M., Hendriks, J., van Erning, L. (eds.) Digital Mammography. Computational Imaging and Vision, vol. 13, pp 457–460. Springer, Dordrecht (1998). https://doi.org/10.1007/978-94-011-5318-8_75

19. Pacheco, A.G., et al.: PAD-UFES-20: a skin lesion dataset composed of patient data and clinical images collected from smartphones. Data Brief **32**, 106221 (2020)

20. He, K., Zhang, X., Ren, S., Sun, J.: Deep residual learning for image recognition. In: Proceedings of the IEEE Conference on Computer Vision and Pattern Recognition, pp. 770–778. IEEE (2016)

21. Falconi, L.G., Pérez, M., Aguilar, W.G., Conci, A.: Transfer learning and fine tuning in breast mammogram abnormalities classification on CBIS-DDSM database. Adv. Sci. Technol. Eng. Syst. J. **5**(2), 154–165 (2020)

22. Cheng, J., et al.: ResGANet: residual group attention network for medical image classification and segmentation. Med. Image Anal. **76**, 102313 (2022)

Hybrid Ensemble-Based Travel Mode Prediction

Paweł Golik[ID], Maciej Grzenda[(✉)][ID], and Elżbieta Sienkiewicz[ID]

Faculty of Mathematics and Information Science, Warsaw University of Technology,
Koszykowa 75, 00-662 Warszawa, Poland
{Pawel.Golik.stud,Maciej.Grzenda,Elzbieta.Sienkiewicz}@pw.edu.pl

Abstract. Travel mode choice (TMC) prediction, which can be formulated as a classification task, helps in understanding what makes citizens choose different modes of transport for individual trips. This is also a major step towards fostering sustainable transportation. As behaviour may evolve over time, we also face the question of detecting concept drift in the data. This necessitates using appropriate methods to address potential concept drift. In particular, it is necessary to decide whether batch or stream mining methods should be used to develop periodically updated TMC models. To address the challenge of the development of TMC models, we propose the novel Incremental Ensemble of Batch and Stream Models (IEBSM) method aimed at adapting travel mode choice classifiers to concept drift possibly occurring in the data. It relies on the combination of drift detectors with batch learning and stream mining models. We compare it against batch and incremental learners, including methods relying on active drift detection. Experiments with varied travel mode data sets representing both city and country levels show that the IEBSM method both detects drift in travel mode data and successfully adapts the models to evolving travel mode choice data. The method has a higher rank than batch and stream learners.

Keywords: Travel mode choice · stream mining · concept drift

1 Introduction

The growth in the volume of data streams has caused data mining methods, which analyze bounded and stationary datasets, to be potentially unable to adapt to shifting data patterns and dynamic phenomena [3]. In addition to the regular fluctuations and random variations in the data, the *concept drift* phenomenon [3,5] is frequently observed due to reasons such as seasonality [4]. More precisely, concept drift arises when the probability $p(x)$ and/or $p(y|x)$ changes between consecutive labelled stream instances $\{x, y\}$ [4].

An example of a phenomenon possibly affected by concept drift is the selection of travel modes by travellers. Predicting travel mode choices (TMCs) [11,12] is an important aspect of transportation planning, as it helps understand under what conditions people are likely to use e.g. private cars or public transport

© The Author(s), under exclusive license to Springer Nature Switzerland AG 2024
I. Miliou et al. (Eds.): IDA 2024, LNCS 14641, pp. 191–202, 2024.
https://doi.org/10.1007/978-3-031-58547-0_16

(PT). This can influence inter alia the development of PT routes, PT schedules, but also cycle lanes. However, travellers' preferences and choices may change with time, which offers a compelling example of changing data streams.

The TMC datasets typically used to model travel mode choices include features such as duration and reason of the trip, and traveller's attributes, e.g., age and gender, but also e.g. the number of PT interchanges [12], represented in vector $\mathbf{x}$. Each trip record includes as well a travel mode selected by the traveller [7,10,11], i.e., a class $y_i \in Y$, which includes using a car, public transport, cycling and walking. As trips are made over time, it is not obvious whether classifiers predicting travel mode should be trained with batch methods assuming stationary settings i.e., fixed $p(y|x)$ and $p(x)$ or stream mining methods addressing concept drift [8]. Stream mining algorithms offer the advantage of continuous incremental model training. This enables a model to adapt to concept drift, but it fundamentally alters the training process and the employed machine learning algorithms by including either passive or active adaptation to concept drift [3]. Alternatively, models can be trained on batches of data, a typical approach. In this scenario, models can effectively detect patterns present in batches due to the ability to iterate over data multiple times, often yielding improved outcomes. Still, their effectiveness might be reduced when the incoming data shifts [6].

Given the formulation of a TMC problem as a classification task, [10,11], a question is posed whether online classifiers should be applied to learn possibly evolving models reflecting the evolving decisions of travellers, or whether the magnitude of concept drift(s) is not sufficient to use online learners rather than batch models. The answer is likely to depend on how stationary the process is in different cities or countries and may even change with time. Hence, we propose an ensemble method combining multiple batch and online methods to reduce the risk of selecting an under-performing method. Furthermore, the method we propose, can utilize multiple batch-learning models. It enables the use of distinct configurations of drift detection and batch model retraining strategies, referred to as *drift handling strategies*, for each batch learner.

Hence, in this work, we propose the novel Incremental Ensemble of Batch and Stream Models (IEBSM) method for predicting the preferred mode of transport. We evaluate both the method and baseline learners using various real datasets of successively recorded trips provided by respondents. The primary contributions of this work are as follows:

- We propose the novel IEBSM ensemble method combining drift detectors with batch and online learners. The method automates the use of multiple batch and online methods, drift detection and the retraining of batch models. The experiments we performed show that the IEBSM method yields performance gains over batch and online methods for various travel mode choice tasks. It provided highest ranked TMC models. We provide the open source implementation of the IEBSM method[1].

---

[1] Supplementary material with detailed results for all experiments and the repository with the source code developed using inter alia `river` (https://riverml.xyz/) and `scikit-learn` libraries are available at https://github.com/Shaveek23/batchstream.

- We investigate whether statistically significant changes occur in travel mode choice data for a number of travel mode choice data sets and confirm that such changes occured in each of the data sets. Moreover, we confirm that the IEBSM method both detected changes and successfully managed the introduction of selected updated batch models.

The remainder of this work is organized as follows. In Sect. 2, we provide an overview of related works. This is followed in Sect. 3 by the proposal of the novel method aiming to automate the use of varied underlying batch and online learners under concept drifting data streams. The results of the evaluation of the method and reference methods are provided in Sect. 4. This is followed by the conclusions and summary of future works in Sect. 5.

2 Related Works

TMC modelling [7,10,11] is concerned with predicting the travel mode most likely used by a person for their trip. Recently, the benefits arising from the use of machine learning methods for TMC tasks were discussed in [7,10]. In [7], random forest was shown to yield the best accuracy and computational cost among the tested classifiers. Over time, aspects such as temporal changes in the environment, seasonality, and evolving human preferences are all likely to affect those choices. Hence, in some TMC datasets such as those used in [10], apart from respondent and trip attributes, such as age, education and distance travelled, the features related to weather conditions were included. Still, batch machine learning methods not considering possible changes in travel mode choice decision boundaries $p(y|x)$ are typically used both in comparative studies [7,10] and surveys of machine learning for TMC modelling [11].

Apart from batch methods, online incremental learning methods have been developed as well, which are also suitable for learning in non-stationary environments [4]. Notable methods include adaptive random forest [8], which builds upon the random forest method to enable learning from non-stationary data streams. This way, real concept drift [4], i.e., a change in $p(y|x)$, can be addressed by changing the ensemble members. In the case of adaptive random forest, ensemble members can be replaced with new base learners better matching shifted class boundaries $p(y|x)$. Frequently, the evaluation of online models relies on first making prediction $\hat{y} = h_i(\mathbf{x}_i)$ with the current model h_i to use the instance to get a new, possibly different model $h_{i+1} = learn(h_i, \{\mathbf{x}_i, y_i\})$ through incremental training. This approach is referred to as test-then-train [3,8]. This illustrates the fact that incremental learning methods respecting stream mining assumptions are constrained by the fact that they can inspect each example at most once [3]. This may result in models of a lower performance than the models built within a batch process relying on the access to the entire data set and the ability to iteratively revisit all instances during the training.

As traditional ML models deployed in production settings might experience performance degradation over time, their application to evolving and potentially infinite data streams called for a new approach. Hence, when Machine Learning solutions are used in IT systems, a growing emphasis on the Machine Learning Operations (MLOps) process is observed. MLOps focuses on addressing data changes through Continuous Monitoring and Continuous Training steps, ensuring models adapt to data and concept shifts to maintain performance [15]. There are multiple methods for concept drift detection, mainly focusing on monitoring data distribution [3]. Similarly, the adaptation of batch models can be performed in a number of ways, such as retraining from scratch on all available data [17] or just the latest instance window.

The combination of online and batch learning methods has been considered before. In [9], neural model training and incremental training of an online learner was proposed in the form of a hybrid model that switches between the multilayer perceptron and stream mining models based on their recent accuracy over a sliding window. In [13], the authors combined initially trained batch models (e.g. decision trees) and gradually converted them into online models. This approach leverages the simultaneous predictions from both online and batch members. However, this approach does not involve concept drift detection or model adaptation. Instead, it adds new batch learners built on recent instances over time.

Hence, the question arises of how to build TMC classifiers, while considering concept drift of unknown magnitude. Importantly, not only human preferences towards different modes can change with time (e.g. depend on the time of the year) causing $p(y|x)$ changes, but also $p(x)$ clearly changes. Examples include travel to schools less likely to happen during school holidays.

3   Ensemble of Batch and Online Learners

The method proposed in this work relies on learning an ensemble of base models including both batch and online models to respond to possible virtual drift, i.e., changes in $p(x)$, and real drift i.e., changes in $p(y|x)$. To evaluate the IEBSM ensembles as well as reference online and batch methods, the test-then-train approach is applied. Online learners are trained the same way and with the same data stream irrespective of whether they are evaluated as standalone reference learners or participate in the ensemble. Similarly, batch learners are retrained in line with Algorithm 1 described below, both when they are evaluated as reference methods and when they are a part of an ensemble.

3.1   Training of Online and Batch Learners with TMC Data Streams

For online learning algorithms, the learners are provided with new labelled examples and updated incrementally, as defined in the test-then-train approach. In the case of batch learning algorithms, newly arriving $\{x_i, y_i\}$ instances are placed in the cache of the most recent instances. Then, drift detectors assess the cache to identify concept drift following the predefined drift handling strategy. If a drift is detected, the batch model undergoes retraining, using data from the cache under the chosen retraining strategy. Subsequently, the newly trained model and the previous model are evaluated on the successive n_{comp} instances. If the retrained model demonstrates superior performance compared to the old model, it replaces the previous model in use. This illustrates the challenge of batch learning adapted to an online setting, i.e., its dependence on hyperparameters such as n_{comp}. The batch learning models are trained for the first time using the initial n_{first_fit} instances. Prior to the collection of all these instances, a majority class model generates prediction output as the label of the class that has been observed most frequently up to that point. In this way, batch and stream models can be evaluated with the same instances $\{\mathbf{x}_i, y_i\}, i = 1, \ldots$ irrespective of the duration of the *warm-up* period of n_{first_fit} instances.

In our approach, the monitoring strategies used to detect drift rely on the analysis of the most recent instances. To achieve this, we partition these instances into two equal batches of s instances i.e., the reference batch and the current batch. Every s instances, we then compare the distributions of these two batches by applying statistical tests. A test is applied to each feature separately to detect possible changes in the distribution of the values of j-th feature x^j and label y. We associate a threshold θ with those tests, the interpretation of which varies depending on the drift detection method. As the tests to be applied depend on feature types, we discuss them in detail in Sect. 4. Besides testing for changes in the input feature/target distribution, we utilise detection techniques that monitor the performance of a model. A performance drop, defined as the F_1 macro score on the current batch falling below α of the F_1 macro score on a reference batch, suggests a concept drift. To initiate retraining, at least one drift detection method identifying a change in the data distribution of some feature or a change in model performance must detect a concept drift.

3.2   Building an Ensemble of Batch and Online Learners

As defined in Algorithm 1, we propose an ensemble-based method that aims to maximise the performance of travel mode choice predictions by combining predictions of both batch and online learners. The method builds an ensemble of N base learners, some of which can be online learners such as adaptive random forest [8], while the remaining ones can be batch learners. In line with the test-then-train approach, for every new instance, each base learners generates a prediction first. In the case of batch models, we propose to rely on majority class prediction prior to the collection of a sufficiently large training data set.

The IEBSM ensemble generates the ultimate prediction by aggregating the outputs from all member models, as shown in lines 11–16 of Algorithm 1. We propose two approaches for combining predictions. In both approaches, we record the value of a performance measure such as F_1 for every ensemble member in the prequential approach, i.e., over a sliding window of instances. Under the **Weighted Voting (WV)** approach we combine predictions using weights w_i assigned to each member m_i of ensemble M according to their recent performance calculated on the sliding window of instances. In the case of the **Dynamic Switching (DS)** approach, the final prediction of the ensemble is the prediction from the recently top-performing ensemble member. This corresponds to assigning $w_i = 1$ to the best model m_i, and $w_j = 0, j \neq i$ otherwise.

Algorithm 1: Training and evaluation of IEBSM models.

Input: $\{x_1, y_2\}, \ldots, \{x_i, y_i\}, \ldots$ - a labelled data stream, c - a method combining predictions of members, e - a method evaluating members, $S = \{S_1, \ldots, S_K\}, K \leq N$ - a set of drift handling strategies (one per each batch base learner), each defined by a vector of hyperparameter values controlling the way drift is detected and a batch model retrained

Data: $M = \{m_1, \ldots, m_N\}$ - an ensemble of base learners

1 **foreach** $\{x_i, y_i\} \in data\_stream$ **do**
2     $\hat{Y}_i \leftarrow [], scores_i \leftarrow []$
3   **foreach** $m \in M$ **do**
4   **if** $m.type == batch$ **and** $i <= n_{first\_fit}$ **then**
5           $\hat{Y}_i[m] \leftarrow get\_majority\_class(m.cache)$
6   **if** $i == n_{first\_fit}$ **then**
7              $m = m.first\_fit()$
8   **else**
9           $\hat{Y}_i[m] \leftarrow m.predict(x_i)$
10       $e.update\_model\_score(y_i, \hat{Y}_i[m])$
11   **if** $c == DS$ **then**
12       $weigths \leftarrow zeros(len(members))$
13       $weights[argmax(scores_i)] \leftarrow 1$
14   **if** $c == WV$ **then**
15       $weights \leftarrow (scores_i/sum(scores_i))$
16    $\hat{y}_i \leftarrow argmax_{c \in classes}\{\Sigma_{j \leftarrow 0, \hat{Y}_i[j]==c}^{mlen-1} weights[j]\}$
17   **foreach** $m \in M$ **do**
18   **if** $m.type == online$ **then**
19          $m = m.update\_model(x_i, y_i)$
20   **if** $m.type == batch$ **then**
21          $m.update\_cache(x_i, y_i)$
22   **if** $m.has\_concept\_drift\_occurred(S(m))$ **then**
23             $m.shadow\_model \leftarrow train\_on\_cached\_instances(m, S(m))$
24          $m \leftarrow evaluate\_shadow\_model(m, m.shadow\_model, x_i, y_i, \hat{Y}_i[m])$
25    $update\_performance\_metrics(y_i, \hat{y}_i)$

Next the training of base learners is considered. In the case of online learners, they are simply provided with the new instance $(\mathbf{x}_i, y_i)$, which may trigger updates of a model m. In the case of batch learners, first a cache of recent instances is updated. This is to store data to be used for potential retraining of a batch model based on a recent data set. Next, drift handling strategy $S(m)$ is used to define the way drift detection is performed, e.g. whether it is focused on virtual drift only and/or the performance of model m, and how sensitive drift detection is, which is defined by the settings of statistical tests. This illustrates the complexity of using batch learners in the case of concept drifting data streams.

In case a drift is detected, a new model is developed and stored as a shadow model to potentially replace the original one. This happens once its performance is found to be actually superior to the performance of the original model. Hence, in line 24, the new pair of true and predicted labels is used to update the performance of the shadow model $m.shadow\_model$, if any, and compare it to the performance of the original model m and decide whether it should replace the model m or not. In this way, drift detection is combined with the checking of the performance of a newly developed shadow model over a window of new instances not used to train it.

Table 1. The summary of data streams

Data stream	Instances	Features	Classes	Description
Ohio (OHI)	122,331	156	12	2001–2003 Ohio survey [14]
London (LON)	81,086	41	4	London Travel Demand Survey [12]
Optima (OPT)	2,265	497	4	Swiss survey data [2]
NTS	230,608	17	4	Dutch National Travel Survey with environment and weather features [7,10]
N-MW	144,905	2,571	21	The National Household Travel
N-NE	145,564	2,437	21	Survey (NHTS) conducted in
N-SE	209,485	2,586	21	2016 and 2017 [16], divided
N-SW	190,279	2,505	21	into five regions of the US
N-W	233,323	2,553	21	

4 Results

4.1 Data Streams and Libraries

The experiments performed with online, batch, and combined methods were assessed on real travel mode data streams, with the overview provided in Table 1. The datasets vary in the number of features, classes, i.e., travel modes, and instances. Each instance corresponds to an actual trip reported by a survey participant, with the employed travel mode designated as the target variable. For a more comprehensive description of data stream preparation, please refer to the supplementary material, where additional description is provided.

To implement the proposed methods and the evaluation framework, we used the River library (online learning methods), and the LightGBM package (the LGBM classifier). For the batch learning methods, except for LGBM, the scikit-learn library was used, while the concept drift detection implementation relied on the EvidentlyAI [1].

4.2 Experiments

We conducted a series of experiments for each data stream, which included online learning experiments, batch learning experiments, and experiments using the IEBSM method. The precise configurations of online and batch learning methods are detailed in the supplementary material accompanying this work.

For online learning, we employed the Hoeffding Adaptive Tree (HAT), Adaptive Random Forest (ARF), Streaming Random Patches (SRP), Online Gaussian Naive Bayes ((O)NB), and Online Logistic Regression ((O)LR) algorithms. The batch learning experiments made use of the Logistic Regression (LR) (with prior standardization of the training batch), Gaussian Naive Bayes (NB), Decision Tree Classifier (DT), LGBM, and Random Forest (RF) algorithms. Furthermore, we applied each batch learning algorithm to the data streams using three distinct drift handling strategies:

- S1: Basic drift detection of changes in input features, target and model performance drift with $\theta = 0.03$, $s = 10,000$, $\alpha = 0.2$
- S2: Performance drift detection only with $s = 10,000$, $\alpha = 0.2$
- S3: Frequent drift detection of changes in input features, target and model performance with $\theta = 0.02$, $s = 5,000$, $\alpha = 0.2$, i.e., relying on smaller windows of $5,000$ instances than in S1.

Batch experiments were equivalent to running Algorithm 1 with one batch base learner and its drift handling strategy. We have chosen the values for the hyper-parameters θ and s through preliminary tests aimed at determining the values resulting in possibly high performance of the models. In the batch-learning experiments, we employed a retraining strategy, which involved training a new model using all historical instances that arrived after the last model replacement. After each retraining, we assessed the performance of a shadow model relative to the old one, over the following $n_{comp} = 500$ instances. Depending on their performance, we would replace the old model with the new one. The first training took place after the initial $n_{first\_fit} = 2500$ examples.

Moreover, we conducted baseline experiments in which we trained each batch algorithm on the initial 2,500 instances (strategy B1) and subsequently used that model for predictions on the remaining data stream. We also repeated the baseline experiments using an initial training set of 25,000 instances (strategy B2) for a more comprehensive analysis.

Finally, in the experiments including batch models and involving drift detection and model adaptation we dynamically selected a specific statistical test based on the input feature/target column. For numerical columns with the number of unique values $n\_unique > 5$ we used Wasserstein Distance when $s > 1,000$; and two-sample Kolmogorov-Smirnov test otherwise. For categorical columns or numerical (with $2 < n\_unique <= 5$), Jensen–Shannon divergence when $s > 1,000$; or chi-squared test were used otherwise. Finally, for binary categorical features ($n\_unique = 2$): Jensen–Shannon divergence was used when $s > 1,000$; and proportion difference test for independent samples based on Z-score otherwise.

Combining Online and Batch Learning with the IEBSM Method. The experiments using IEBSM included an ensemble of four instances of the same batch learning classifier, each utilizing a distinct drift handling (DH) strategy, along with three online learning classifiers: (O)NB, HAT, and ARF. We tested the LGBM and RF as the batch learning algorithms. We employed a single batch learning algorithm for all batch members within each IEBSM experiment to reduce variation arising from diverse algorithms. This allowed us to single out the impact of distinct drift handling strategies, namely:

- S4: investigating changes in input features, target, and model performance with $\theta = 0.02$, $s = 2,500$, $\alpha = 0.2$
- S5: investigating changes in model performance only with $s = 2,500$, $\alpha = 0.2$
- S6: investigating changes in input features, target, and model performance with $\theta = 0.03$, $s = 10,000$, $\alpha = 0.2$

- S7: investigating changes in input features, target, and model performance with $\theta = 0.02$, $s = 10,000$, $\alpha = 0.2$

In the S5 and S7 settings, the retraining batch corresponds to the window of last s instances. In contrast, for S4 and S6, all instances since the last model replacement are considered for retraining. The other hyperparameter values (e.g., n_{first_fit}, n_{comp}) were the same as in the single online/batch learning experiments. The seven ensemble members described above were combined using two methods c outlined in Sect. 3.2.

After conducting experiments, it became evident that online learning methods exhibited significantly inferior performance compared to batch learning methods. To demonstrate the effect of model combination with the IEBSM method and eliminate the impact of under-performing online models, we conducted additional IEBSM experiments: DS-BATCH and WV-BATCH experiments utilizing only LGBM S4, S5, S6, and S7 models. Moreover, we performed DS-ONLINE and WV-ONLINE experiments combining HAT, ARF, and (O)NB models solely. For the nine data streams, we calculated each method's average ranking position based on the value of F_1 macro score. Table 2 shows the obtained results for the selected experiments (all experiment results are provided in the supplementary material).

Table 2. Ranks of selected methods across all streams and F_1 macro score for each data stream. Data streams were arranged in order based on the increasing number of features. A ranking score combined with the corresponding position in the overall ranking (in brackets). † - For the OPT data stream, hyperparameters values set to $s = 100$ and $s = 250$ (instead of 2,500 and 10,000), $n_{first\_fit} = 150$, and $n_{comp} = 50$. For N-* data streams, the input feature drift detection disabled in DH strategies S1, S3, S4, S6, S7 due to the performance issues caused by a large number of features.

Method	Rank	Data stream								
	(pos.)	NTS	LON	OHI	OPT†	N-NE	N-SW	N-W	N-MW	N-SE
DS-RF	4.33(1)	0.532	0.544	0.206	0.393	**0.464**	**0.464**	**0.453**	**0.476**	0.453
WV-RF	5.78(2)	0.534	0.546	0.197	0.362	0.460	0.460	0.435	0.458	**0.456**
DS-LGBM	7.78(3)	0.540	**0.549**	**0.224**	0.436	0.377	0.358	0.320	0.368	0.316
RF-S3	8.11(4)	0.520	0.530	0.205	0.382	0.414	0.421	0.417	0.403	0.428
DS-BATCH	8.44(5)	0.541	0.538	**0.224**	0.446	0.379	0.357	0.320	0.365	0.314
WV-LGBM	9.00(6)	**0.542**	0.546	0.215	0.438	0.359	0.357	0.306	0.352	0.316
WV-BATCH	10.83(12)	0.539	0.531	0.216	**0.459**	0.358	0.356	0.307	0.349	0.314
LGBM-S3	12.39(13)	0.530	0.533	0.213	0.453	0.358	0.339	0.302	0.349	0.314
LGBM-B1	18.61(19)	0.355	0.512	0.213	0.307	0.358	0.339	0.302	0.349	0.314
RF-B2	23.11(23)	0.486	0.426	0.196	0.274	0.279	0.229	0.197	0.232	0.220
LGBM-B2	25.00(24)	0.506	0.433	0.162	0.299	0.167	0.179	0.183	0.118	0.142
WV-ONLINE	25.44(26)	0.481	0.504	0.166	0.338	0.077	0.071	0.066	0.081	0.069
SRP	26.22(27)	0.375	0.430	0.177	0.234	0.231	0.161	0.210	0.210	0.195
RF-B1	26.33(28)	0.356	0.516	0.164	0.199	0.253	0.211	0.146	0.189	0.167
DS-ONLINE	26.78(29)	0.375	0.432	0.162	0.312	0.084	0.080	0.074	0.090	0.078

4.3   Discussion

It follows from Table 2 that batch-learning experiments RF-S3 and LGBM-S3 employing DH strategies to possibly adapt batch models, outperformed the baseline experiments RF-B* and LGBM-B* in which RF and LGBM models trained once on initial batch of data were used next to predict travel modes for all the remaining instances (B1 and B2 strategies, sample results in Table 2 provided inter alia for RF as RF-B1 results). This finding demonstrates that implementing the aforementioned DH strategies significantly benefits batch-learning models when faced with travel mode choice data. Among different ways the RF and LGBM models can be updated, strategy S3 stood out as the most effective.

Surprisingly, the online learning methods yielded the poorest performance results, as illustrated by the SRP results. One potential explanation could be the abundance of features in our data streams, among which many might be irrelevant. The SRP classifier emerged as the most effective online learning method, albeit with the trade-off of longer execution times.

The use of IEBSM, evaluated in the DS-* and WV-* experiments, enhanced performance compared to individual batch and online model setups, and provided the best overall results. This combination successfully mitigated the challenges caused by the need to choose online and batch learning methods. Moreover, utilising multiple batch members with varied DH strategies and varied hyperparameters in turn reduced the need to pre-select the optimal strategy. While the Random Forest algorithm used to develop batch ensemble members yielded superior results, as shown by the DS-RF outcome, the DS-LGBM demonstrated faster operation despite the inferior performance.

Interestingly, the IEBSM ensembles comprised solely of batch-learning LGBM classifiers ([DS/WV]-BATCH) resulted in a worse global rank than the corresponding experiments that combined both LGBM and online classifiers ([DS/WV]-LGBM). These findings underscore that, even in cases where online learning demonstrates suboptimal performance, IEBSM ensembles can benefit from the diversity their members offer. Similarly, when using combining ensembles solely with online learners ([DS/WV]-ONLINE), a notable enhancement in performance, compared to the performance of the experiments utilizing single instances of HAT, ARF, and O(NB), was observed.

Finally, the role of drift detections and shadow models in the IEBSM approach can be analyzed. Figure 1 presents the total number of drift detections and actual model replacements for all batch members for the highest-ranked method, i.e., the IEBSM-based DS-RF[2] ensemble, which notably included both online and random forest models. It follows from the figure, that detections of statistically significant changes in data have occurred in all data streams. These were followed by the replacement of batch models. Hence, the shadow models built with more recent data were found to be superior to the original models they replaced. Newly developed models were only in some cases found to yield better

---

[2] Due to the extensive number of experiments conducted, detailed results for all experiments are provided in the supplementary materials.

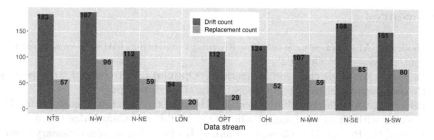

Fig. 1. The number of drift detections and model replacements. DS-RF experiments.

performance, as the drift count significantly exceeds the actual model replacement count. This confirms that both detection and the evaluation of shadow models are vital components of the highest-ranked approach to building TMC models i.e., the DS-RF approach.

5 Conclusions

In the prevailing majority of cases, modelling of travel mode choices is performed with batch learning methods. However, factors such as seasonality suggest that when predicting TMC decisions incorporating concept drift detection and adaptation could be justified. On the other hand, change detection could occur too frequently and reduce the potential of newly developed models in turn. A possible solution to the problem can rely on the use of both online and batch learners. Our experiments performed with multiple travel mode choice data sets confirm the need for continuous monitoring and retraining of TMC models. Combining batch and online learning clearly yields improved performance of the models. Furthermore, the IEBSM method eases the challenge of choosing a learning method and drift detection settings by employing multiple base members including both online and batch learners with different drift handling strategies. This resulted in the best rank of the IEBSM approach.

Future works entail exploring various combining approaches, such as different ways of assigning member weights. Furthermore, travel mode choice data sets can be used to foster the development of future stream mining methods.

Acknowledgements. Paweł Golik: The research has been conducted as part of the third edition of the CyberSummer@WUT-3 competition at the Warsaw University of Technology as part of the Research Centre POB for Cybersecurity and Data Science (CB POB Cyber&DS). Maciej Grzenda: The research leading to these results has received funding from the EEA/Norway Grants 2014–2021 through the National Centre for Research and Development. CoMobility benefits from a 2.05 million €grant from Iceland, Liechtenstein and Norway through the EEA Grants. The aim of the project is to provide a package of tools and methods for the co-creation of sustainable mobility in urban spaces.

References

1. Evidently AI - open-source machine learning monitoring. https://www.evidentlyai. com/. Accessed 05 Oct 2023
2. Bierlaire, M.: Mode choice in Switzerland (Optima) (2018). https://transp-or.epfl. ch/documents/technicalReports/CS_OptimaDescription.pdf
3. Bifet, A., Gavaldà, R., Pfahringer, B., Holmes, G.: Machine Learning for Data Streams with Practical Examples in MOA. MIT Press, Cambridge (2018)
4. Ditzler, G., Roveri, M., Alippi, C., Polikar, R.: Learning in nonstationary environments: a survey. IEEE Comput. Intell. Mag. **10**(4), 12–25 (2015)
5. Gama, J.: Knowledge Discovery from Data Streams. CRC Press, New York (2010)
6. Gandomani, T.J., Tavakoli, Z., Zulzalil, H., Farsani, H.K.: The role of project manager in agile software teams: a systematic literature review. IEEE Access **8**, 117109–117121 (2020)
7. García-García, J.C., García-Ródenas, R., López-Gómez, J.A., Ángel Martín-Baos, J.: A comparative study of machine learning, deep neural networks and random utility maximization models for travel mode choice modelling. Transp. Res. Procedia **62**, 374–382 (2022)
8. Gomes, H.M., et al.: Adaptive random forests for evolving data stream classification. Mach. Learn. **106**(9), 1469–1495 (2017)
9. Grzenda, M., Kwasiborska, K., Zaremba, T.: Hybrid short term prediction to address limited timeliness of public transport data streams. Neurocomputing **391**, 305–317 (2020)
10. Hagenauer, J., Helbich, M.: A comparative study of machine learning classifiers for modeling travel mode choice. Expert Syst. Appl. **78**, 273–282 (2017)
11. Hillel, T., Bierlaire, M., Elshafie, M.Z., Jin, Y.: A systematic review of machine learning classification methodologies for modelling passenger mode choice. J. Choice Model. **38**, 100221 (2021)
12. Hillel, T., Elshafie, M.Z., Jin, Y.: Recreating passenger mode choice-sets for transport simulation: a case study of London, UK. Proc. Inst. Civil Eng. Smart Infrastruct. Construct. **171**(1), 29–42 (2018)
13. Pishgoo, B., Azirani, A.A., Raahemi, B.: A dynamic feature selection and intelligent model serving for hybrid batch-stream processing. Knowl.-Based Syst. **256**, 109749 (2022)
14. Team, T.: 2001–2003 Ohio statewide household travel survey. Technical report, Livewire Data Platform; NREL; Pacific Northwest National Lab (PNNL) (2023)
15. Testi, M., et al.: MLOps: a taxonomy and a methodology. IEEE Access **10**, 63606–63618 (2022)
16. U.S. Department of Transportation, Federal Highway Administration: 2017 national household travel survey (2017). http://nhts.ornl.gov
17. Yang, J., Rivard, H., Zmeureanu, R.: On-line building energy prediction using adaptive artificial neural networks. Energy Build. **37**(12), 1250–1259 (2005)

Natural Language Processing

Beyond Words: A Comparative Analysis of LLM Embeddings for Effective Clustering

Imed Keraghel[1,2](✉), Stanislas Morbieu[2], and Mohamed Nadif[1]

[1] Centre Borelli UMR9010, Université Paris Cité, 75006 Paris, France
imed.keraghel@u-paris.fr
[2] Kernix Software, 75014 Paris, France

Abstract. The document clustering process involves the grouping of similar unlabeled textual documents. This task relies on the use of document embedding techniques, which can be derived from various models, including traditional and neural network-based approaches. The emergence of Large Language Models (LLMs) has provided a new method of capturing information from texts through customized numerical representations, potentially enhancing text clustering by identifying subtle semantic connections. The objective of this paper is to demonstrate the impact of LLMs of different sizes on text clustering. To accomplish this, we select five different LLMs and compare them with three less resource-intensive embedding methods. Additionally, we utilize six clustering algorithms. We simultaneously assess the performance of the embedding models and clustering algorithms in terms of clustering quality, and highlight the strengths and limitations of the models under investigation.

Keywords: Large Language Models · Embeddings · Clustering

1 Introduction

In the rapidly evolving area of natural language processing (NLP), the advent of Large Language Models (LLMs) such as the GPT series [1,2] has significantly enhanced our ability to process and analyze large volumes of texts. These models function by transforming texts into high-dimensional vectors called embeddings, which are commonly used for tasks such as translating or answering questions. Using them for clustering is quite new and has not yet been extensively explored.

The objective of this article is to explore the emerging field of embeddings generated by Large Language Models (LLMs) through a thorough comparative analysis. Our study encompasses various models, including both smaller-scale models and larger architectures like the GPT series. The main goal is to examine the impact of a language model's size on the quality of its embeddings for clustering similar texts.

Clustering, categorizing text based on similarity, is crucial in NLP. LLM-generated embeddings could enhance this by capturing text's semantic nuances. Our core hypothesis posits that there might be a direct correlation between the size of an LLM and its proficiency in creating effective embeddings. This could imply that larger models yield more accurate embeddings, thereby improving clustering results. On the other hand, there is a potential for a point of diminishing returns where increases in model size no longer contribute to significant enhancements in clustering performance. In fact, overly large models may even detract from clustering quality due to overfitting on training tasks. This study compares small and large language models to find out if the larger ones create better text groupings or if the smaller ones are equally effective while using fewer resources. This research will help inform future work and real-world uses of these models to sort and understand large textual datasets.

2   Related Work

The use of textual embeddings in NLP tasks has been extensively studied, with a primary focus on supervised tasks such as classification [3]. However, the use of textual embeddings, especially those derived from LLMs, in unsupervised tasks such as clustering remains underexplored. Despite the proven capabilities of LLMs in capturing nuanced semantic relationships within text, their potential for clustering has not been fully exploited or understood within the academic domain. Previous research, including studies conducted by [4,5], has focused on investigating the effect of different embedding techniques on clustering results. These studies have shown that the choice of embedding significantly affects the performance of algorithms. Research in this area has mainly focused on comparing traditional embeddings, which are static and contextualized embeddings. The contextualized embeddings are often limited to BERT-based transformer models. The results of these studies have been diverse, with some suggesting that traditional methods perform similarly or even better than newer methods in specific situations. Recent studies such as [6] acknowledge the strength of LLMs in enhancing clustering but stop short of employing their embeddings directly, opting instead for keyword enrichment strategies. This highlights a gap in the field, indicating that the full capabilities of LLM embeddings have not been entirely harnessed for unsupervised clustering applications. In [7], the authors benchmarked numerous models, including LLMs, for clustering, providing a broad overview of performance but without comparing clustering algorithms or delving into details.

This paper addresses this lacuna by comparing the clustering efficacy of embeddings from variously sized LLMs. Our main objective is to determine whether these advanced embeddings can greatly enhance clustering tasks.

3   Models and Algorithms

Textual embeddings, a fundamental component in NLP for converting texts into numerical vectors, are varied in type and derivation method. Traditional embeddings use simple techniques like TF-IDF. Static embeddings, exemplified by

Table 1. Description of models.

Model	#Parameters	#Layers	#Embedding size
JoSE	–	–	100
MiniLM-L12-v2	33 million	12	384
BERT	110 million	12	768
BLOOMZ-560m	560 million	24	1024
BLOOMZ-3B	3 billion	30	2560
Mistral	7.3 billion	32	4096
Llama 2	13 billion	40	5120
text-embedding-ada-002	–	–	1536

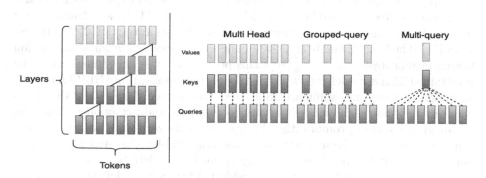

Fig. 1. Dual attention mechanisms: Grouped-Query and Sliding Window.

Word2Vec [8], maintain a consistent vector per word regardless of context. In contrast, contextual embeddings, such as those of BERT [9] and GPT [2], adjust vectors based on the context of word usage, allowing nuanced language model training. In our contribution, we aim to evaluate the effectiveness of textual embeddings derived from five LLMs, namely BLOOMZ-560m, BLOOMZ-3B [10], Mistral-7B [11], Llama-2-13B [12], and GPT besides a compact transformer-based model, MiniLM [13], which serves as a standard for comparison. BERT and JoSE [14] have also been considered in this study to determine whether the use of energy-intensive models (LLMs) is justified or not. We study this effectiveness in the context of unsupervised learning, particularly in the objective of document clustering.

The properties of the studied models are given in Table 1, with a comprehensive examination presented in the following sections.

Mistral 7B is a LLM developed by Mistral AI[1] with 7.3 billion parameters. It outperforms competitors like Llama-2-13B on various evaluation criteria [11]. This model employs innovative attention mechanisms such as Grouped Query Attention (GQA) for faster inferences [15] and Sliding Window Attention (SWA) to handle longer sequences more efficiently [16].

---

[1] https://mistral.ai/.

The SWA mechanism enhances efficiency by allowing each model layer to concentrate on a data segment within a sliding window. This approach conserves computing resources, scaling with window size and data length. Additionally, SWA aids in retaining and utilizing information beyond the current window through data layering.

The GQA mechanism allows for simultaneous focus on critical text parts. It organizes question words into groups, guiding attention to text areas essential for answering. Per Fig. 1, GQA can process several questions at once by forming distinct groups of vectors of the question word. This multigroup approach applies attention to input tokens, enabling the model to efficiently address multiple questions by concentrating on relevant text portions simultaneously. **BLOOMZ** is an enhanced iteration of the BLOOM model [17]. The original Bloom model was developed through the collaboration of over a thousand AI experts with the aim of creating a freely available and widely accessible LLM. It was trained on a diverse dataset and its architecture is based on the Megatron-LM GPT-2 framework [18]. The BLOOMZ family of models undergoes a process known as instruction tuning or multitask fine-tuning [10]. This process involves the refinement of the pre-trained Bloom model with a diverse datasets across various NLP tasks such as question answering, summarization, and translation. The datasets serve as a means to enhance the model's ability to interpret and act upon instructions encapsulated within prompts, thereby improving its zero-shot task generalization capabilities. In essence, zero-shot learning enables the model to adeptly perform tasks it has not been explicitly trained on, solely based on its understanding derived from the prompts provided. **Llama 2** model, part of Meta's Llama family of open-source LLMs, is an autoregressive language model that uses an optimized transformer architecture and has parameter counts ranging from 7 billion to 70 billion [12]. The largest variant, which incorporates GQA (like Mistral), enhances inference speed without compromising quality. LLama models have exhibited strong competitiveness compared to existing open-source models and have achieved performance levels comparable to some proprietary models. However, it is important to note that they still fall behind more advanced models like GPT-4. **text-embedding-ada-002** is a specialized variant of the GPT-3 architecture developed by OpenAI for generating text embeddings[2]. It uses the transformer framework to create vector representations of text that capture semantic meaning. This model is pre-trained on a vast corpus of text data and subsequently fine-tuned for the specific purpose of text embedding. Although it has fewer parameters than the more extensive GPT-3 models, it retains the capability of the GPT framework, making it accessible to a wide range of applications via API access. **MiniLM** is a different kind of language model that emphasizes a smaller and more efficient design compared to LLMs [13]. The main innovation in MiniLM is in its training approach. It involves distilling the knowledge of a powerful, large teacher-language model into a much smaller student model. This process, known as self-attention distillation, involves training the student to imitate the behavior of the teacher as closely as possible. The

---

[2] https://platform.openai.com/docs/guides/embeddings.

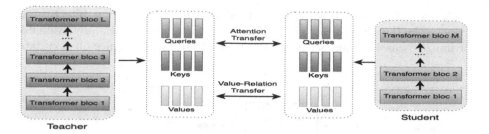

Fig. 2. Deep self-attention distillation: student model training through advanced imitation of teacher model's final layer with enhanced value-relation transfer for self-attention mechanisms, as implemented in MiniLM models.

student model does this by learning to predict the teacher's outputs from its intermediate layers rather than just its final output. Despite its smaller size, MiniLM manages to maintain a high level of language understanding and generation capabilities (Fig. 2).

3.1   Clustering Algorithms

In unsupervised learning, clustering methods are routinely employed in data embedding to make new discoveries from large and complex data sets. Several clustering algorithms exist; in our study, we use five algorithms in order to evaluate the different models. These algorithms were chosen for their popularity in the community or for their use when combining them with dimensionality reduction methods. Specifically, we focus on deep clustering variants, which have demonstrated impressive results in image clustering [19]. Therefore, the selected clustering algorithms are the following. K-means from Scikit-learn [20], Spherical K-means from the coclust framework [21], CAEclust from [22] Deep K-means and Deep Clustering Network from [23]. **K-means** is a widely used algorithm for clustering. It aims to partition a dataset into k non-overlapping clusters, minimizing the sum of squared distances between data points and their respective cluster centroids. The process involves initializing centroids, assigning points to the nearest centroid, updating centroids based on mean values, and iterating until convergence. **Spherical K-means (SK-means)** is a variant of K-means designed for data residing on a unit hypersphere. Unlike K-means that uses Euclidean distance, Spherical K-means employs cosine distance to measure dissimilarity between data points considering the angle between them. This makes it suitable for scenarios where only direction or relative orientation matters, such as text clustering.

A survey [24] describes a wide range of different deep representation learning methods. Here, we focus on the AEs and its variants because combined with clustering methods, they can lead to interesting results in various fields. **CAEclust** is a Python package developed to implement consensus clustering [22]. It relies on Autoencoders (AEs) and spectral clustering. It aims to facilitate deep clustering by merging representations from various Denoising Autoencoders (DAEs), as shown in Fig. 3 (left), without the need to define a specific architecture, as proposed by the same authors in [19]. CAEclust has been successfully evaluated for image datasets. Thus, we plan to test CAEclust's performance on text data.

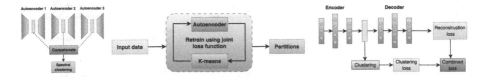

Fig. 3. Left: Overview of CAEclust. Center: Overview of methods based on retraining using an AE, K-means, and a joint loss function. Right: A typical network architecture for a deep clustering algorithm. Such an algorithm uses an optimization objective that combines the reconstruction error and the clustering error (joint optimization approach).

Representation learning followed by cluster analysis is often helpful in data science. The K-means algorithm applied to data embedding derived from UMAP [25] or AEs, for example, is a popular approach. This procedure is performed sequentially and is referred to as the tandem approach. However, AE may sometimes be an unsuitable method to reduce the dimension before clustering; it can fail to retain information that could be valuable for the clustering task. Thus, jointly optimizing for both tasks –RL and clustering– is a good alternative. Thus, we propose two popular algorithms. **Deep K-means (DKM)** combines jointly representation learning and clustering [23]. Its aim is to minimize both the reconstruction error and the clustering error in the learned embedded space, by iteratively updating the autoencoder parameters and cluster representatives. Figure 3 (center) shows the steps of this algorithm. **Deep Clustering Network (DCN)** algorithm [26] also performs unsupervised clustering using a deep autoencoder, but does not use an auxiliary distribution. Instead, it uses an optimization objective that is a direct combination of a reconstruction error and a clustering error, as depicted in Fig. 3 (right).

4   Numerical Experiments

In this section, we elucidate the various components of our experimental setup. In the remainder of our experiments, we use the following datasets:

– BBC News[3]: a dataset sourced from the BBC News, encompasses a collection of 2,225 articles labeled across five categories: business, entertainment, politics, sport, and tech.

---

[3] http://mlg.ucd.ie/datasets/bbc.html.

Table 2. Description of datasets. The balance represents the ratio between the smallest and largest class. #Tokens indicates the mean token count.

Datasets	Characteristics			
	#Documents	#Clusters	Balance	#Tokens
BBC News	2,225	5	0.75	390
20 Newsgroup	18,846	20	0.63	284
IMDb	50,000	2	1	231
Web Content	1,408	16	0.14	747

- 20 Newsgroups[4]: consists of 18,846 newsgroup documents, distributed across 20 different topical newsgroups. Originating from Usenet newsgroups in the late 1990s, it encompasses a broad spectrum of themes such as politics, religion, and science.
- IMDb: a collection of movie reviews retrieved from the IMDb website used for binary sentiment classification [27]. The dataset serves as a cornerstone for NLP research and machine learning studies, designed to provide a substantial and balanced collection of positive and negative reviews.
- Web Content[5]: a compilation of data created through an extensive scraping of various websites. By extracting text from a myriad of web pages, this dataset offers a diverse and rich set of information from different domains and types of content. This dataset comprises a collection of 1,408 samples, categorized into 16 distinct classifications reflecting a broad spectrum of Web Content. The categories span a variety of sectors and interests, from education and news to e-commerce and sport.

Data Preprocessing. We select four datasets originating from various fields, each with its own distinct attributes, thus necessitating customized preprocessing strategies. The 20 Newsgroups and IMDb datasets undergo pre-processing, including removing apostrophes, HTML tags, special characters, punctuation, URLs, and converting texts to lowercase. In contrast, the BBC News dataset, which needs less cleaning, undergoes only a conversion to lowercase. The Web Content dataset, already preprocessed, receives no additional treatment. The characteristics of these datasets are detailed in Table 2.

4.1   Evaluation Metrics

Using labeled datasets, we evaluate clustering algorithms performance with external indices: Accuracy (ACC), Normalized Mutual Information (NMI) [28], and Adjusted Rand Index (ARI) [29]. The ACC quantifies the degree to which each cluster contains data points corresponding to their respective classes. NMI,

---

[4] http://qwone.com/~jason/20Newsgroups/.
[5] https://www.kaggle.com/datasets/hetulmehta/website-classification.

which ranges from 0 to 1, evaluates the information commonality between the suggested clustering and the ground truth labels. Finally, ARI measures the similarity between two clustering, taking into account both the cluster assignments and the ground truth labels, when available. The ARI metric ranges from -1 to 1, where a higher value indicates better agreement between the true labels and the clustering. Intuitively, NMI quantifies how much the estimated clustering is informative about the true clustering, while the ARI measures the degree of agreement between the estimated clustering and the reference partition. Both NMI and ARI are equal to 1 if the resulting clustering partition is identical to the ground truth.

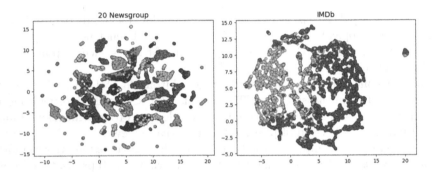

Fig. 4. UMAP: 2D Plot of obtained clusters by K-means using GPT embeddings.

4.2   Experimental Settings

The experiments were conducted on a professional workstation with specific hardware specifications: an Intel® Core™ i9-12950HX CPU running at 2.6 GHz and 64 GB of DDR5 memory operating at 4,800 MHz. It is important to note that the extraction of embeddings for LLMs required GPU usage, which was carried out on the Pro version of Google Colab equipped with an Nvidia A100 40 GB.

When it comes to clustering algorithms, K-means initialize using *the K-means++* technique, which chooses the starting centroids by sampling based on their contribution to total inertia. We cap the iterations at 300 and run 10 initial setups to strengthen the clustering stability. The same parameters, 300 iterations and 10 starts, are applied to Spherical K-means. As we are in an unsupervised setting, we employ default values for Deep K-means, DCN, and CAEclust to maintain consistency. The performances of the algorithms are detailed in Table 3.

For each experiment, the model is launched 10 times, and the best result according to the objective function to optimize was selected (Inertia for KMeans, combined loss for DCN, etc.). For the embeddings, a maximum size of 2,000 tokens was set for all models except for BERT and MiniLM, where it was fixed at 512. Finally, as baselines, we introduce two other models, namely BERT [9] and Joint

Table 3. For each dataset, the best score of each clustering algorithm applied with all embedding models, is highlighted in bold, and second-best is underlined. For instance for the Website content dataset, concerning `K-means`, the best score in term of ACC is **83.85** obtained with `GPT` and the second 73.08 is obtained with `MiniLM`.

Emb.	Clus.	Website content			20 Newsgroup			IMDb			BBC News		
		ACC	NMI	ARI	ACC	NMI	ARI	ACC	NMI	ARI	ACC	NMI	ARI
JoSE	K-means	72.87	71.56	60.27	56.71	57.05	41.98	68.84	10.57	14.18	95.46	85.55	89.26
	SK-means	**77.41**	73.30	65.12	53.73	55.41	40.59	69.08	11.34	14.55	69.17	66.98	60.81
	UMAP K-means	77.63	76.26	68.00	52.81	61.03	46.16	50.54	00.01	00.01	95.70	86.98	89.99
	DKM	76.28	70.60	62.91	56.73	56.07	42.87	55.96	01.06	01.40	90.52	74.77	77.90
	DCN	66.62	59.60	48.83	56.78	54.57	41.90	61.08	03.69	04.89	81.35	60.68	61.12
	CAEclust	72.80	70.04	60.63	44.68	51.90	33.53	51.36	00.04	00.05	95.60	86.59	89.60
BERT	K-means	41.05	40.08	22.21	34.94	38.64	20.45	54.08	00.51	00.65	94.29	83.43	86.82
	SK-means	42.12	39.26	23.49	35.88	40.05	21.24	54.18	00.54	00.68	93.57	81.78	85.03
	UMAP K-means	54.55	55.52	37.49	50.49	52.02	35.59	55.24	00.83	01.08	93.30	82.53	84.36
	DKM	52.20	48.97	32.73	46.06	50.17	33.28	53.02	00.33	00.35	93.03	81.16	84.22
	DCN	52.06	47.93	30.59	48.73	49.16	33.72	52.44	00.23	00.22	**91.73**	77.70	**80.68**
	CAEclust	54.97	51.28	34.77	44.6	51.88	33.49	54.97	51.28	34.77	92.90	80.50	83.73
MiniLM	K-means	73.08	71.41	60.43	59.34	60.94	45.14	57.28	01.54	02.10	95.87	87.54	90.31
	SK-means	66.48	67.91	55.33	58.09	60.12	43.61	57.30	01.62	02.11	72.09	64.47	60.37
	UMAP K-means	82.60	77.91	72.34	64.56	66.29	50.29	60.52	03.27	04.41	95.64	86.94	89.55
	DKM	77.34	72.30	63.77	61.72	61.06	47.40	58.82	02.26	03.09	91.28	78.41	80.73
	DCN	74.01	67.91	59.24	60.20	59.48	45.89	64.54	06.77	08.44	77.21	53.47	55.16
	CAEclust	74.36	72.42	63.68	60.72	60.68	44.52	60.46	03.19	04.36	95.06	85.71	88.47
BLOOMZ-560M	K-means	24.64	27.43	13.48	24.54	27.90	12.22	60.48	03.20	04.37	89.12	73.84	74.81
	SK-means	30.33	30.55	15.46	25.02	28.53	12.50	60.42	03.16	04.32	89.21	74.04	75.01
	UMAP K-means	56.32	55.34	38.86	34.32	38.84	20.50	64.40	06.07	08.28	94.97	84.82	88.18
	DKM	17.68	12.62	04.20	07.38	00.93	00.28	53.00	00.26	00.34	31.96	04.57	04.03
	DCN	15.77	12.43	04.50	25.64	29.60	13.72	63.74	05.56	07.53	70.56	54.58	47.35
	CAEclust	63.35	65.03	49.55	55.86	56.03	38.70	**86.72**	**45.04**	**53.93**	95.69	87.39	89.82
BLOOMZ-3B	K-means	43.75	40.23	23.38	29.25	33.76	15.00	58.72	02.30	03.02	89.62	74.06	75.72
	SK-means	45.03	41.35	26.38	28.98	33.83	15.13	58.76	02.32	03.05	89.57	73.97	75.61
	UMAP K-means	55.68	53.68	37.41	38.45	43.94	24.33	58.40	02.09	02.80	95.42	86.77	89.38
	DKM	47.94	46.42	28.84	15.14	23.18	07.78	57.12	01.66	02.01	92.27	78.83	81.77
	DCN	39.63	36.43	20.40	34.54	39.72	20.57	58.80	02.33	03.08	89.48	72.32	75.82
	CAEclust	63.42	63.40	48.78	44.30	50.50	31.30	83.72	36.24	45.47	70.88	72.88	62.70
Mistral-7B	K-means	56.46	56.42	41.05	44.53	48.24	26.62	63.04	05.02	06.78	96.36	88.51	91.49
	SK-means	58.10	59.60	43.96	42.98	47.30	26.10	63.00	04.99	06.74	71.15	69.94	64.03
	UMAP K-means	72.94	72.75	61.32	54.75	60.65	41.44	69.08	10.77	14.54	**97.12**	**91.23**	**93.13**
	DKM	65.70	64.85	51.39	56.34	59.34	41.00	64.10	08.41	07.94	87.33	77.94	73.80
	DCN	55.18	54.89	38.11	53.92	55.82	36.23	62.02	04.49	05.76	89.84	77.01	77.23
	CAEclust	59.80	60.56	44.10	40.73	48.64	29.22	77.18	22.59	29.54	94.65	87.19	88.42
Llama-2-13B	K-means	55.33	54.12	36.62	45.60	47.98	27.13	55.82	00.99	01.34	**97.35**	**91.30**	**93.76**
	SK-means	47.73	53.12	37.40	43.39	47.66	25.79	55.76	00.97	01.31	88.63	80.68	76.97
	UMAP K-means	72.16	70.54	58.73	54.18	61.45	42.52	54.30	00.55	00.72	71.37	80.33	66.29
	DKM	70.10	68.17	55.84	56.85	60.69	39.74	52.64	01.15	00.27	88.09	79.78	75.62
	DCN	55.61	56.43	41.24	51.40	56.82	37.39	60.66	03.36	04.53	87.37	78.23	74.14
	CAEclust	56.61	58.82	41.56	53.19	55.25	35.00	74.48	18.04	23.96	97.35	92.04	93.70
GPT	K-means	**83.85**	**82.67**	**76.69**	65.33	67.63	53.45	80.12	29.69	36.28	97.08	90.88	93.19
	SK-means	74.34	79.09	67.81	64.28	68.04	52.22	80.58	30.71	37.39	96.99	90.82	92.99
	UMAP K-means	**84.20**	**83.78**	**78.18**	70.77	72.17	60.02	71.74	14.20	18.89	96.99	90.49	92.79
	DKM	**82.63**	**83.24**	**76.09**	67.68	69.07	56.48	74.30	17.93	23.60	96.04	89.42	90.93
	DCN	**78.98**	**78.32**	**70.09**	63.93	66.84	52.58	75.54	19.76	26.08	88.90	75.37	75.80
	CAEclust	**75.41**	**75.79**	**63.96**	**73.32**	**74.83**	**62.75**	76.66	22.27	28.42	96.67	90.05	92.21

Spherical Embedding (JoSE) [14]. JoSE has been chosen for its parsimony and efficiency, outperforming popular models such as Word2Vec, GloVe and BERT in tasks of clustering and textual similarity while being simple and energy-efficient. To represent documents, an average of the vectors produced by the embedding models was computed for each token. To access the GPT-3 embeddings, we use the text-embedding-ada-002 model through OpenAI's API, which charges $0.0001 for every 1,000 tokens.

4.3   Results and Discussion

Below are the essential points that result from numerous experiments:

- Undeniably, the GPT model is the best regardless of the nature of the datasets. This is verified regardless of the clustering algorithms used. Furthermore, note that whatever datasets used, with GPT, K-means appears a simple and effective solution for clustering. This was confirmed by a ranking test, which revealed that K-means is the most efficient algorithm on GPT embeddings with an average rank of 2.4. It is followed by UMAP K-means and SK-means, with ranks of 2.9 and 3.2, respectively.
- Unlike the BLOOMZ-560M, BLOOMZ-3B, Mistral-7B and Llama-2-13B models that are sensitive to class imbalance, GPT and also MiniLM remain robust. This can be seen on the Website Content dataset whose classes are very unbalanced, the performance of the clustering methods remains very modest compared to GPT and MiniLM. Furthermore, note that for BLOOMZ-560M, BLOOMZ-3B, Mistral-7B and Llama-2-13B, all clustering algorithms fail, except for CAEclust.
- The performance of the BLOOMZ-560M, BLOOMZ-3B, Mistral-7B and Llama-2-13B models fluctuates depending on the datasets, unlike the GPT model and MiniLM model to a lesser extent. This is probably due to the imbalance and the degree of overlap of the clusters. Indeed, these models seem only suitable when the classes are easily distinguishable, as with BBC News.
- The developers of the Mistral-7B model claim that it compares well to Llama-2-13B, although it has fewer parameters. This claim is substantiated by comparing the performance of clustering algorithms using embeddings from both models.
- It does not appear that there is a relationship or correlation between the *size* of the model and the effectiveness of these embeddings. Indeed, it's observed that Mistral-7B exhibits performance comparable to that of Llama-2-13B and that the *smaller* model MiniLM outperforms models that are larger.
- One might wonder about the need to use all the dimensions generated by the models. Indeed, this represents a cost which can be reduced by a non-linear dimensionality reduction technique such as UMAP [25]; for all models, we fixed the dimension at 20 and the number of neighbors equal to 50. Thus, we can measure the impact of K-means applied on the reduced matrix. In Table 3 we observe that for all models except GPT and Llama-2-13B, the

use of UMAP often leads to improved performance, as illustrated in Fig. 4. However, we noted that by increasing the number of neighbors to 90, we can improve the clustering for GPT and Llama-2-13B. Such a suitable number of neighbors was detected based on the within-cluster sum of squares minimized by K-means as a function of the neighbours. In this way, we respect the context of unsupervised learning.

- Note that the performance of the JoSE Model is interesting; it is often better than BERT while embeddings do not require large resources.
- Finally, except for the GPT model, the deep clustering methods DKM, DCN and CAEclust are not always more efficient than a simple K-means.

5   Conclusion and Perspectives

In this paper, our objective was to evaluate several LLM models in an unsupervised learning context. We evaluated six clustering algorithms from benchmarks with different characteristics in terms of the degree of mixing (class distribution), the number of classes that can be balanced or not. Unquestionably, the GPT model is the best followed by Mistral-7B and MiniLM; all clustering algorithms record very good scores compared to the other models. Even if our objective is not to compare clustering algorithms, we can note that despite its simplicity, K-means can be used without concern for such embeddings. Furthermore, we noted that the deep CAEclust algorithm also proves interesting on such data. On the other hand, since we chose BERT and JoSE as Baselines models, it is interesting to emphasize the good performance of the JoSE model, which is a parsimonious model.

References

1. Radford, A., Narasimhan, K., Salimans, T., Sutskever, I., et al.: Improving language understanding by generative pre-training (2018)
2. Brown, T., et al.: Language models are few-shot learners. NeurIPS **33**, 1877–1901 (2020)
3. Sun, C., Qiu, X., Xu, Y., Huang, X.: How to fine-tune BERT for text classification? In: Sun, M., Huang, X., Ji, H., Liu, Z., Liu, Y. (eds.) CCL 2019. LNCS (LNAI), vol. 11856, pp. 194–206. Springer, Cham (2019). https://doi.org/10.1007/978-3-030-32381-3_16
4. Saha, R.: Influence of various text embeddings on clustering performance in nlp. arXiv:2305.03144 (2023)
5. Ravi, J., Kulkarni, S.: Text embedding techniques for efficient clustering of twitter data. In: Evolutionary Intelligence, pp. 1–11 (2023)
6. Viswanathan, V., Gashteovski, K., Lawrence, C., Wu, T., Neubig, G.: Large language models enable few-shot clustering. arXiv:2307.00524 (2023)
7. Muennighoff, M., Tazi, N., Magne, L., Reimers, N.: Mteb: massive text embedding benchmark. arXiv:2210.07316 (2022)
8. Mikolov, T., Chen, K., Corrado, G., Dean, J.: Efficient estimation of word representations in vector space. arXiv:1301.3781 (2013)

9. Devlin, J., Chang, M.W., Lee, K., Toutanova, K.: Bert: pre-training of deep bidirectional transformers for language understanding. arXiv:1810.04805 (2018)
10. Muennighoff, N., et al.: Crosslingual generalization through multitask finetuning. arXiv:2211.01786 (2022)
11. Jiang, A.Q., et al.: Mistral 7b. arXiv:2310.06825 (2023)
12. Touvron, H., et al.: Llama 2: open foundation and fine-tuned chat models. arXiv:2307.09288 (2023
13. Wang, W., Wei, F., Dong, L., Bao, H., Yang, N., Zhou, M.: Minilm: deep self-attention distillation for task-agnostic compression of pre-trained transformers. Neurips **33**, 5776–5788 (2020)
14. Meng, Y., et al.: Spherical text embedding. NeurIPS **32** (2019)
15. Ainslie, J., Lee-Thorp, J., de Jong, M., Zemlyanskiy, Y., Lebrón, F., Sanghai, S.: GQA: training generalized multi-query transformer models from multi-head checkpoints. arXiv:2305.13245 (2023)
16. Beltagy, I., Peters, M.E., Cohan, A.: Longformer: the long-document transformer. arXiv:2004.05150 (2020)
17. Le Scao, T., et al.: Bloom: a 176b-parameter open-access multilingual language model. arXiv:2211.05100 (2022)
18. Shoeybi, M., Patwary, M., Puri, R., LeGresley, P., Casper, J., Catanzaro, B.: Megatron-lm: training multi-billion parameter language models using model parallelism. arXiv:1909.08053 (2019)
19. Affeldt, S., Labiod, L., Nadif, M.: Spectral clustering via ensemble deep autoencoder learning (SC-EDAE). Pattern Recogn. **108**, 107522 (2020)
20. Buitinck, L., et al.: API design for machine learning software: experiences from the scikit-learn project. arXiv:1309.0238 (2013)
21. Role, F., Morbieu, S., Nadif, M.: Coclust: a python package for co-clustering. J. Stat. Softw. **88**(7), 1–29 (2019)
22. Affeldt, S., Labiod, L., Nadif, M.: Caeclust: a consensus of autoencoders representations for clustering. Image Process. Line **12**, 590–603 (2022)
23. Fard, M.M., Thonet, T., Gaussier, E.: Deep k-means: jointly clustering with k-means and learning representations. Pattern Recogn. Lett. **138**, 185–192 (2020)
24. Karim, M.R., et al.: Deep learning-based clustering approaches for bioinformatics. Brief. Bioinf. 1–23 (2020)
25. McInnes, L., Healy, J., Melville, J.: Umap: uniform manifold approximation and projection for dimension reduction. arXiv:1802.03426 (2018)
26. Yang, B., Fu, X., Sidiropoulos, N.D., Hong, M.: Towards k-means-friendly spaces: simultaneous deep learning and clustering. In: ICML, pp. 3861–3870 (2017)
27. Maas, A.L., Daly, R.E., Pham, P.T., Huang, D., Ng, A.Y., Potts, C.: Learning word vectors for sentiment analysis. In: Annual Meeting of the Association for Computational Linguistics: Human Language Technologies, pp. 142–150 (2011)
28. Strehl, A., Ghosh, J.: Cluster ensembles–a knowledge reuse framework for combining multiple partitions. J. Mach. Learn. Res. **3**, 583–617 (2002)
29. Steinley, D.: Properties of the hubert-arable adjusted rand index. Psychol. Methods **9**(3), 386 (2004)

Data Quality in NLP: Metrics and a Comprehensive Taxonomy

Vu Minh Hoang Dang$^{(\boxtimes)}$ ⓘ and Rakesh M. Verma ⓘ

University of Houston, Houston, USA
`vdang9@uh.edu, rmverma2@central.uh.edu`

Abstract. Data quality is a crucial factor for the success of natural language processing (NLP) models. However, there is a lack of a standard taxonomy for data quality in NLP, which makes it difficult to assess and improve the quality of NLP datasets. In this work, we propose a comprehensive taxonomy for data quality in NLP, covering various aspects such as linguistic, semantic, anomaly, classifier performance, and diversity. We also introduce a novel metric to measure the difficulty of a dataset, which reflects the inherent challenges of the data. We evaluate our taxonomy using a wide variety of NLP datasets that span multiple domains and tasks. The results show that our taxonomy can effectively capture the changes in data quality and provide valuable insights for data creators and users. We believe that our work is a significant contribution to the field of NLP, as it provides a systematic and holistic approach to data quality assessment.

Keywords: Data Distance Measure · Data Quality Index · Exploratory Data Analysis · Deception Datasets

1 Introduction

With the rapid development of machine learning (ML), it is astonishing that neural language models have achieved human-level performance across several natural language processing (NLP) datasets. However, a series of recent works have shown that many popular NLP datasets, such as SQUAD and SNLI have unwanted biases, resulting from the annotation process [24,25]. Low data quality leads to the failure of models to generalize the gained knowledge. As a consequence, their performance drops when tested with real word data. These errors during dataset creation happened due to the lack of a standard taxonomy for data quality in NLP. Dataset creators often rely on their own defined metrics or perform literature reviews to explore suitable metrics for data quality.

This is a hard problem because of multiple factors. Languages are diverse and not created equally. This means that each language may have different properties, and therefore, is hard to be generalized. A property of data quality can be important for a set of languages, while it may be irrelevant for others. Secondly, the relationship between different aspects of data quality can be challenging to

I. Miliou et al. (Eds.): IDA 2024, LNCS 14641, pp. 217–229, 2024.
https://doi.org/10.1007/978-3-031-58547-0_18

define. Moreover, different tasks or models may give more weight to different aspects of quality.

In this work, we first explore different existing metrics for data quality in NLP through literature reviews. During this process, we dig deep into the definition, formula, and usefulness of each metric. Then, we find the relationship between the metrics and categorize them. Using this knowledge, we will construct the taxonomy. It will be evaluated using a set of high-quality datasets and a set of low-quality datasets. This acts as proof that the taxonomy works well in any setting. Below are the contributions of this work:

- We propose a comprehensive taxonomy for data quality in NLP.
- We propose a novel metric to measure the difficulty of a dataset.
- We evaluate our proposed solution using a wide variety of NLP datasets.

1.1 Data Quality

In this work, we define data as a nonempty collection D of facts and/or information with a nonempty set of well-defined objectives O and a set of constraints C. We take inspiration from [3,9,18], which are ML works that also explore data quality. Because the authors of these works want to cover as many types of data as possible, their works are generic and do not explore deeper aspects of data quality in NLP. However, they provide some excellent dimensions for data quality. We can then define data quality with respect to objectives O and constraints C as having the following dimensions:

- **Consistency** with respect to constraints C - Consistency means that all data entries in the dataset adhere to a specific format, standard, or rule without exceptions. If the constraints in C are logical and consistent, then we can formalize consistency as $D \in Mod\ C$, where $Mod\ C$ denotes the class of all models of C, i.e., the dataset D is a model of the formulas (constraints) in C. An example of a constraint is that a certain attribute of each data instance should be a nonnegative real number.
- **Appropriateness** in terms of objectives O - Appropriateness implies that the collected data are directly relevant and suitable for the tasks or goals defined by O. For example, the objective is to analyze the reading habits of teenagers. In this scenario, appropriate data would be "book genres preferred by teenagers," while an example of inappropriate data would be "favorite food of adults." This could be formalized using relevance logic [1].
- **Representativeness** with respect to O - Representativeness indicates that the dataset encompasses a comprehensive range of data instances that provide a holistic view of the subject as defined by the objectives O. For example, the objective is to analyze dietary preference across the US. Then, a representative dataset would need to include preferences from people of various ages, ethnicities, regions, and socio-economic backgrounds, and ideally, the number of samples for each scenario should be the same.

Table 1. Comparison of different EDA tools, other NLP approaches, and our approach in relation to data quality dimensions: Consistency, Appropriateness, Representativeness, and Completeness. ✓ indicates the approach supports the respective dimension, while ✗ indicates that it does not.

Approaches	Consistency	Appropriateness	Representativeness	Completeness
EDA Toolkits	✓	✗	✗	✗
Dataset Pruning	✓	✓	✗	✗
Stopping Model Bias	✗	✓	✗	✗
Adversarial Dataset Creation	✗	✗	✓	✓
Counterfactual Data Augmentation	✗	✗	✓	✓
Our Approach	✓	✓	✓	✓

– **Completeness** with respect to O - Completeness denotes that all the information necessary to achieve or analyze the objectives O is present in the dataset. For example, the objective is to analyze monthly sales of a product throughout the year. Completeness here would mean that the dataset contains data for all 12 months without missing any month.

In NLP, these dimensions are tailored to ensure that the data will perform well with language models and algorithms. For instance, consistency is crucial in NLP because language data often needs to be in a specific format for computational processes. Appropriateness plays a role since irrelevant data can skew the model's learning. Representativeness is important due to the diverse ways language is used across different groups, and completeness is critical because missing linguistic data can lead to misinterpretations or biases in language understanding.

2 Related Work

Exploratory Data Analysis (EDA) is an important first step in any work involving a new dataset. EDA involves understanding the characteristics of the dataset, detecting anomalies, and obtaining insights to inform subsequent data processing steps. Several toolkits such as TDDA,[1] YdataProfiling,[2] and TDFV[3] have emerged to facilitate this process, each with its unique features and limitations. These EDA toolkits offer valuable features for exploring and assessing data quality, but they fall short of addressing the unique challenges posed by NLP datasets. In sum, while these EDA toolkits provide essential tools for data quality exploration, many focus heavily on consistency and are less comprehensive in evaluating NLP data's appropriateness, representativeness, and completeness. Our proposed taxonomy provides a comprehensive view of the data quality of an NLP dataset, including meaningful metrics for all defined dimensions of data quality in this work. Our approach focuses solely on textual data, which directly

---

[1] https://github.com/tdda/tdda.
[2] https://github.com/ydataai/ydata-profiling.
[3] https://github.com/tensorflow/metadata.

addresses the limitation of most EDA toolkits. Furthermore, the metrics we use will address the lack of mechanisms for appropriateness, representativeness, and completeness in EDA toolkits for textual data (Table 1).

Data Quality in NLP. In the specific domain of NLP, ensuring data quality presents its own set of challenges and requires tailored approaches. To address the challenges of data quality in NLP, researchers have proposed various approaches. Dataset Pruning methods [14,21] improve appropriateness and consistency by eliminating irrelevant samples but risk losing representativeness and completeness, impacting model performance. Stopping model bias techniques [4,10] prevent model biases using ensemble techniques that avoid biased data patterns, enhancing appropriateness without addressing other data quality dimensions. Adversarial dataset creation approaches [2,15,17] construct bias-free datasets using classifiers to strengthen model robustness against adversarial examples, which may compromise data appropriateness, consistency, or completeness. Counterfactual data augmentation methods [8,11] add counterfactual examples to datasets, aiming to boost fairness and generalization, focusing on representativeness and completeness but neglecting consistency and appropriateness. Recently, efforts have been made to generalize metrics for evaluating data quality in NLP, with the introduction of the Data Quality Index (DQI) [16]. However, DQI has its own limitations, as it primarily focuses on assessing data quality from a linguistic perspective, often overlooking other important factors such as diversity or label quality. Moreover, some of its components are not generic enough and can only be applied to a specific text format. Our approach takes advantage of DQI, using some of its indices that are valuable to our taxonomy. We use other metrics as well as propose a new metric to address the limitation of DQI, viz., the lack of inter-class metrics.

Dimensions of Data Quality. There exist many frameworks for data quality such as [12,19,26], with each framework proposing many dimensions. However, not all dimensions are suitable for our purpose. For example, integrity, reputation, and accessibility may be more related to the process of data collection and management than the inherent quality of the data with respect to its utility in NLP. Therefore, they cannot be applied directly to NLP without modifications. In our work, we selected the most relevant dimensions that can aid in measuring data quality for textual data. The reasoning behind our choices is given in Sect. 1.1.

3   Taxonomy for Data Quality in NLP

Ensuring data quality is paramount in Natural Language Processing (NLP) to achieve high performance and generalizability in models. We propose a taxonomy (Fig. 1) that delineates various aspects of data quality with a brief definition and their significance in NLP. The taxonomy's categories encompass various metrics that correspond to the data quality dimensions outlined in Sect. 1.1.

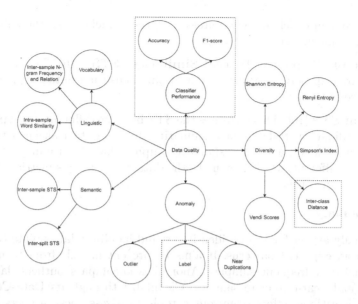

Fig. 1. The proposed taxonomy for data quality in NLP. Parts that are inside dashed boxes require a labeled dataset.

3.1   Linguistic

Linguistic quality emphasizes the structural and grammatical aspects of a dataset. Linguistically sound datasets ensure that models are trained on well-structured and coherent data, fostering better understanding and generalization.

Vocabulary [16]. This metric takes into account the following characteristics of the data: the size of the vocabulary versus the size of the dataset and the distribution of sentence lengths. The higher the average vocabulary, the better the completeness and the representativeness of the dataset and the consistency of the sentence length.

Inter-sample N-gram Frequency and Relation [16]. The terms are constructed to analyze the distribution of the granularity (e.g. 2-gram, 3-gram) considered and impose an acceptable range of values for each granularity. It is used to analyze the consistency of word frequency distribution across the dataset.

Intra-sample Word Similarity [16]. This measures the similarity between words within a sample, aiding in assessing the data's appropriateness for diverse linguistic tasks. The closer the mean of all similarities is to the hyperparameter value, the higher the data quality, since words with a low similarity may imply noisy data and a high similarity may imply high pair-wise bias in data.

3.2   Semantic

This dimension addresses the meaningful content of the data. Semantic quality is concerned with the clarity, relevance, and depth of meaning in the content.

Ensuring semantic richness guarantees that the models grasp the subtleties of language, enhancing their performance.

Inter-sample Semantic Textual Similarity (STS) [16]. This assesses the semantic overlap between different data points, ensuring data's representativeness and completeness.

Inter-split STS [16]. This measures the STS between different data splits (e.g., training, validation, test). A high similarity implies data leakage between the training and test set, and a low similarity implies that the training set and test set are very different, which may make the dataset hard. This metric aligns with the principle of consistency.

3.3   Anomaly

The anomaly aspect focuses on identifying and handling data points that differ greatly from expected patterns. It aims to prevent models from being misled by unusual or infrequent samples. Anomalies encompass outliers, label inaccuracies, and near-duplicate samples. Cleanlab,[4] through its Data-Centric AI (DCAI) algorithms, offers open-source tools to assess these aspects. Outliers signify inconsistency, while label accuracy ensures data completeness and suitability for supervised learning. Near-duplicate detection helps identify highly similar samples, aiding in enhancing dataset consistency and representativeness.

3.4   Classifier Performance

This is related to how well the data can train a model for its intended task. A dataset that supports high classifier performance is indicative of its appropriate, representative, and complete nature. We chose F1-score and accuracy for this category. Accuracy represents the fraction of predictions our model got right. It is a popular metric but can be misleading for imbalanced datasets. F1-score is the harmonic mean of precision and recall. It provides a balance between the two, especially important in imbalanced datasets.

3.5   Diversity

This category addresses the variety and range covered by the dataset. Ensuring data diversity helps models to be versatile, adaptable, and resilient to overfitting. A diverse dataset is a precondition for high representativeness. To achieve this, we use a wide variety of entropy-related metrics, such as entropy [22], Renyi entropy [20], and Simpson's Index [23]. We also use Vendi Scores [7], which is defined as the exponential of the entropy of the eigenvalues of a user-defined similarity matrix. The Vendi Score can help measure data quality by capturing the effective number of unique and dissimilar elements in a sample and accounting for correlations between features.

---

[4] https://docs.cleanlab.ai/stable/index.html.

Inter-class Distance. We propose a new metric to measure the difficulty of the data set. Inter-class Distance is defined as the average Jensen-Shannon distance between the distributions of classes within a dataset. The main motivation of this metric is due to the lack of an inter-class metric in [16]. As the field evolves, it is imperative to understand the inherent challenges of the datasets. We chose Jenshen-Shannon distance because we wanted a bounded, symmetric metric that is computationally manageable and suitable for discrete probability distributions. If the class distribution remains similar across classes, then the dataset will be harder to learn. A class distribution is defined by the user. The use of different distributions can lead to different results.

In this work, we chose the Inverse Document Frequency (IDF) distribution. IDF is a statistical measure used to assess how common or uncommon a word is amongst the corpus. The idea is that if a word occurs in too many documents, it may not be as important as a word that occurs less frequently. Using IDF distributions to evaluate the difficulty of a dataset aligns with the goal of understanding the uniqueness and commonality of words (or features) within and across classes. By focusing on words that are unique to particular classes and less common across the dataset, IDF can highlight the features that are most discriminative. Classes that have high IDF scores for a different set of words are likely to be more distinguishable from each other, potentially reducing the difficulty of the dataset. The difficulty of the dataset provides information on how challenging a dataset might be for a model to learn, allowing researchers to tailor their approaches more effectively. Formally, given a dataset with n classes, with class i having the probability distribution P_i and the probability distribution of the rest of the dataset is $P_{\neq i}$, and $JSDist$ denotes Jenshen-Shannon distance, the inter-class distance ICD is computed as:

$$ICD = \frac{1}{n} \sum_{i=1}^{n} JSDist(P_i || P_{\neq i}). \tag{1}$$

4 Experimental Setup

We selected a suite of datasets that are exemplary in illustrating the range of deception that can be encountered. The cornerstone of our dataset collection is the Generalized Deception Dataset (GDD) [27], which comprises datasets from diverse domains such as job scams, fake news, phishing, political statements, and product reviews. Each dataset has two versions: new and old. The old version was constructed using a variety of sources with some cleaning [27]. The new version was constructed using the old version but with another round of cleaning. The cleaning process for the new version includes removing non-English documents, duplications, outliers, as well as noisy data points. For political statements, the creators also changed the label scheme. Additionally, we incorporate the TwiBot datasets [5,6], which are specifically tailored to detect bot-driven behaviors and anomalies on the Twitter platform. The TwiBot datasets have many features other than textual data. However, our work will use only their textual data.

We chose these datasets because we were aiming to construct a quality dataset of deceptive attacks for domain-independent deception detection. However, we expect that this can be applied to other NLP datasets. The goal of this experiment is to observe how different aspects of data quality change between the old and new version of a dataset.

The datasets were preprocessed so that they were in the appropriate format. For the experiments, the datasets need to have two features: text and label. The metrics for the linguistic component and the semantic component are implemented based on the formulas from [16]. All similarity metrics are computed using cosine similarity as suggested by [16]. The metrics for the anomaly component are implemented using CleanLab. The entropy, renyi entropy, and simpsons' index, which are metrics in the diversity component, are implemented based on their definitions [20, 22, 23]. Another metric in the diversity component, Vendi Score, is implemented following the GitHub[5] of Vendi Score. We implement inter-class distance based on its definition provided in Eq. 1. For classifier performance, we employ a wide range of classical classifiers from the sklearn library [13], including logistic regression, decision tree, naive bayes, AdaBoost, k-nearest neighbors (knn), random forest, and support vector machine (SVM). Due to the time constraint and resource constraint, we only use the default hyperparameters for all classifiers.

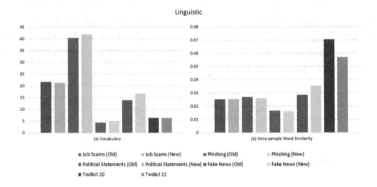

Fig. 2. Linguistic analysis of different datasets in NLP. Figure (a) is the 'Vocabulary' chart showcasing an increase in Phishing and Fake News datasets. Figure (b), the 'Intra-sample Word Similarity' chart, emphasizes the drop in quality in TwiBot, which indicates that noises and biases have been introduced.

---

[5] https://github.com/vertaix/Vendi-Score.

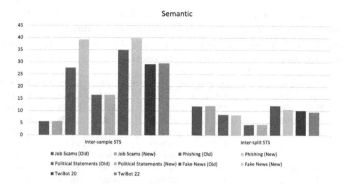

Fig. 3. The chart illustrates the inter-sample and inter-split semantic textual similarity (STS) scores. Phishing and Fake News exhibit an increase in inter-sample STS. Inter-split STS remains mostly the same, with only Fake News having a noticeable decrease

Table 2. Comparative assessment of diversity metrics across multiple NLP datasets. ↑ indicates higher is better, and ↓ indicates lower is better. We observe consistently higher entropy and Renyi entropy values for the TwiBot datasets, pointing towards a greater level of unpredictability. The Fake News dataset shows a decrease in the new version across most metrics, except for Simpson's Index.

		Entropy ↑	Renyi Entropy ↑	Simpson's Index ↓	Vendi Score ↑	Inter-class Distance ↑
Job Scams	Old	7.7972	7.8164	0.001366	1.5047	0.5160
	New	7.7981	7.8174	0.001375	1.4540	0.5169
Phishing	Old	8.8932	8.9201	0.000630	1.6614	0.5845
	New	8.8886	8.9155	0.000631	1.6427	0.5841
Political Statements	Old	7.7734	7.7906	0.001488	1.5566	0.3026
	New	7.7835	7.8009	0.001480	1.5703	0.3171
Fake News	Old	8.7075	8.7328	0.000789	1.7700	0.4549
	New	8.6939	8.7178	0.000785	1.7401	0.4169
TwiBot20		9.9849	10.0487	0.002828	1.5989	0.5748
TwiBot22		10.4437	10.5113	0.002420	1.5822	0.5420

5 Results and Discussion

Linguistic. From Fig. 2(a), we see a large increase in the quality of the vocabulary for Phishing and Fake News. This is the result of the removal of duplicates, as explained in Sect. 4. The higher vocabulary quality indicates that the new datasets have better representativeness than the old datasets. Other datasets remain mostly the same. In Fig. 2(b), there is a drop for TwiBot. This implies that there is more noise or bias in the new version of TwiBot. On the other hand, Fake News dataset shows an improvement, supporting that the cleaning of outliers improves the appropriateness of the dataset.

Semantic. Figure 3 indicates robust inter-sample coherence for the new Phishing and Fake News datasets. The process of removing HTML tag and some

Table 3. Performance comparison of various machine learning classifiers in terms of f1-score, evaluated on both old and new versions of the datasets. The Naive Bayes (Multinomial) classifier yielded a score of 0 for the Job Scams dataset, indicating the extreme imbalance issue of the Job Scams dataset.

Datasets		Logistic Regression	Decision Tree	Naive Bayes	AdaBoost	K Neighbors	Random Forest	SVM
Job Scams	Old	0.1791	0.5965	0	0.2706	0.4946	0.5122	0.3841
	New	0.1955	0.5232	0	0.3049	0.5729	0.4713	0.4106
Phishing	Old	0.9476	0.9213	0.9070	0.9261	0.9666	0.9518	0.9659
	New	0.9441	0.9190	0.8995	0.9169	0.9611	0.9428	0.9638
Political Statements	Old	0.5087	0.4957	0.3904	0.3007	0.4986	0.4753	0.4875
	New	0.7691	0.6733	0.7843	0.7606	0.7229	0.7604	0.7810
Fake News	Old	0.9328	0.8230	0.7585	0.8781	0.7168	0.8814	0.9432
	New	0.9082	0.7895	0.6501	0.8662	0.6778	0.8581	0.9190
TwiBot20		0.6838	0.5823	0.7292	0.6704	0.6794	0.6768	0.6902
TwiBot22		0.6343	0.5695	0.6817	0.6808	0.7017	0.6371	0.6521

irrelevant metadata, and, therefore, removing noises from Phishing dataset contributes mostly to this improvement, while the process of refining samples and removing outliers and noise explains the Fake News dataset. Removing outliers and noises eliminates samples that have abnormally low similarity with most sentences, while removing duplications eliminates samples that have abnormally high similarity with some other sentences. Inter-split STS remains mostly the same for most datasets, with the exception of Fake News. However, we expect this due to the difference in size between the old Fake New dataset (72, 134 articles) and the new Fake New dataset (20, 456 articles). With a size gap, it can lead to more inconsistent splits, where splits can be too dissimilar or too similar to each other.

Diversity and Classifier Performance. Referring to Table 2, the diversity metrics largely maintain consistency between old and new dataset versions, except for the Fake News dataset which exhibits a decrease in diversity in several indicators. We suspect that the extensive removal of samples led to this problem. When a sample is removed, there is often a trade-off between appropriateness and representativeness. For example, let us have a simple dataset with four data points, two female and two male, and an objective O that is to analyze food preferences across the country. If we determine that one of the male data points is not appropriate for the objective (e.g. food preference has the value of "beer" which is a beverage), we may want to remove it. However, after removing it, we have two female data points and one male data point, which leads to under-representativeness for male. We also see a decrease in the inter-class distance in TwiBot dataset. This indicates that the classes' distributions are closer together, which makes it difficult for the model to learn. We can observe this behavior in Table 3, where the f1-scores of TwiBot22 are lower than TwiBot20 across the board. From Table 3, we observe that Phishing dataset exhibits consistently high f1 scores in most classifiers. Political Statements dataset shows a significant improvement in f1-score in its new version compared to the old version across all classifiers. The change in the label scheme of Political Statements dataset is likely the main contribution to this improvement. F1-scores also reveal the imbalance issue with Job Scams dataset, with some classifiers having very low f1-scores compared to other datasets. Performance degradation in the new Fake News dataset signals a possible over-cleaning issue, where the data's representativeness and thus its quality may have been compromised, as explained above.

Results Summary. For Job Scams, the data quality maintains the same level, with the extreme imbalance issue remaining the same. For Phishing, the new version has better linguistic and semantic quality. For Political Statements, the quality remains the same, but the change in label scheme results in better model's performance. For Fake News the quality correlated with appropriateness increases, while the quality correlated with representativeness decreases. For TwiBot, the quality in most categories decreases when using only textual data.

Limitation. Although we proposed a comprehensive taxonomy for data quality in NLP and demonstrated its effectiveness with multiple datasets, our work is not free of limitations. Firstly, some of the metrics in our taxonomy require labeled datasets. Secondly, there are many more existing metrics that are tailored to specific tasks that we did not consider due to the generic nature of our taxonomy. Moreover, many of the metrics have hyperparameters that can be tuned depending on the use case of the users. While this provides the necessary flexibility, it can be overwhelming for new users. Furthermore, using different hyperparameters can lead to widely different results. Therefore, there is a learning curve to use our taxonomy. Also, we do not consider some dimensions of data quality that are related to data management or the source of the data.

6 Conclusion and Future Works

In this work, we have proposed a comprehensive taxonomy for data quality in NLP based on defined dimensions of data quality. Our taxonomy encompasses various aspects of data quality, including linguistic quality, representational quality, and contextual quality. We have also introduced a novel metric to measure the difficulty of a dataset, which can be used to assess the overall quality of a dataset. We have evaluated our proposed taxonomy using a wide variety of NLP datasets that encompass multiple domains. The results demonstrate that our taxonomy can effectively distinguish between changes in the quality of the datasets at multiple angles. We believe that our work is a significant contribution to the field of NLP. By providing a comprehensive and systematic approach to data quality assessment, we can help improve the reliability and performance of NLP models. There are many directions for future work, such as: Extending the taxonomy to support more metrics, developing new methods or tools to automatically measure or improve the data quality metrics in our taxonomy, or applying the taxonomy to more diverse and challenging NLP datasets, such as multilingual, multimodal, or low-resource data.

Acknowledgments. Research partly supported by NSF grants 2210198 and 2244279, and ARO grants W911NF-20-1-0254 and W911NF-23-1-0191. Verma is the founder of Everest Cyber Security and Analytics, Inc.

References

1. Anderson, A.R., Belnap Jr, N.D., Dunn, J.M.: Entailment, Vol. II: The Logic of Relevance and Necessity, vol. 5009. Princeton University Press, Princeton (2017)
2. Bras, R.L., et al.: Adversarial filters of dataset biases. CoRR arxiv:2002.04108 (2020)
3. Chen, H., Chen, J., Ding, J.: Data evaluation and enhancement for quality improvement of machine learning. IEEE Trans. Reliab. **70**(2), 831–847 (2021). https://doi.org/10.1109/TR.2021.3070863
4. Clark, C., Yatskar, M., Zettlemoyer, L.: Don't take the easy way out: ensemble based methods for avoiding known dataset biases. CoRR arxiv:1909.03683 (2019)
5. Feng, S., et al.: Twibot-22: towards graph-based twitter bot detection. In: Thirty-sixth Conference on Neural Information Processing Systems Datasets and Benchmarks Track (2022)
6. Feng, S., Wan, H., Wang, N., Li, J., Luo, M.: Twibot-20: a comprehensive twitter bot detection benchmark. In: ACM International Conference on Information & Knowledge Management (2021)
7. Friedman, D., Dieng, A.B.: The vendi score: a diversity evaluation metric for machine learning (2023)
8. Gardner, M., et al.: Evaluating NLP models via contrast sets. CoRR arxiv:2004.02709 (2020)
9. Gupta, N., et al.: Data quality toolkit: automatic assessment of data quality and remediation for machine learning datasets (2021)
10. He, H., Zha, S., Wang, H.: Unlearn dataset bias in natural language inference by fitting the residual. CoRR arxiv:1908.10763 (2019)
11. Kaushik, D., Hovy, E.H., Lipton, Z.C.: Learning the difference that makes a difference with counterfactually-augmented data. CoRR arxiv:1909.12434 (2019)
12. Krogstie, J.: A semiotic approach to data quality. In: Nurcan, S., et al. (eds.) BPMDS/EMMSAD -2013. LNBIP, vol. 147, pp. 395–410. Springer, Heidelberg (2013). https://doi.org/10.1007/978-3-642-38484-4_28
13. scikit learn: (2023). https://scikit-learn.org/
14. Li, Y., Vasconcelos, N.: REPAIR: removing representation bias by dataset resampling. CoRR arxiv:1904.07911 (2019)
15. Liu, N.F., Schwartz, R., Smith, N.A.: Inoculation by fine-tuning: a method for analyzing challenge datasets. CoRR arxiv:1904.02668 (2019)
16. Mishra, S., Arunkumar, A., Sachdeva, B.S., Bryan, C., Baral, C.: DQI: measuring data quality in NLP. CoRR arxiv:2005.00816 (2020)
17. Nie, Y., Williams, A., Dinan, E., Bansal, M., et al.: Adversarial NLI: a new benchmark for natural language understanding. CoRR arxiv:1910.14599 (2019)
18. Picard, S., et al.: Ensuring dataset quality for machine learning certification (2020)

19. Pipino, L.L., Lee, Y.W., Wang, R.Y.: Data quality assessment. Commun. ACM **45**(4), 211–218 (2002). https://doi.org/10.1145/505248.506010
20. Rényi, A.: On measures of entropy and information. In: Berkeley Symposium on Mathematical Statistics and Probability, vol. 1: Contributions to the Theory of Statistics, vol. 4, pp. 547–562. UC Press (1961)
21. Sakaguchi, K., Bras, R.L., Bhagavatula, C., Choi, Y.: WINOGRANDE: an adversarial Winograd schema challenge at scale. CoRR arxiv:1907.10641 (2019)
22. Shannon, C.E.: A mathematical theory of communication. Bell Syst. Tech. J. **27**(3), 379–423 (1948)
23. Simpson, E.H.: Measurement of diversity. Nature **163**(4148), 688–688 (1949)
24. Tan, S., Shen, Y., Huang, C., Courville, A.C.: Investigating biases in textual entailment datasets. CoRR arxiv:1906.09635 (2019)
25. Torralba, A., Efros, A.A.: Unbiased look at dataset bias. In: CVPR 2011, pp. 1521–1528 (2011). https://doi.org/10.1109/CVPR.2011.5995347
26. Wand, Y., Wang, R.Y.: Anchoring data quality dimensions in ontological foundations. Commun. ACM **39**(11), 86–95 (1996). https://doi.org/10.1145/240455.240479
27. Zeng, V., Liu, X., Verma, R.M.: Does deception leave a content independent stylistic trace? In: ACM Conference on Data and Application Security and Privacy, pp. 349-351. ACM, New York (2022). https://doi.org/10.1145/3508398.3519358

Building Brownian Bridges to Learn Dynamic Author Representations from Texts

Enzo Terreau[✉] and Julien Velcin[ID]

Université de Lyon, Lyon 2, ERIC UR 3083, Bron, France
{enzo.terreau,julien.velcin}@univ-lyon2.fr

Abstract. Authors writing habits fluctuate throughout their lives. This evolution may stem from engaging in new topics, new genres or by the variation of their writing style. However, most representation models aiming at building meaningful authors embedding focus on static representations. They skip the precious time information useful to build more powerful and versatile representations. Only a limited number of methods learn dynamic representations, each dedicated to a time bin. Here we propose a new representation learning model called BARL (Brownian Bridges for Author Representation Learning). BARL uses Brownian Bridges, a Gaussian process, to embed authors as continuous trajectories through time. Leveraging the Variational Information Bottleneck (VIB) framework, it integrates a pre-trained temporal text encoder to encode authors and documents into the same space, learning a distinct dynamic for each author along with a customized variance. We evaluate BARL on several tasks: authorship attribution, document dating and author classification on two datasets from the literature. BARL outperforms baselines and existing dynamic author embedding models while learning a continuous temporal representation space.

Keywords: Author Representation · Dynamic Author Embedding · Variational Information Bottleneck · Brownian Bridges

1 Introduction

Significant work has been made in learning to encode textual information, spanning from representations at subword levels to sentences and documents of various lengths. This encompasses classic methods developed in information retrieval, such as the TF and TFxIDF vector representation, to more recent techniques involving word embedding and contextual representations based on Transformer-based architectures [6,8]. However, less attention has been dedicated to working at the author level, problem which introduces new features such as topic or style [10]. These representations can be used to solve several downstream tasks, such as author classification or identification, link prediction, and in recommendation systems. Building time-sensitive representations is a must to solve these tasks in a better way.

I. Miliou et al. (Eds.): IDA 2024, LNCS 14641, pp. 230–241, 2024.
https://doi.org/10.1007/978-3-031-58547-0_19

In this paper, we propose a novel method to construct a latent space that captures the dynamic of authors over time. While previous works have addressed this problem before [7,9], they do not explicitly build a unique space applicable not only to one but to several sub-tasks. The latent space we aim to build is shown in Fig. 1, showcasing the trajectories of several authors through the evolution of their publications over time.

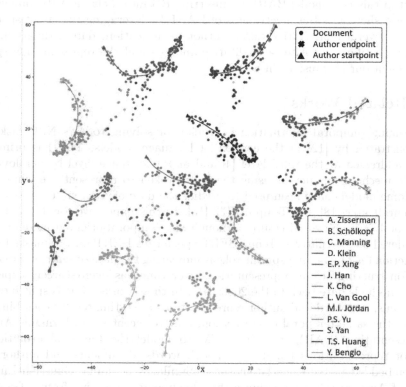

Fig. 1. 2D T-SNE projection of selected author trajectories from the S2G corpus. The color gradient corresponds to the publication date to observe the evolution of its publication over time. Each trajectory goes from its start point ▲ to its end point ✖ (Color figure online)

Our new method, named BARL (Brownian Bridges for Author Representation Learning), is a representation learning method that uses the concept of Brownian Bridges (BB), a Gaussian process previously used for problems such as video features disentangling [5]. BB is a continuous stochastic process that models the transition between two points in the space (start and end points, illustrated by the triangle and cross symbols in Fig. 1).

While BB have already been used for dealing with textual content, it has been done at the document level only to increase consistency of long text generation [19]. In summary, the contribution of this paper is threefold:

– We are the first to use BB to address the dynamic of author representations.

- We introduce the novel BARL model in which we adapt the Variational Information Bottleneck (VIB) framework to deal with author dynamics through the BB mechanism.
- We demonstrate the effectiveness of BARL through quantitative and qualitative experiments, comparing it to the existing literature.

Section 2 delves deeper into related work while Sect. 3 provides details on how we setup this new model BARL by inserting BB into a classic VIB framework. Section 4 showcases the effectiveness of BARL by leveraging the latent space to address several quantitative tasks: author identification, date estimation, and author classification. After a qualitative analysis and the representation space, we present our conclusions in Sect. 5.

2   Related Works

Leveraging temporal information is crucial for solving today's NLP tasks, as demonstrated by [12] in the context of language models. This has primarily been addressed at the word level [4] and at the sentence level by employing a simple mechanism, such as using the [CLS] token to represent the whole text [2]. Some models have been used to predict the date of a given document, such as NeuralDater [18] and TempoBERT [15]. In this paper, we use TempoBERT as a module in our architecture, although any temporal-oriented model can be considered as alternatives. TempoBERT specializes BERT on the masked word prediction task by adding specific tokens indicating the date of sentence creation.

Constructing author representations over time has been scarcely explored. Early methods date back to the 2000s [16]. In these works, time is split into discrete bins, and word and author representations are estimated for each bin. Following the same temporal discretization, a more recent work, Dynamic Author Representation (DAR), uses an LSTM to model the temporal evolution of author representations [7]. In this model, words, documents and authors are not embedded in the same latent space. Similarly, Dynamic Gaussian Embedding of Authors (DGEA) assumes that documents are drawn from a Gaussian distribution depending on authors. Two implementations, one based on Kalman filters and one based on a deep learning architecture, are tested, but as DAR they both rely on a discretization of the time span. Our model, BARL, stands out as the only one capable of embedding both documents and authors into a unique continuous latent space. This feature is a crucial prerequisite for capturing author trajectories over time. This constitutes the main contribution of our work.

3   BARL: Brownian Bridges for Author Representation Learning

3.1   Background

We note D the set of all documents and A the set of authors. In this work, we assume that each document $d_a^t \in D$ is written by one author $a \in A$ at time

$t \in \mathbb{R}^+$. We restrict t to an interval that corresponds to the first timestamp (set to 0, by default) and the maximum timestamp $T \in \mathbb{R}^+$, so $t \in [0, T]$. Our objective is then to build a latent space that includes both document representations z_t^a (i.e., the latent representation of the document d_a^t) and author representations h_t^a (which can differ from z_t^a because there is an interpolation, see below). We assume that there is only one document written by a at time t. This is a mild assumption because the scale of t can be adjusted accordingly.

3.2   Using the Brownian Bridges

In BARL, we use the Brownian Bridge (BB) to build author trajectories in the latent space. In this framework, the trajectory of one author is fully characterized by its initial coordinate $h_0^a \in \mathbb{R}^r$ and its terminal coordinate $h_T^a \in \mathbb{R}^r$, which can both be noted by $H^a = (h_0^a, h_T^a)$, where r is the dimension of the latent space. All the intermediate vector representations h_t^a are interpolated between these two points, assuming a Gaussian noise that increases with the temporal distance from the two endpoints:

$$p(h_t^a | h_0^a, h_T^a) = \mathcal{N}\left((1 - \frac{t}{T})h_0^a + \frac{t}{T}h_T^a, \frac{t(T-t)}{T}\right) \tag{1}$$

where $\mathcal{N}(\mu, \sigma^2)$ is the normal distribution with μ mean and σ^2 variance with diagonal given in Eq. 1.

In the following, we aim to make latent representations of documents z_t^a close to their related author representation h_t^a. This mechanism, inspired by the Time Control model of [19], is illustrated in Fig. 2.

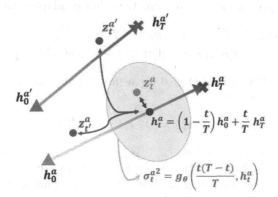

Fig. 2. Illustration of two authors' trajectories (a and a'). We would like to make z_t^a, in green, close to h_t^a, while we make documents farther from the other authors (such as $z_t^{a'}$) or documents written at another time (such as $z_{t'}^a$), in red. h_t^a is the interpolation between the start point h_0^a (▲) and the end point h_T^a (✖) (Color figure online)

3.3 Variational Information Bottleneck

In the Time Control theory [19], the initial and terminal states are fixed for all documents, which is not compatible with our aim that is to *learn* the author representation. This leads us to propose a new model that integrates the BB mechanism into the VIB framework, which precisely aims to learn the latent representations z (in our case, both H^a and z_t^a). The general object if of VIB is the following [1]:

$$\arg \max_z I(z, y) - \beta.I(z, x) \tag{2}$$

where (x, y) are the input and output data (y can be the labels in a classification problem), z is the latent representation, $\beta \in \mathbb{R}^+$ is the trade-off hyper-parameter and I is the Mutual Information. VIB is based on the Information Bottleneck principle [17] that aims at maximally compressing the information in z, such that z is highly informative regarding the labels (i.e., z can be used to predict the labels y). It boils down to minimizing the following loss using the lower bound introduced in [1]:

$$\mathcal{L}_{vib} = -\mathbb{E}[\log q(y|z)] + \beta.KL(p(z|x)||q(z)) \tag{3}$$

where $\mathbb{E}$ is the usual expectation, $q(y|z)$ is the variational approximation of $p(y|z)$, $q(z)$ is for $p(z)$ and $KL(\cdot||\cdot)$ stands for the Kullback-Leibler divergence.

3.4 Learning Author Representations

In our context, we have naturally built author/document (z_t^a, z_t^d) pairs as positive examples (i.e., $y = 1$), which are complemented by sampling negative pairs (i.e., $y = 0$): $(z_{t'}^a, z_t^d)$ where the time differs, $(z_t^{a'}, z_t^d)$ where author a' differs, or both. In this work, we propose a new loss function that is adapted to our problem:

$$\mathcal{L}_{BARL} = -\mathbb{E}_{p(z_t^a|d_t^a), p(h_t^a|a)}[\log q(y|z_t^a, h_t^a)] \\ + \beta[KL(p(h_t^a|a)||q(h_t^a)) + KL(p(z_t^a|d_t^a)||q(z_t^a))] \tag{4}$$

where the probability of a label y (0 for negative pairs and 1 for positive pairs, see above) is given by [13]:

$$q(y = 1|z_t^a, z_t^d) = \sigma(-c_a||z_t^a - z_t^d||_2 + e_a) \tag{5}$$

where σ is the sigmoid function, $c_a \in \mathbb{R}_*^+$ and $e_a \in \mathbb{R}$ are learnable parameters. Expectation in Eq. 4 is intractable for most deep encoders. However, we can approximate it by sampling L observations by training example using the reparameterization trick of VAE [11]:

$$z_t^a = \mu_t^a + \eta_t^a \odot \epsilon, \ h_t^a = (1 - \frac{t}{T})h_0^a + \frac{t}{T}h_T^a + \frac{t(T-t)}{T}\epsilon \ \text{with} \ \epsilon \sim \mathcal{N}(0,1) \tag{6}$$

where z_t^a is fully determined by its mean $\mu_t^a = f_\theta(d_t^a)$ and variance $\eta_t^{a2} = g_\Theta(d_t^a)$, f_θ and g_Θ two functions being learnt as in standard VAE. We can point out that the VIB framework perfectly fits the Gaussian process of BB.

3.5 Model Architecture of BARL

A schematic representation of our model is proposed in Fig. 3.

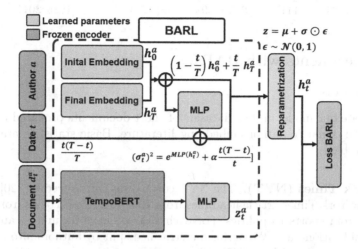

Fig. 3. Architecture of our model **BARL**. At the learning stage, we provide as input the document along a timestamp (a different timestamps for negative examples) and author (a different author for negative examples). During the inference, we can encode any unseen document and interpolate known author representation for any date.

For each author we learn their start and end points (h_0^a and h_T^a) through two embedding layers. Each point matches each author first and last work date. In Sect. 4 we evaluate the extrapolation capacity of BARL by setting a maximum date T greater than the last known writing date of every author.

In the BB modeling, the variance is only time-based ($\sigma_t^{a\,2} = \frac{t(T-t)}{T}$). Each author has its own writing dynamic, with drastic topic or stylistic changes, thus following [9] we compute a log-variance using a 2-layers MLP with $\texttt{tanh}$ and $\texttt{LeakyReLU}$ activation functions on the author representation h_t^a. Our final variance is given by: $(\sigma_t^a)^2 = e^{\mathrm{MLP}(h_t^a)} + \alpha\frac{t(T-t)}{T}$. α is a learnable parameter we add which weights the relative importance of both variance, initialized to one.

The entering block of our model for documents maps a document in natural language to a vector. We propose to use a pre-trained text encoder. Models that are pre-trained on large datasets are now easily available online. They have been proved successful on many NLP tasks with a simple fine-tuning phase. The VIB framework allows to naturally introduce a pre-trained text encoder. Here we use a frozen TempoBERT [15], a BERT model fine-tuned on temporal masking, and a 3-layers MLP with $\texttt{LeakyReLU}$ activation. We do not compute any document variance as experiments give better results without it. Note that any text encoder (ideally time-based) can be used and even fine-tuned during the training stage, making our model language agnostic to the very choice of the encoder.

Table 1. Basic statistics of both datasets NYT and S2G

Dataset	Authors	Avg. Tokens	Texts by Author	Period
NYT	546	8.4($\pm$2.5)	76($\pm$51)	$[1990, 2015]$
S2G	1117	8.7($\pm$2.9)	48($\pm$27)	$[1985, 2017]$

4   Experiments with BARL

4.1   Datasets

We evaluate our model on two datasets of short documents prepared by [7] and used in the dynamic author embedding literature. Basic statistics are summarized in Table 1.

New York Times (NYT). The NYT corpus was introduced by [20] and is a set of New York Times article headlines from 1990 to 2015. It comes from various sections, from sports to politics. For each title we have its unique author and its full publication date. We perform a stratified split by authors into train, test and validation sets with a 70/20/10 ratio. We use it in an imputation scheme to evaluate the modeling capability of BARL, each set coming from the whole time span.

Semantic Scholar (S2G). The S2G corpus was introduced by [3]. It is a set of scientific article titles published in machine learning conference from 1985 to 2017. For each article we only have its year publication date (corresponding to one timestep) and it may have multiple authors. Each distinct author of a document yields a unique (author, document) pair. We perform the same imputation scheme as for NYT but we also use it to evaluate the ability of BARL to extrapolate. For each author we split into train, test and validation sets with a 70/20/10 ratio following the chronological order. Thus, in the prediction scheme the last timesteps (which differs for every author) are not seen during the training stage.

4.2   Parameter Settings and Competitors

We select hyper-parameters using a grid search on each validation set. We use a $5e^{-4}$ learning rate with linear decay for 100 epochs and apply early stopping. To fasten the learning stage, we add a writing year token at the beginning of each training example for the first 5 epochs, with an increasing masking probability. We also compare BARL to its variations with only a time-based variance (BARL_t), without any variance (BARL_{novar}) and with an L_2 loss rather than the VIB framework (BARL_{L_2}). The results of these ablation studies are discussed in a dedicated section (see Sect. 4.6). Our code is publicly available[1].

---

[1] https://github.com/EnzoFleur/barl.

We compare our model against several baselines. First, we consider the Universal Sentence Encoder (USE) [6], a very powerful sentence encoder. Then, BERT models [8] fine-tuned on the authorship attribution task ($BERT_A$), on the document dating task ($BERT_T$) and on both tasks ($BERT_{T+A}$). We also evaluate TempoBERT [15] alone. Finally, we compare BARL with two recent dynamic author embedding methods detailed in Sect. 2: DAR[2] (Dynamic Author Representation) [7] and DGEA (Dynamic Gaussian Embedding of Authors) [9]. We choose the K-DGEA version of DGEA as it is the fastest and obtain similar results than R-DGEA. For both models we use the configuration given by the authors as they were both trained on the same datasets. We evaluate our model on three tasks, with 10 repetitions for each competitors corresponding to the standard deviation in each result table.

4.3   Results in Authorship Attribution

Authorship attribution consists in assigning to each document its author(s). We use accuracy and coverage error (CE) as metrics. For S2G, we use the usual metric for multi-author corpus: Label Ranking Average Precision (LRAP). For each document embedding we choose its closest author embeddings using cosine similarity. Results are shown in Table 2. Our model BARL ranks first among author embedding methods and third overall (second on S2G in prediction), with at least a two-points increase in coverage error against K-DGEA. DAR cannot produce meaningful document embedding without the author information which is fatal here. $BERT_A$ performs the best even if USE is better in coverage error on NYT. USE confirms its huge modeling capability as an on-the-shelf sentence encoder, challenging fine-tuned models.

Authorship attribution is harder on S2G than on NYT, with more authors and a more specific vocabulary. The time information is more important in a context of scientific publication: this explains why TempoBERT and $BERT_{T+A}$ both achieve better ranks on S2G. Our model is competitive with fine-tuned encoder on a unique specific task while its objectives are multiple: grasp the link between an author and its production with its evolution through time.

4.4   Results in Document Dating

Document dating aims to predict the publication year of each document. The two associated metrics are accuracy and mean absolute error (MAE). We use a K-Nearest-Neighbors as classifier to predict each date, except for baselines with a classification head and TempoBERT. This task is not suitable for prediction as we cannot predict unseen years. Results are shown in Table 3.

Our model BARL achieves the best results on every axis, except MAE on S2G. BARL outperforms fine-tuned model on the document dating task. $BERT_{T+A}$ gets better results than $BERT_T$ which shows the contribution of the author information to predict a document writing period. BARL seems to make

---

[2] https://github.com/edouardelasalles/dar.

Table 2. Authorship attribution on NYT in imputation and on S2G in imputation and prediction. Best models in bold, second underlined, std in parentheses.

Méthodes	NYT (546 authors)		S2G (1117 authors)			
	Imputation		Imputation		Prediction	
	CE ↓	Accuracy ↑	CE ↓	Accuracy ↑	CE ↓	Accuracy ↑
USE	**11.1 (0.0)**	12.7 (0.0)	19.9 (0.0)	10.4 (0.0)	22.3 (0.0)	7.6 (0.0)
$BERT_A$	11.4 (0.8)	**14.3 (0.4)**	**12.8 (0.8)**	**13.4 (0.7)**	**17.9 (1.3)**	**8.9 (0.8)**
$BERT_T$	88.2 (1.0)	0.5 (0.2)	90.2 (1.9)	0.8 (0.2)	94.0 (2.3)	1.1 (0.2)
$BERT_{T+A}$	33.4 (0.8)	3.7 (0.6)	15.3 (1.2)	10.6 (0.7)	25.9 (1.2)	6.7 (0.5)
TempoBERT	12.1 (1.1)	10.2 (0.9)	15.9 (1.5)	10.5 (1.1)	21.6 (0.7)	6.9 (0.4)
DAR	47.1 (1.1)	1.1 (0.2)	38.7 (2.0)	1.2 (0.4)	41.8 (1.2)	0.8 (0.4)
K-DGEA	14.5 (1.2)	9.8 (0.8)	20.4 (1.3)	10.1 (0.8)	28.2 (1.1)	6.6 (0.5)
BARL	11.9 (0.8)	12.2 (1.0)	15.5 (1.4)	10.4 (1.0)	20.6 (1.1)	7.3 (0.8)
$BARL_{L_2}$	27.2 (1.1)	10.1 (0.9)	24.2 (1.2)	9.7 (1.0)	35.5 (1.3)	4.3 (0.9)
$BARL_t$	12.5 (0.7)	11.9 (1.1)	15.9 (1.6)	10.3 (1.2)	21.1 (1.2)	7.1 (0.8)
$BARL_{novar}$	13.1 (0.4)	11.3 (0.6)	19.9 (1.1)	10.1 (0.6)	22.3 (1.1)	7.0 (0.9)

the best of it. Topics and vocabulary in S2G is more time-related as the average MAEs are smaller, even with a bigger global time span.

The two specialized author embedding models are far behind BARL, as they use a discrete representation of time which allows less smoothness and precision in dynamic document representation. The results in both tasks, authorship attribution and document dating, confirm that the main objective of our model has been reached.

4.5   Results in Author Classification

The last task focuses is author classification on S2G corpus in prediction. We associate each author a to the conference (IJCAI, ACL, EMNLP, ...) in which he or she published the most at each timestep $(t_0^a, t_1^a, ..., t_k^a)$ of the training set. We aim to predict their conference for the last timesteps $(t_{k+1}^a, ..., T^a)$. Here we only evaluate the author embedding method in their ability to extrapolate authors' dynamics. We use a linear SVM optimized with grid-search as classifier and accuracy as metric. Results are given in Table 4. In prediction, we aim at evaluating the extrapolation ability of representation models for unseen future timesteps. As DAR produces static and dynamic representations, we show the results for both representations and their concatenation. Here, author embedding models easily outperform the two language models USE and TempoBERT. Even if S2G only provides the publication year, canceling one of the main asset of BARL, our model shares the first rank with DAR. Most of the information in DAR representations are hold in the static representations of each author. Our model is able to interpolate as well as extrapolate authors' dynamics while producing continuous representations.

Table 3. Results in document dating in imputation. Best model in bold, second underlined, std in parentheses.

Méthodes	NYT (26 years)		S2G (33 years)	
	Accuracy ↑	MAE ↓	Accuracy ↑	MAE ↓
USE	10.4 (0.0)	6.8 (0.0)	9.81 (0.0)	5.1 (0.0)
$BERT_A$	10.2 (0.1)	6.3 (0.1)	8.9 (0.5)	5.5 (0.3)
$BERT_T$	13.0 (0.1)	5.5 (0.2)	12.2 (0.4)	4.4 (0.2)
$BERT_{T+A}$	12.0 (0.1)	5.4 (0.2)	12.3 (0.3)	**4.2 (0.2)**
TempoBERT	13.1 (0.2)	5.2 (0.1)	12.1 (0.3)	4.7 (0.5)
DAR	5.6 (0.2)	7.5 (0.3)	4.8 (0.4)	10.0 (1.1)
K-DGEA	10.6 (0.4)	6.7 (0.3)	7.4 (0.3)	6.8 (0.5)
BARL	**13.3 (0.3)**	**4.7 (0.3)**	**12.6 (0.5)**	4.3 (0.4)
$BARL_{L_2}$	10.5 (0.2)	5.3 (0.3)	8.9 (0.2)	5.4 (0.3)
$BARL_t$	12.8 (0.4)	5.0 (0.3)	12.2 (0.4)	4.6 (0.2)
$BARL_{novar}$	11.7 (0.3)	5.1 (0.2)	11.8 (0.3)	4.8 (0.3)

4.6 Ablation Study

We now discuss the results of BARL with its variations (see Tables 2 and 3, bottom lines). The worst results are obtained with the simple L_2 loss. The variational framework of VIB allows more versatility which is key to build such a complex representation space. The same conclusion goes when we look at BARL variations without any variance. Using only a temporal variance is enough to get sufficient results on both authorship attribution and document dating tasks. But it restrains the ability to grasp each author dynamic. Adding an author-related variance allows to capture for each author a distinct evolution, either sticking to the same topic or going through a lot of different topics during their career.

4.7 Qualitative Analysis

We propose a qualitative analysis of authors' paths learned by BARL on S2G (Fig. 1). We represent the ten most prolific authors and some of their co-authors. Authors that have a short publication time span (K. Cho, 4 years, S. Yan, 13 years) are represented by shorter paths. Our model represents locally each author dynamic. Each point cloud is also more spread at the end of each author trajectory, as the quantity of topics an author works on tends to diversify overs time. It is even more the case for authors with publications in lots of various conferences (e.g., Y. Bengio, P. Yu). Finally, co-authorships clearly emerge with close trajectories (e.g., C. Manning and D. Klein, K. Cho and Y. Bengio). These few examples illustrate the modeling power of BARL and its potential to represent authors evolution through time, which takes the co-authorship features into account.

Table 4. Accuracy on author classification for S2G in prediction. Each headline show the portion of training set used.

Methods	S2G Prédiction Accuracy (22 classes)			
	100 %	75%	50%	25%
USE	28.8 (1.5)	28.3 (1.7)	25.8 (2.4)	24.8 (1.6)
TempoBERT	29.2 (2.3)	28.6 (2.5)	27.5 (1.6)	26.1 (1.6)
DAR (dynamic)	17.5 (1.3)	17.4 (1.4)	17.4 (2.3)	17.2 (1.5)
DAR (static)	34.5 (1.1)	34.2 (1.8)	34.2 (2.0)	33.4 (1.6)
DAR (concat)	35.3 (1.6)	35.0 (1.4)	**34.9 (0.9)**	**34.7 (1.2)**
K-DGEA	35.0 (2.1)	34.7 (2.1)	34.2 (1.8)	33.0 (1.9)
BARL	**35.7 (2.0)**	**35.5 (2.0)**	34.8 (1.8)	34.5 (1.9)

5 Conclusion

We presented BARL, a dynamic author and document embedding model based on Brownian Bridges to represent authors as continuous trajectories through time. To fit this Gaussian modeling, BARL integrates BB into the VIB framework, which brings more versatility and smoothness when capturing author evolution. Our model outperforms existing works in dynamic author representation and it is competitive with specifically fine-tuned encoders. It encodes document and author into the same space and it can integrate any pre-trained temporal text encoder. In future works, we will incorporate more advanced encoders trained on event extraction to process longer and more complex texts (see [18]). We will also test other Gaussian processes [5] and add more trajectory points to grasp more complex evolution. We also plan to use this framework at the document scale to embed each document as a trajectory of sentences and see if specific stories schemes arise for example, following [14].

Acknowledgements. This work was granted access to the HPC/AI resources of IDRIS under the allocation 2023-AD011012369R2 made by GENCI. The author(s) acknowledge(s) the support of the French Agence Nationale de la Recherche (ANR), under grant ANR-19-CE38-0007 (project LIFRANUM).

References

1. Alemi, A.A., Fischer, I., Dillon, J.V., Murphy, K.: Deep variational information bottleneck. In: ICLR (2017)
2. Amba Hombaiah, S., Chen, T., Zhang, M., Bendersky, M., Najork, M.: Dynamic language models for continuously evolving content. In: ACM SIGKDD on Knowledge Discovery and Data Mining, p. 2514 (2021)
3. Ammar, W., Groeneveld, D., Bhagavatula, C., Beltagy, I., Crawford, M., et al.: Construction of the literature graph in semantic scholar. In: Proceedings of the Conference of the ACL: Human Language Technologies, vol. 3, p. 84 (2018)

4. Bamler, R., Mandt, S.: Dynamic word embeddings. In: ICML, pp. 380–389. PMLR (2017)
5. Bhagat, S., Uppal, S., Yin, Z., Lim, N.: Disentangling multiple features in video sequences using gaussian processes in variational autoencoders. In: Vedaldi, A., Bischof, H., Brox, T., Frahm, J.-M. (eds.) ECCV 2020. LNCS, vol. 12368, pp. 102–117. Springer, Cham (2020). https://doi.org/10.1007/978-3-030-58592-1_7
6. Cer, D., Yang, Y., Kong, S., Hua, N., Limtiaco, N., et al.: Universal sentence encoder for English. In: EMNLP: System Demonstrations, pp. 169–174 (2018)
7. Delasalles, E., Lamprier, S., Denoyer, L.: Learning dynamic author representations with temporal language models. In: 2019 IEEE ICDM, pp. 120–129. IEEE (2019)
8. Devlin, J., Chang, M.W., Lee, K., Toutanova, K.: BERT: pre-training of deep bidirectional transformers for language understanding. In: Conference of the North American Chapter of the ACL, pp. 4171–4186 (2019)
9. Gourru, A., Velcin, J., Gravier, C., Jacques, J.: Dynamic Gaussian embedding of authors. In: WWW 2022, pp. 2109–2119. Association for Computing Machinery (2022)
10. Jawahar, G., Ganguly, S., Gupta, M., Varma, V., Pudi, V.: Author2vec: learning author representations by combining content and link information. In: WWW 2016 Companion, pp. 49–50, April 2016
11. Kingma, D.P., Welling, M.: Auto-encoding variational Bayes. In: ICLR Conference (2014)
12. Lazaridou, A., et al.: Mind the gap: assessing temporal generalization in neural language models. In: NeurIPS, pp. 29348–29363 (2021)
13. Oh, S.J., Murphy, K., Pan, J., Roth, J., Schroff, F., Gallagher, A.: Modeling uncertainty with hedged instance embedding. In: Proceedings of ICLR (2019)
14. Reagan, A.J., Mitchell, L., Kiley, D., Danforth, C.M., Dodds, P.S.: The emotional arcs of stories are dominated by six basic shapes. EPJ Data Sci. 5, 31 (2016)
15. Rosin, G.D., Guy, I., Radinsky, K.: Time masking for temporal language models. In: ACM WSDM, pp. 833–841 (2022)
16. Sarkar, P., Siddiqi, S.M., Gordon, G.J.: Approximate Kalman filters for embedding author-word co-occurrence data over time. In: Airoldi, E., Blei, D.M., Fienberg, S.E., Goldenberg, A., Xing, E.P., Zheng, A.X. (eds.) ICML 2006. LNCS, vol. 4503, pp. 126–139. Springer, Heidelberg (2007). https://doi.org/10.1007/978-3-540-73133-7_10
17. Tishby, N., Pereira, F.C., Bialek, W.: The information bottleneck method. In: 37th Allerton Conference on Communication, Control, and Computing, p. 368 (1999)
18. Vashishth, S., Dasgupta, S.S., Ray, S.N., Talukdar, P.: Dating documents using graph convolution networks. In: Annual Meeting of the ACL, pp. 1605–1615. Association for Computational Linguistics, Australia (2018)
19. Wang, R.E., Durmus, E., Goodman, N., Hashimoto, T.: Language modeling via stochastic processes. In: ICLR (2022)
20. Yao, Z., Sun, Y., Ding, W., Rao, N., Xiong, H.: Dynamic word embeddings for evolving semantic discovery. In: ACM WSDM, p. 673-681. (2018)

Automatically Detecting Political Viewpoints in Norwegian Text

Tu My Doan$^{(\boxtimes)}$ ⓘ, David Baumgartner ⓘ, Benjamin Kille ⓘ,
and Jon Atle Gulla ⓘ

Norwegian University of Science and Technology, Trondheim, Norway
{tu.m.doan,david.baumgartner,benjamin.u.kille,jon.atle.gulla}@ntnu.no

Abstract. We introduce three resources to support research on political texts in Scandinavia. The encoder-decoder transformer models *sp-t5* and *sp-t5-keyword* were trained on political texts. The *nor-pvi* (available at https://tinyurl.com/nor-pvi) data set comprises political viewpoints, stances, and summaries for Norwegian. Experiments with four distinct tasks show that large-scale models, such as *nort5* perform slightly better. Still, *sp-t5* and *sp-t5-keyword* perform almost on par and require much less data and computation.

Keywords: Political Viewpoints · Political Dataset · LLMs

1 Introduction

Citizens struggle to stay informed as more and more political information is published. Identifying, analyzing, and presenting political information demands system support. The emergence of the internet and social media has affected the way we perceive politics, creating a vast amount of online data and diverse viewpoints. Large Language Models (LLMs) have become essential tools in navigating this complexity, capturing the nuances of language in a layered network. They are quite effective in assisting the analysis and interpretation of the extensive range of information available online. LLMs have been trained and used widely in various domains such as health, finance, and educations. Given that politics plays an important role in shaping our society and touches every aspect of our lives, it is crucial to have tools like LLMs to help us better understand political texts. These models can provide deeper insights into the complex language and concepts used in political discussions. Due to the specialized requirements of this domain, the need for tailored resources, including domain-specific LLMs and relevant data sets, becomes crucial.

Norway has a constitutional monarchy with the Stortinget (Parliament) [21, Chapter 4]. Elections are held every four years. As of January 2024, there are 169 members of parliament representing ten parties[1]. However, research in this area, particularly for the Norwegian language, is limited. To bridge this research gap, we present a Norwegian dataset annotated with political viewpoints, stances,

---

[1] https://www.stortinget.no/ (Accessed on 28 January 2024).

I. Miliou et al. (Eds.): IDA 2024, LNCS 14641, pp. 242–253, 2024.
https://doi.org/10.1007/978-3-031-58547-0_20

and summaries namely *nor-pvi*. We also introduce political-domain LLMs: *sp-t5* and *sp-t5-keyword* (encoder-decoder). The models were trained on parliamentary speeches covering four languages: Norwegian, Swedish, Danish, and Icelandic. We investigate a task that is challenging even for humans: can we (automatically) identify political viewpoints in speeches? We define a political viewpoint as an actionable opinion expressed in relation to a political task or problem (see [4] for a formal definition). We ask:

RQ1 Does fine-tuning help LLMs to succeed in identifying political viewpoints in Norwegian?

RQ2 How to effectively evaluate the results of political viewpoint identification (PVI) task?

RQ3 How well do the identified viewpoints align with human annotations?

2   Related Work

We present work related to political text analysis, domain- and language-specific LLMs, and masking techniques.

2.1   Political Text Analysis

Political text analysis can be divided into three categories: (i) political ideologies/leaning/party detection, (ii) political stance/framing detection, and (iii) political viewpoints extractions [4]. Extensive work has been conducted related to (i), specifically classifying political party affiliation [5,13,15] or ideology detection [3,12]. Similarly, several papers investigate stance detection [8,9,35]. Viewpoint extraction breaks down into different tasks, such as identifying topics [25], opinions [33], perspectives [18], or comparing viewpoints [23][2]. These studies touch on viewpoints, yet automatic political viewpoint identification (PVI) has plenty of space to explore. The lack of data resources and tools holds the field back.

2.2   Domain- and Language-Specific LLMs

Recent years have seen the development of domain- and language-specific LLMs. We focus on politics and Scandinavian languages. English encoder models for politics include a BERT model for stances and ideologies [20], and ConfliBERT [10] focusing on political conflict and violence. In Scandinavian languages, we find a variety of encoder models including NB-BERT [15], NorBERT [16,28], SP-BERT [6] (all Norwegian), as well as Swedish [22], Danish [11], Finnish [36], and Icelandic [30] models. In addition, there are some encoder-decoder models that are either multilingual—mT5 [37], mBART [19]—or language-specific such as NorT5 [28], or North-T5[3] for Norwegian. Our models, *sp-t5* and *sp-t5-keywords* address the intersection between politics and Scandinavian languages directly.

---

[2] For a more comprehensive treatment, see [4].

[3] https://huggingface.co/north/t5_base_NCC.

2.3   Masking Techniques

Transformers use different masking strategies to learn to represent language. Recently, more effort has been invested into masking keywords to advance standard Masked Language Modeling (MLM). For instance, researchers applied keyword masking for a BERT model [7], or NER/NEL for biomedical texts [1]. Another study [38] examined the influence in LLM training of Masking Ratio Decay (MRD), or POS-Tagging Weighted (PTW) masking.

In this work, we focus on PVI. We observe that few work has attempted to automatically extract viewpoints from political texts. We introduce a dataset and two LLMs to provide political analysts with resources to deal with the increasing amount of political texts in Scandinavian languages. Besides, we explore the use of opinion keyword masking to guide the encoder-decoder model to identify political viewpoints more reliably.

3   The nor-pvi Dataset

Political Viewpoint Identification requires annotated political texts. We introduce the *nor-pvi* dataset comprising 4027 speeches from the Talk of Norway[4] [17]. The dataset was annotated by three native speakers with 1232 viewpoints, 4027 summaries, and 1220 stances (For: 800, Neutral:110, Against:310). Some speeches lacked viewpoints whereas other speeches contained multiple viewpoints. We keep all distinct viewpoints. Viewpoints come in the form of phrase(s) or sentence(s). For every viewpoint, annotators also attached a stance. Stance labels are decided by majority. Conflicting stances are discarded. To create summaries, we use ChatGPT[5]. Subsequently, four native speakers verified and corrected the summaries. Summary length has the average of 56 words.

4   Encoder-Decoder Models

We introduce two new encoder-decoder models for Scandinavian political texts: *sp-t5* and *sp-t5-keyword*. First, we discuss the training data. Then, we describe the setup and training process.

4.1   Training Datasets

Norwegian Datasets: We obtain parliamentary speeches from different sources.

– **The Talk of Norway (ToN)** [17] comprises 250 373 speeches from the Norwegian Parliament (1998–2016). ToN was annotated with 83 metadata variables, including sentence and token boundaries, lemmas, parts-of-speech, and morphological features. Data were collected from the Storting API, the official API for data from the Norwegian Parliament.

---

[4] For evaluation purposes, these texts are excluded from the training data.
[5] We use both GPT-3.5 and GPT-4 version at https://chat.openai.com/.

- **Norwegian Colossal Corpus (NCC)** [14] was created by National Library of Norway (NLN)[6]. This is a large-scale general text corpus, primarily in Norwegian (Bokmål and Nynorsk), but also includes Scandinavian languages and others like English, Spanish, French and more. NCC is 49GB in size with 7B words, comprising newspapers, books, government documents, etc.
- **Norwegian Parliamentary Speech Corpus (NPSC)** [31][7] was developed by Norwegian Language Bank. NPSC has 140 h of Norwegian Parliament audio meetings from 2017 and 2018, with 65 000 sentences (1.2M words) in transcripts (Bokmål and Nynorsk) with speakers' metadata.
- To increase the number of Norwegian political speeches, we crawled additional data from the Norwegian Parliamentary website's API[8] from January 2019 to October 2023, yielding 6887 speeches.

Swedish and Danish Dataset: We use Parl Speech (v2) [27] comprising 6.3M parliamentary speeches from major legislatures in Austria, Czech Republic, Germany, Denmark, UK, Sweden, Spain, Netherlands and New Zealand, spanning 21 to 32 years until 2020[9]. These speeches are primarily sourced from Parliament websites. The dataset features 11 variables such as date, speaker, party, and text. There are 355 059 and 455 076 speeches for Swedish and Danish respectively. We also crawled more recent data from the Swedish Parliament website[10].

Icelandic Dataset: IGC-Parl corpus [32][11] has 404K speeches (totalling 209M words) from the Althingi, Iceland's Parliament, spanning 1911 to mid-2019. It offers extensive metadata and automatic linguistic annotations (POS tags and lemmas). However, the annotation accuracy has not been human-verified.

We pre-processed texts from all sources, using regular expressions to eliminate references to the parliament's president and markup. We also removed redundant white spaces. Speeches with fewer than 60 tokens, often questions or answers, were excluded. We acquired a dataset of about 1.44 million speeches: 16% Norwegian, 32% Danish, 25% Swedish, and 27% Icelandic.

4.2   Setup and Training

We group the corpora into 2 groups for training language models: (i) *general data* (NCC dataset) and (ii) *political data* (ToN, NCC, NPSC, crawled speeches, Swedish, Danish and Icelandic parliament speeches). We only train *base-size* models (580M parameters, vocab size 250K) due to limited computing resources.

---

[6] https://huggingface.co/datasets/NbAiLab/NCC.
[7] https://huggingface.co/datasets/NbAiLab/norwegian_parliament.
[8] https://data.stortinget.no/om-datatjenesten/bruksvilkar/.
[9] Details at https://doi.org/10.7910/DVN/L4OAKN.
[10] https://data.riksdagen.se/data/anforanden/.
[11] https://repository.clarin.is/repository/xmlui/handle/20.500.12537/14.

Scandinavian Politics T5 (SP-T5): For T5 model, we pre-trained from northT5-base[12] on political dataset. We trained model for a total of 2.69M steps with Adafactor optimizer [29] and sequence length 512. First 2.16M steps was done on batch size 128 and 0.53M steps on batch size 96 on TPU[13] v3-8 and v2-8 respectively.

SP-T5 with Opinion Keywords Masking (SP-T5-kw) We introduce another version of SP-T5 using opinion keywords masking for Norwegian. The motivation is evaluate whether model can better identify political viewpoints in Norwegian text. We created a list of 1822 opinion keywords for Norwegian. We scanned political speeches for verbs indicating expressing viewpoints and collected them. In addition, we collected a set of adjectives that could be combined with "være" (English: to be) to express viewpoints. For instance, "det er nødvendig" (English: it is necessary) forms such an expression. Then, we extracted the conjugated forms of the verbs from Ordbokene[14]. We also manually compiled a list of adverbs that could be placed between the personal pronoun and the verb. Finally, we combined all forms with the three pronouns "jeg" (English: I), "vi" (English: we), and "det" (English: it). SP-T5-kw was trained using similar settings (training steps, batch size, sequence length) as SP-T5. The only difference is in the last 0.53M steps, we apply keyword masking strategy.

5   Experiments and Evaluations

We use various language models for the evaluation: norT5-base [28][15], northT5-base[16], mT5-base [37][17] and our SP-models. We fine-tuned those models on four tasks. We report results (mean and standard deviation) over three runs. These are considered as baselines. The idea is to have a fair comparison among various LLMs, not to achieve state-of-the-art results. We format data as below:

> input: prefix + text
>
> target: label

Political Leaning Classification: This task focuses on identifying political leaning of political speeches in both Norwegian and Swedish. We use part of the language model training data for the political leaning classification task. Labels are annotated by consulting with experts. There are 46 387 Swedish speeches (left: 23 348, right: 23 039) and 6465 Norwegian speeches (left: 2916,

---

[12] The model was trained from mT5 checkpoint for 500K steps mainly on NCC dataset. See https://huggingface.co/north/t5_base_NCC.

[13] TPUs are special computing nodes operated by Google Cloud.

[14] https://ordbokene.no.

[15] https://huggingface.co/ltg/nort5-base.

[16] https://huggingface.co/north/t5_base_NCC.

[17] https://huggingface.co/google/mt5-base.

right: 3549). The labels are left/right (Norwegian: venstre/høyre and Swedish: vänster/höger). We use same prefix "find political leaning:" which is translated to corresponding languages ("hitta politisk tillhörighet" in Swedish, "finn politisk tilhørighet" in Norwegian). For evaluation metrics, we use Accuracy and F_{1macro}.

Translating European Parliament Speeches: We use Europarl bilingual dataset [34]. Experiments were conducted both ways for Danish and Swedish. A manual inspection of a sample of documents revealed that occasionally the length of two lines differed markedly. To prevent training the model on different content, we removed all cases where the difference exceeded 80 characters. This has resulted into a dataset of 892 727 items. We translate the prefix "translate from [source language] to [target language]:" into corresponding source language ("oversæt fra dansk til svensk" in Danish and "översätt från svenska till danska" in Swedish). The [source/target language] can be Danish/Swedish or Swedish/Danish. We evaluate results using BLEU score [26].

Political Speeches Summarization: In this task, we summarize political speeches in Norwegian (*nor-pvi* dataset). We use prefix "summarize:" ("oppsummer" in Norwegian). We adopt ROGUE score [2] as evaluation metric.

Political Viewpoint Identification (PVI): The task focuses on identifying political viewpoints from *nor-pvi* dataset described in Sect. 3. We use prefix: "find viewpoint:" ("finn synspunkt" in Norwegian). We consider ROUGE metric [2] and human evaluation. We manually checked whether the generated texts from models are viewpoints (Yes: 1/ No: 0 answer) and report mean and standard deviation results.

6 Results and Discussions

Table 1 shows experimental results of four tasks: (Table 1a) political leaning classification, (Table 1b) political speeches summarization, (1c) political viewpoint identification and (Table 1d) EU parliament speeches translation. Across all tasks, we notice that *nort5* performs better in most cases. *sp-t5* and its variation performance is the best in some cases. Our keyword masking strategy improves model in the task of PVI for all ROUGE metric. The performance of *north-t5* is consistently moderate, showing average results across various tasks. *mt5* indicates notably lower performance in most cases.

For task (Table 1a), *nort5* performs the best among four models for Norwegian at 75% accuracy, a gap of 21% comparing to *mt5*. For Swedish, our models are at most 2% better than *nort5*. All models show less than 0.1 standard deviation.

Table 1b shows results for summarization task. We observe similar trend across all models. Our SP-models perform quite well in this task. However, the

Table 1. Experimental results. We only compare *base-size* models. Orange color shows highest value and blue color for second highest. (∗) denotes our models.

(a) Political Leaning

Model	Norwegian		Swedish	
	Acc.	F_{1macro}	Acc.	F_{1macro}
mt5	$0.54^{\pm0.0}$	$0.44^{\pm0.0}$	$0.73^{\pm0.0}$	$0.72^{\pm0.0}$
north-t5	$0.57^{\pm0.0}$	$0.53^{\pm0.0}$	$0.79^{\pm0.0}$	$0.79^{\pm0.0}$
nort5	$0.75^{\pm0.0}$	$0.75^{\pm0.0}$	$0.84^{\pm0.0}$	$0.84^{\pm0.0}$
sp-t5$^{(*)}$	$0.64^{\pm0.0}$	$0.61^{\pm0.0}$	$0.87^{\pm0.0}$	$0.87^{\pm0.0}$
sp-t5-kw$^{(*)}$	$0.66^{\pm0.0}$	$0.65^{\pm0.0}$	$0.86^{\pm0.0}$	$0.86^{\pm0.0}$

(b) Summarization

Model	ROUGE		
	R-1	R-2	R-L
mt5	$38.59^{\pm0.1}$	$13.82^{\pm0.1}$	$24.97^{\pm0.1}$
north-t5	$39.13^{\pm0.2}$	$14.13^{\pm0.1}$	$25.28^{\pm0.6}$
nort5	$39.80^{\pm0.0}$	$15.14^{\pm0.1}$	$26.98^{\pm0.2}$
sp-t5$^{(*)}$	$40.42^{\pm0.3}$	$14.94^{\pm0.2}$	$27.59^{\pm0.3}$
sp-t5-kw$^{(*)}$	$40.38^{\pm0.1}$	$15.05^{\pm0.1}$	$27.50^{\pm0.1}$

(c) PVI

Model	ROUGE			Human (%)
	R-1	R-2	R-L	
mt5	$40.49^{\pm1.5}$	$27.84^{\pm2.2}$	$34.44^{\pm1.8}$	$33.48^{\pm47.3}$
north-t5	$39.52^{\pm1.9}$	$26.30^{\pm1.9}$	$33.04^{\pm2.1}$	$32.17^{\pm46.8}$
nort5	$48.99^{\pm0.9}$	$39.77^{\pm1.1}$	$43.89^{\pm0.6}$	$56.84^{\pm49.6}$
sp-t5$^{(*)}$	$41.35^{\pm0.1}$	$28.89^{\pm0.5}$	$35.79^{\pm0.2}$	$48.29^{\pm50.1}$
sp-t5-kw$^{(*)}$	$45.66^{\pm0.9}$	$34.18^{\pm1.0}$	$40.32^{\pm0.9}$	$46.58^{\pm50.0}$

(d) Translation

Model	BLEU	
	DA-SV	SV-DA
mt5	$43.22^{\pm0.2}$	$50.59^{\pm0.3}$
north-t5	$43.88^{\pm0.4}$	$51.02^{\pm0.7}$
nort5	$46.02^{\pm0.4}$	$55.11^{\pm0.0}$
sp-t5$^{(*)}$	$45.35^{\pm0.6}$	$52.02^{\pm0.2}$
sp-t5-kw$^{(*)}$	$45.49^{\pm0.3}$	$51.68^{\pm0.8}$

differences among ROUGE metrics are quite high, for example, a gap of about 25 between R-1 and R-2 in *sp-t5-kw*. High R-1 scores indicate that models are able to capture the gist of content fairly well. However, they struggle more in maintaining the structure and complex relationship in the original text. The results suggest that there is more room for improvements.

In translation task (Table 1d), all models perform somewhat similar in each sub-task. We observe that the BLEU score [26] for translation from Swedish to Danish is better than the reverse direction. Both *nort5* and *sp-t5* perform better than other models in both directions. Standard deviation is also low for this task across all models.

In PVI task (Table 1c)— main focus of our work— results are comparable between *nort5* and *sp-t5-kw*, with a difference of 3.3/5.6/3.6 in ROUGE-1/2/L respectively and a variance of about 10% for human evaluation. In Table 2, we have an example of (translated) speech with ground-truth and models' outputs. Looking into model test outputs, we notice that in most cases, models are struggling to generate the correct viewpoints. For different viewpoints that come from the same speech, models tend to generate the same outputs. This suggests that they are not very effective in identifying various viewpoints coming from the same speech. For human evaluation, we presented two annotators with predictions from all five models for a sample of 83 political speeches where a different group of annotators assured that these outputs contained viewpoints. They decided whether each predicted viewpoint was accurate, thus we obtained scores that were either 1 or 0. The column labeled *Human* presents the average scores for each model and their standard deviations. The *nort5* achieves the highest score

Table 2. Translated sample speech from Norwegian and model outputs for PVI task. Viewpoints are highlighted in pink color. The ground-truth is annotated viewpoint from the speech. There are five models' generated viewpoints.

Sample speech

In my view, the Right party's proposal is a gross exaggeration, where the Right party are blinded by the Confederation of Norwegian Enterprise's (NHO) and the business sector's need for labor. At the same time, the Right party overlook the societal consequences this could have in the form of increased immigration to Norway. Does the Right party not believe that in the event of, for example, an economic downturn with a subsequent increase in unemployment, it will primarily be all the unskilled, new labor immigrants who will be affected and first go into unemployment? An far too large proportion of today's immigrants are already outside the regular workforce. Wouldn't it be better to include these people in the workforce before taking such a strong initiative to bring in more labor immigrants, as the Right party has done here? And besides: In her speech, Erna Solberg mentioned that there was a need to bring in 10,000 labor immigrants per year. Is this in addition to all the asylum seekers and refugees - with subsequent family reunification - that otherwise come to Norwegian society every year? Is this really all that the Right party actually wants in this context?

Ground-truth	mt5 output
A far too large proportion of today's immigrants are already outside the regular workforce. Wouldn't it be better to include these people in the workforce before taking such a strong initiative to admit more labor immigrants, as the Right Party has done here?	Moreover, in her speech, Erna Solberg mentioned that there was a need to admit 10,000 labor immigrants per year. Is this to be in addition to all the asylum seekers and refugees - with subsequent family reunification - that otherwise come to Norwegian society each year?

north-t5 output	nort5 output
In my opinion, the Right party's proposal is a gross exaggeration, where the party is blinded by the needs of NHO (Confederation of Norwegian Enterprise) and the business sector for labor. At the same time, they overlook the societal consequences this could have in terms of increased immigration to Norway. Don't they believe that, for example, during an economic downturn with a subsequent increase among asylum seekers and refugees	There was a need to admit 10,000 labor immigrants per year. Should this be in addition to all the asylum seekers and refugees - with subsequent family reunification - that otherwise come to the Norwegian society every year? Wouldn't it be better to get these included in the workforce before taking such a strong initiative to bring in more labor immigrants, as the Right party here has done? Moreover, in her speech, Erna Solberg mentioned that there was a need to admit 10,000 labor immigrants.

sp-t5 output	sp-t5-kw output
Does the Right party not believe that in the event of, for example, an economic downturn with a subsequent increase in unemployment, it will primarily be all the unskilled, new labor immigrants who will be affected and first go into unemployment? An excessively large proportion of todayŠs immigrants are already outside the regular workforce.	At the same time, the Right party overlooks the societal consequences this could have in the form of increased immigration to Norway. Don't they think that, for instance, during an economic downturn with a subsequent rise in unemployment, it will primarily be all the unskilled, new labor immigrants who will be affected and first to leave when taking such a strong initiative to bring in more?

followed by *sp-t5*. All models fail in almost half of the cases. Besides, judging whether a text passage constitutes a viewpoint can be subjective. A different set of annotators can come to other conclusions. Overall, these results suggests that the task is challenging for both human and machine learning models. Nevertheless, enhancements can be made through strategies like expanding the training data set, extending the training duration of LLMs, refining fine-tuning parameters, and employing more advanced evaluation techniques.

Three out of four tasks focus mainly on Norwegian language. This is more beneficial for *nort5* model which was intensively trained on a lot of Norwegian texts, using many 128 GB GPUs and large global batch size 8192. Our models (*sp-t5* and *sp-t5-kw*) were trained on less data. As they are domain-specific language models, the amount of political text for Norwegian is quite limited. SP-models were trained with fewer steps due to limited computing resources (one TPU v3-8 or v2-8) compared to *nort5*. However, having substantial computing resources to train a language model like *nort5* may not be feasible for all, particularly for those with limited funding. Our experiments illustrate the possibility of training with reduced resources, accepting a minor compromise in performance. This approach could be advantageous for those aiming to train their own models on domain- and/or language-specific texts, where the availability of extensive computing power is not a prerequisite.

7   Conclusion and Future Work

In this work, we introduce our *sp-t5* and *sp-t5-keyword* models—encoder-decoder architecture—for political texts in Scandinavian languages. We also annotate a political dataset for Norwegian language namely *nor-pvi* with political viewpoints, stances, and summaries. This is our effort to bridge the gap in low-resource languages and under-represented domains such as politics. Experiments are conducted on various *T5-base-size* language models (our *sp-models*, *nort5*, *north-t5*, and *mt5*) on four tasks: political leaning classification, political speeches summarization, EU parliament speeches translation and political viewpoint identification. Results indicate that LLMs can be helpful tools to identify political viewpoints if they are fine-tuned (RQ_1). We relied on ROUGE metrics, human evaluation and discussed their shortcomings (RQ_2). Exploring differences between texts generated with LLMs and human annotators, we found that all models failed to identify about half the viewpoints contained in a sample of political speeches (RQ_3). More research is necessary to create models that can reliably help to educate people about political decision making.

Due to the lack of language and domain experts in Swedish, Danish and Icelandic, we could not perform complete evaluation for all tasks. Manual annotation is labor-intensive and expensive, we were able to focus only on Norwegian language. For future work, we would like to extend our dataset to other languages and conduct more complete experiments. We also want to extend our evaluation metric beyond the standard ones (BLEU, ROUGE) in future experiments to better understand model performances such as HOLMS [24]. Conducting model explanations is also future work for PVI. We will explore further strategies to use keywords for more targeted training for viewpoint identification. Training *sp-gpt2* decoder model is also our goal for future.

Acknowledgements. This work is done as part of Trondheim Analytica project and funded under Digital Transformation program at Norwegian University of Science and Technology (NTNU), 7034 Trondheim, Norway. This work has been partly funded by the SFI NorwAI, (Center for Research-based Innovation, 309834). Model training was supported by Cloud TPUs from Google's TPU Research Cloud program.

References

1. Borovikova, M., Ferré, A., Bossy, R., Roche, M., Nédellec, C.: Could Keyword Masking Strategy Improve Language Model? In: Métais, E., Meziane, F., Sugumaran, V., Manning, W., Reiff-Marganiec, S. (eds.) NLDB 2023. LNCS, vol. 13913, pp. 271–284. Springer, Cham (2023). https://doi.org/10.1007/978-3-031-35320-8_19
2. Chin-Yew, L.: Looking for a Few Good Metrics: ROUGE and its Evaluation. In: Proceedings of the 4th NTCIR Workshops (2004)
3. Djemili, S., Longhi, J., Marinica, C., Kotzinos, D., Sarfati, G.E.: What does Twitter have to say about Ideology? In: NLP 4 CMC: Natural Language Processing for Computer-Mediated Communication/Social Media-Pre-conference Workshop at Konvens 2014. vol. 1. Universitätsverlag Hildesheim (2014)
4. Doan, T.M., Gulla, J.A.: A survey on political viewpoints identification. Online Soc. Networks Media **30** (2022). https://doi.org/10.1016/j.osnem.2022.100208
5. Doan, T.M., Kille, B., Gulla, J.A.: Using language models for classifying the party affiliation of political texts. In: Rosso, P., Basile, V., Martínez, R., Métais, E., Meziane, F. (eds.) NLDB. LNCS, pp. 382–393. Springer, Cham (2022). https://doi.org/10.1007/978-3-031-08473-7_35
6. Doan, T.M., Kille, B., Gulla, J.A.: SP-BERT: a language model for political text in scandinavian languages. In: Metais, E., Meziane, F., Sugumaran, V., Manning, W., Reiff-Marganiec, S. (eds.) NLDB 2023. LNCS, vol. 13913, pp. 467–477. Springer, Cham (2023). https://doi.org/10.1007/978-3-031-35320-8_34
7. Golchin, S., Surdeanu, M., Tavabi, N., Kiapour, A.: Do not mask randomly: effective domain-adaptive pre-training by masking in-domain keywords. In: Can, B., et al. (eds.) RepL4NLP. ACL (2023). https://doi.org/10.18653/v1/2023.repl4nlp-1.2
8. Hardalov, M., Arora, A., Nakov, P., Augenstein, I.: Cross-domain label-adaptive stance detection. In: Moens, M.F., Huang, X., Specia, L., Yih, S.W.T. (eds.) CEMNLP. ACL (2021). https://doi.org/10.18653/v1/2021.emnlp-main.710
9. Hardalov, M., Arora, A., Nakov, P., Augenstein, I.: Few-shot cross-lingual stance detection with sentiment-based pre-training. In: Proceedings of the AAAI Conference on Artificial Intelligence, vol. 36 (2022)
10. Hu, Y., et al.: ConfliBERT: a pre-trained language model for political conflict and violence. In: NAACL. ACL (2022). https://doi.org/10.18653/v1/2022.naacl-main.400
11. Hvingelby, R., Pauli, A.B., Barrett, M., Rosted, C., Lidegaard, L.M., Søgaard, A.: DaNE: a named entity resource for Danish. In: Proceedings of the 12th Language Resources and Evaluation Conference, pp. 4597–4604 (2020)

12. Iyyer, M., Enns, P., Boyd-Graber, J., Resnik, P.: Political ideology detection using recursive neural networks. ACL **1** (2014). https://doi.org/10.3115/v1/P14-1105
13. Kannangara, S.: Mining Twitter for fine-grained political opinion polarity classification, ideology detection and sarcasm detection. In: WSDM. ACM (2018). https://doi.org/10.1145/3159652.3170461
14. Kummervold, P.E., Wetjen, F., De la Rosa, J.: The NORWEGIAN colossal corpus: a text corpus for training large norwegian language models. In: LREC. European Language Resources Association (2022)
15. Kummervold, P.E., De la Rosa, J., Wetjen, F., Brygfjeld, S.A.: Operationalizing a national digital library: the case for a Norwegian transformer model. In: NoDaLiDa (2021)
16. Kutuzov, A., Barnes, J., Velldal, E., Øvrelid, L., Oepen, S.: Large-scale contextualised language modelling for Norwegian. In: NoDaLiDa. Linköping University Electronic Press, Sweden (2021)
17. Lapponi, E., Søyland, M.G., Velldal, E., Oepen, S.: The Talk of Norway: a richly annotated corpus of the Norwegian parliament, 1998–2016. LREC, pp. 1–21 (2018). https://doi.org/10.1007/s10579-018-9411-5
18. Lin, W.H., Wilson, T., Wiebe, J., Hauptmann, A.: Which side are you on? IDENTIFYING perspectives at the document and sentence Levels. In: CoNLL-X. ACL (2006)
19. Liu, Y., et al.: Multilingual denoising pre-training for neural machine translation. Trans. Assoc. Comput. Linguistics **8**, 726–742 (2020)
20. Liu, Y., Zhang, X.F., Wegsman, D., Beauchamp, N., Wang, L.: POLITICS: pre-training with same-story article comparison for ideology prediction and stance detection. In: Findings of the Association for Computational Linguistics: NAACL 2022. ACL (2022). https://doi.org/10.18653/v1/2022.findings-naacl.101
21. Maagerø, E. and Simonsen, B.: Norway: Society and Culture. Cappelen Damm Akademisk, 3rd edn. (2022)
22. Malmsten, M., Börjeson, L., Haffenden, C.: Playing with Words at the National Library of Sweden - Making a Swedish BERT. CoRR **abs/2007.01658** (2020). https://arxiv.org/abs/2007.01658
23. Menini, S., Tonelli, S.: Agreement and disagreement: comparison of points of view in the political domain. In: COLING 2016, the 26th International Conference on Computational Linguistics, pp. 2461–2470 (2016)
24. M'rabet, Y., Demner-Fushman, D.: HOLMS: alternative summary evaluation with large language models. In: Proceedings of the 28th International Conference on Computational Linguistics, pp. 5679–5688 (2020)
25. Paul, M., Girju, R.: A two-dimensional topic-aspect model for discovering multifaceted topics. In: Proceedings of the Twenty-Fourth AAAI Conference on Artificial Intelligence, pp. 545–550. AAAI 2010, AAAI Press (2010)
26. Post, M.: A call for clarity in reporting BLEU scores. In: Proceedings of the Third Conference on Machine Translation: Research Papers, pp. 186–191. ACL (2018). https://www.aclweb.org/anthology/W18-6319
27. Rauh, C., Schwalbach, J.: The ParlSpeech V2 data set: full-text corpora of 6.3 million parliamentary speeches in the key legislative chambers of nine representative democracies (2020). https://doi.org/10.7910/DVN/L4OAKN
28. Samuel, D., et al.: NorBench – a benchmark for Norwegian language models. In: NoDaLiDa. University of Tartu Library (2023)
29. Shazeer, N., Stern, M.: Adafactor: adaptive learning rates with sublinear memory cost. In: ICML, pp. 4596–4604. PMLR (2018)

30. Snæbjarnarson, V., et al.: A warm start and a clean crawled corpus - a recipe for good language models. In: LREC, pp. 4356–4366. ELRA, Marseille, France (2022)
31. Solberg, P.E., Ortiz, P.: The Norwegian Parliamentary Speech Corpus. arXiv preprint arXiv:2201.10881 (2022)
32. Steingrímsson, S., Barkarson, S., Örnólfsson, G.T.: IGC-parl: Icelandic corpus of parliamentary proceedings. In: Proceedings of the Second ParlaCLARIN Workshop. pp. 11–17. ELRA, Marseille, France (2020)
33. Thonet, T., Cabanac, G., Boughanem, M., Pinel-Sauvagnat, K.: VODUM: a topic model unifying viewpoint, topic and opinion discovery. In: Ferro, N., et al. (eds.) ECIR 2016. LNCS, vol. 9626, pp. 533–545. Springer, Cham (2016). https://doi.org/10.1007/978-3-319-30671-1_39
34. Tiedemann, J.: Parallel data, tools and interfaces in OPUS. In: Proceedings of the Eighth International Conference on Language Resources and Evaluation (LREC'12). ELRA (2012)
35. Vamvas, J., Sennrich, R.: X-Stance: A Multilingual Multi-Target Dataset for Stance Detection. CoRR **abs/2003.08385** (2020). https://arxiv.org/abs/2003.08385
36. Virtanen, A., et al.: Multilingual is not enough: BERT for Finnish. arXiv preprint arXiv:1912.07076 (2019)
37. Xue, L., et al.: mT5: a massively multilingual pre-trained text-to-text transformer. In: NAACL. ACL (2021). https://doi.org/10.18653/v1/2021.naacl-main.41
38. Yang, D., Zhang, Z., Zhao, H.: Learning better masking for better language model pre-training. arXiv preprint arXiv:2208.10806 (2022)

AHAM: Adapt, Help, Ask, Model Harvesting LLMs for Literature Mining

Boshko Koloski[1,2(✉)], Nada Lavrač[1,3], Bojan Cestnik[1,4], Senja Pollak[1], Blaž Škrlj[1], and Andrej Kastrin[5]

[1] Jožef Stefan Institute, Ljubljana, Slovenia
boshko.koloski@ijs.si
[2] International Postgraduate School Jožef Stefan, Ljubljana, Slovenia
[3] University of Nova Gorica, Vipava, Slovenia
[4] Temida d.o.o, Ljubljana, Slovenia
[5] University of Ljubljana, Institute for Biostatistics and Medical Informatics, Ljubljana, Slovenia
andrej.kastrin@mf.uni-lj.si

Abstract. In an era of rapidly increasing numbers of scientific publications, researchers face the challenge of keeping pace with field-specific advances. This paper presents methodological advancements in topic modeling by utilizing state-of-the-art language models. We introduce the AHAM methodology and a score for domain-specific **adapt**ation of the BERTopic framework to enhance scientific text analysis. Utilizing the LLaMa2 model, we generate topic definitions through one-shot learning, with **help** from domain experts to craft prompts that guide literature mining by **asking** the model to label topics. We employ language generation and translation scores for inter-topic similarity assessment, aiming to minimize outlier topics and overlap between topic definitions. AHAM has been validated on a new corpus of scientific papers, proving effective in revealing novel insights across research areas. We also examine the impact of sentence-transformer domain adaptation on topic **model**ing precision, using datasets from arXiv, focusing on data size, adaptation niche, and the role of domain adaptation. Our findings indicate a significant interaction between domain adaptation and topic modeling accuracy, especially regarding outliers and topic clarity. We release our code at https://github.com/bkolosk1/aham

Keywords: topic modeling · domain adaptation · sentence-transformers · literature-based discovery

1 Introduction

The large number of publications that appear every day makes it almost impossible for researchers to keep up with the latest findings, even within their narrow scientific field. In a flood of information, researchers can overlook valuable segments of knowledge. Text mining is an effective technology that not only supports the analysis of existing knowledge, but also enables researchers to infer new

© The Author(s), under exclusive license to Springer Nature Switzerland AG 2024
I. Miliou et al. (Eds.): IDA 2024, LNCS 14641, pp. 254–265, 2024.
https://doi.org/10.1007/978-3-031-58547-0_21

knowledge facts from the information hidden in the data. Topic modeling is a text-mining technique that enables researchers to explore the thematic landscape of a collection of documents from a higher level perspective [25]. Although topic modeling is an established field of research, the recent advent of Large Language Models (LLMs) has led to a qualitative leap in their development. This paper presents methodological advances in domain adaptation for topic modeling by utilizing state-of-the-art language models. The approach is demonstrated using a corpus of scientific papers from the field of literature-based discovery (LBD), a vibrant research field of our research interest [6,9]. The specific contributions of this paper are the following:

- A topic modeling objective function AHAM that on the one hand evaluates the quality of topic modeling by measuring the semantic and lexical similarity in the topic names generated by BERTopic modeling, and on the other hand takes into account domain outliers while adapting to a new domain.
- Two specialized sentence transformers fine-tuned for distinct arXiv domains: one for general science and one for biomedical science.
- Quantitative assessment of domain-adaptation effects on two publicly available domains, and qualitative assessment of topic modeling on publications from the literature-based discovery (LBD) research domain.

The remainder of this paper is organized as follows: Sect. 2 discusses related work, Sect. 3 describes the data used, Sect. 4 outlines the proposed methodology, Sect. 5 explains the research questions and presents the results. Section 6 presents the qualitative evaluation of the proposed AHAM approach using the collected data set of LBD publications, and Sect. 7 concludes with a summary and directions for further work.

2 Related Work

The field of natural language processing has witnessed a remarkable transformation with the advent of **Large Language Models** (LLMs), which can be divided into two groups: Masked Language Models (MLMs) such as BERT [1] and generative Causal Language Models (CLMs) such as LLaMa2 [24]. These foundational models have set new standards for understanding and generating text with human-level precision [12]. Building on this groundwork, the sentence transformers [16] have emerged as a specialized evolution. These models, which are tailored to the task of learning sentence representations, ensure that semantically similar sentences are closely aligned in the vector space. One notable application of sentence transformers is BERTopic [4], which has revolutionized topic modeling with its unique approach. BERTopic clusters sentences based on semantic similarity, providing a refined and context-sensitive thematic analysis that outperforms conventional methods. In a similar vein, KeyBERT [3] advances the field of keyword extraction. Utilizing sentence-transformer technologies, it effectively extracts key terms and phrases from extensive texts [7,20]. A pivotal area of research in the use of these models is domain adaptation. Wang et al. [26]

proposed an approach for unsupervised domain adaptation, employing sequential denoising auto-encoders to learn from corrupted data. Another approach to domain adaptation involves generative pseudo-labeling (GPL) [27], where researchers use a surrogate generative model, such as T5 [15] that is trained to generate queries for specific passages [22]. These queries are then ranked by a cross-encoder [17] and used as downstream fine-tuning data for the sentence transformer. The development of prompting techniques in LLMs [29], particularly in-context one-shot learning [8], represents a significant stride in model interaction. This approach involves crafting specific prompts that enable models to learn from a single example within the prompt context, thereby generating more relevant and contextually nuanced responses. This technique is crucial in eliciting accurate and specific outputs from models like LLaMa2 [24], demonstrating a high level of understanding and flexibility in language generation [14].

3   Experimental Data: Literature-Based Discovery Publications

Literature-based discovery (LBD) is a vibrant area of research, with the first approaches reported in the mid-1980s. LBD postulates that knowledge in one domain may be related to knowledge in another domain, but without the relationship being explicit. In his pioneering work [21], by reading separate literature on fish oil and Raynaud's disease, Swanson observed that some knowledge concepts were shared between the two sets of documents. This led to the discovery that fish oil could be used in the treatment of Raynaud's disease, a hypothesis that was later clinically confirmed. Over the last four decades, researchers have proposed several different approaches, ranging from basic latent semantic indexing [2] to techniques based on knowledge graphs [18] and state-of-the-art large language models [28]. A detailed overview of LBD approaches can be found in recent studies [19,23].

We have compiled a novel comprehensive corpus of LBD publications by merging lists of representative publications on LBD that were manually compiled by the authors of two recent surveys of the LBD field [6,23]. To ensure that the latest advances are included, we also added 11 papers published in the last two years that were not included in the previous surveys. Our corpus comprises 389 publications spanning from 1986 to 2023. We concatenated the title and abstract fields to eliminate empty features because six papers did not contain an abstract.

We used this corpus to quantitatively evaluate the proposed methodology in Sect. 5, followed by its qualitative analysis in Sect. 7.

Independent of our methodology, we continue this section by preforming exploratory corpus analysis by aligning the temporal trend of the LBD field with the latent categorization of the articles by the LLaMa2 language model bellow.

The publication frequency of these articles peaked in 2015, with variations observed over the years. We prompted the LLaMa2 model to semantically categorize these articles into two groups (by one-shot in-context learning): those

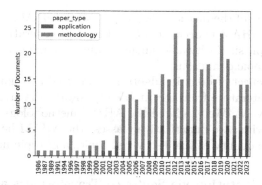

Fig. 1. Documents distribution per year, semantically labeled by LLaMa2 prompting.

proposing 'methodology' and those applying already developed methodologies 'application'. We utilize the one-shot learning capability of the LLaMa2 model. Our goal in this pseudo-categorization was to trace the evolution of the field, assuming that the 'methodology' articles are those that introduce new techniques, while the 'application' articles are likely to focus on the use or refinement of existing methods. We report the results in Fig. 1. The field of LBD began in 1986 with a paper reporting Swanson's [21] discovery of a link between fish oil and Raynaud disease, categorized here as a 'methodology' paper. In the early years, the literature mostly explored methodological issues. Over time, the focus of the work expanded to include both 'methodology' and 'application', with the latter gaining traction from 1999 onwards. A notable surge in 'methodology' articles occurred around 2008, with a peak in 2015, followed by some fluctuation. On the other hand, 'application' articles saw a steady increase, underscoring the sustained interest and necessity for LBD research. By 2023, the production of both 'methodology' and 'application' articles appears to have stabilized, indicating a balanced advancement in these two areas.

4   Methodology

This section presents the AHAM (Adapt, Help, Ask, Model) methodology and score, where by utilizing the LLaMa2 model we generate topic definitions through one-shot learning, with **help** from domain experts to craft prompts that guide literature mining by **asking** the model to label topics discovered by the BERTopic topic **model**ing framework. In summary, the proposed methodology consists of the following steps:

1. **Adapt:** Domain adaptation (see Sect. 4.1).
2. **Help:** Crafting a prompt by the help of domain experts for topic modeling (see Sect. 4.2).
3. **Ask:** Prompting the model with the crafted prompt (see Sect. 4.2).
4. **Model:** Modeling the topics by BERTopic (see Sect. 4.3).

An integral part of the methodology is the novel AHAM heuristic (presented in Sect. 4.4). The AHAM heuristic employs language generation and translation scores for inter-topic similarity assessment, aiming to minimize outlier topics and overlap between topic definitions.

To assess the effect of domain adaptation, we used two domain-specific corpora of scientific literature from the ArXiV repository (the arXiv and medarXiv datasets, respectively). We validated the AHAM methodology on the LBD corpus (see Sect. 3) where we quantitatively assess the AHAM heuristic and qualitatively assess the quality of the discovered topics and topic names.

4.1   Domain-Adaptation via Sentence-Transformers and BERTopic

Sentence transformers [16] represent document representation learning methods that aims to ensure that semantically similar documents should have closer embeddings through contrastive learning. For a query q, a positive p^+ (similar to q), and a negative p^- (different from q) sentence, the loss is: $L(q, p^+, p^-) = \max(0, f(q, p^+) - f(q, p^-) + \text{margin})$. The function f measures sentence similarity, often with cosine similarity, and margin is a hyperparameter for the minimum positive-negative pair distance. However, sentence transformers are initially pre-trained on general domains and may lack specific tuning for a particular domain of interest. To circumvent this limitation, Wang et al. (2021) [27] proposed **Domain Adaptation via Generative Pseudo Labeling (GPL)** introducing a multi-step adaptation process. Query Generation with T5 uses T5 variant [15] to generate synthetic queries Q_{synth} from target domain D_t, leading to $Q_{\text{synth}} = \text{T5}(D_s)$. Mining Negative Passages then finds diverse negative sentences P_{neg} for each query q_i using pretrained models and nucleus sampling. Training triplets $t = (q_i, p_i^+, p_i^-)$ are formed, and a cross encoder CE scores each pair (q_i, p_i^+) and (q_i, p_i^-) as $\delta_{t_i} = CE(q_i, p_i^+) - CE(q_i, p_i^-)$, creating a training corpus for tuning the model S to D_t with MarginMSE loss [5]. In our study we use the $S = all\text{-}mini\text{-}LM\text{-}v12$ sentence-transformer model [16].

BERTopic Modeling. The initial step in BERTopic is to reduce the dimensionality of the adapted sentence-transformer embeddings using UMAP [11] to facilitate more efficient clustering, as clustering in high-dimensional spaces can be challenging. This is followed by the application of the HDBSCAN clustering algorithm [10] to identify distinct clusters. The method then progresses to extracting document-level features to characterize the vicinity of each topic, initially employing class-based cTFIDF to assess word significance at the corpus level, thereby accentuating their importance to specific topics. Subsequently, for each cluster, the KeyBERT keyword extraction technique, which utilizes the S sentence-transformer representation, is applied to retrieve and rank relevant keywords.

Hyperparameters. The UMAP configuration involves setting the number of nearest neighbors and components to 5, with a minimum distance of 0.0, and adopting cosine similarity as the score. For HDBSCAN, the minimum cluster size is set to 5, with all other parameters left at their default values. To ensure reproducibility we seeded all of our experiments with random seed of 42.

Final Outputs. At the end of the processing pipeline for a set of N documents, we identify k distinct clusters. Among these, a unique cluster is designated as $k_{outliers}$, representing the outliers, which include documents that did not closely associate with any other cluster.

4.2 Prompt Engineering of LLMs to Design Topic Names

We utilize the LLaMa2 language model, specifically the *Llama-2-13b-chat-hf* variant from the HuggingFace[1] repository, setting the temperature to 0, the context window to 500 tokens, and the repetition penalty to 1.1. The objective of this phase is to semantically utilize the extracted keywords and the most central documents in each topical cluster. We aim to leverage the LLM's capabilities to derive meaningful semantic labels. This is achieved through training via one-example in-context learning, guided by prompt designs crafted by domain experts of the meta-literature analysis application of our interest. Following related work, we have engineered a three-level prompt structure:

– System prompt, to give personality to the LLM for labeling topics as a domain expert:

> You are a helpful, respectful, and honest research assistant for labeling topics.

– One shot, example prompt, from which we **help** the model with guidance for in-context learning. Notice that this prompt has been crafted in a specific manner, reflecting our contributions to the field, but it could have been crafted differently by the end user.

> I have a topic that contains the following documents:
> - Bisociative Knowledge Discovery by Literature Outlier Detection.
> - Evaluating Outliers for Cross-Context Link Discovery.
> - Exploring the Power of Outliers for Cross-Domain Literature Mining.
> The topic is described by the following keywords: bisociative, knowledge discovery, outlier detection, data mining, cross-context, link discovery, cross-domain, machine learning'.
> Based on the information about the topic above, please create a simple, short, and concise computer science label for this topic. Make sure you only return the label and nothing more.
> [INST]: Outlier-based knowledge discovery

– The query prompt, which **asks** the model to label topics from the keywords and the most central documents for a given topic, is personalized. This personalization ensures that the model returns computer-science topics from the field of Literature-Based Discovery without explicitly mentioning the topics:

---

[1] https://huggingface.co/.

> I have a topic that contains the following documents [DOCUMENTS]
> The topic is described by the following keywords: [KEYWORDS]
> Based on the information about the topic above, please create a simple, short
> and concise computer science label for this topic. Make sure you only return the
> label and nothing more.

4.3 Assessing Adaptation Through Evaluation of Topic Naming

To assess the effect of domain adaptation on topic modeling, we used the idea that, on average, the names of topic labels tend to be dissimilar. To measure the numerical similarities between topics, we have chosen to use three distinct similarity scores TopicSimilarity to evaluate the similarity between topic names of Topic A and Topic B:

- **Levenstein fuzzy matching**: We employ the *fuzzywuzzy*[2] Python implementation which utilizes normalization based on the length to provide a normalized *lexical* similarity score between the topic names.
- **BERTscore**: Uses BERT model embeddings to evaluate the similarity between A and B, by comparing the semantic embedding similarity of present n-grams w BERTscore $= \frac{1}{|A|} \sum_{w \in A} \max_{w' \in B} \cos(w, w')$
- **Semantic similarity (using all-mini-LM-v12)**: Evaluates the semantic closeness of A and B using the cosine of the angle between their vector representations, S(A) and S(B), as: $\cos(A, B) = \frac{S(A) \cdot S(B)}{\|S(A)\| \cdot \|S(B)\|}$

4.4 AHAM Heuristic

We propose a heuristic that jointly combines the number of outliers $|k_{\text{outliers}}|$, the number of clusters $|k|$, and the similarity across the generated topic names with the TopicSimilarity score to the number of steps n_{steps} in the domain-adaptation. We define the AHAM$_{\text{objective}}$ as:

$$\text{AHAM}_{\text{objective}} = 2 \cdot \frac{|k_{\text{outliers}}|}{|k|} \cdot \sum_{\substack{i=1 \\ j=i+1}}^{|k|} \frac{\text{TopicSimilarity}(k_i, k_j)}{k(k-1)}$$

We evaluate at every $10,000$ step of the GPL and select the topic modeling that minimizes the AHAM$_{\text{objective}}$ score within some evaluation budget of n_{steps}.

5 Quantitative Exploration of the AHAM Objective

To our knowledge, no prior research has yet investigated the impact of further domain adaptation (using methods like the aforementioned GPL) of sentence transformers within a specific domain for topic discovery in the BERTopic framework. We selected two domain datasets from the ArXiv repository [13]:

---

[2] https://github.com/seatgeek/fuzzywuzzy

arXiv, which includes 25,000 entries from a wide range of scientific disciplines, and *medarXiv*, containing 8,500 entries from the medical science field. The general domain dataset, *arXiv*, with its larger data volume, allowed us to develop an experimental suite to assess the impact of size and domain granularity. In contrast, the smaller but domain-specific *medarXiv* enabled us to examine the effects of a smaller, more specialized dataset. To address this gap, this experimental setting concentrates on four research questions:

RQ1. Does the domain-specific adaptation of the sentence transformer lead to a more precise topic differentiation? Table 1 and Fig. 2 provide insight into the impact of domain-specific adaptation of a sentence transformer on topic modeling within arXiv and medarXiv datasets. The results indicate a significant decrease in both the number of outliers and the number of topics at step 20k (improving the baseline results of 43 outliers to 29 for the arXiv adaption and to 10 for the medarXiv adaption), which suggests that the adaptation process likely improves the transformer's ability to discern more relevant topics and reduces the identification of outlier topics. The benefit of domain adaptation is further evidenced by the sharp reduction in topics, pointing to a more focused topic differentiation.

Table 1. Comparison of Topic Modeling Performance Post-Adaptation for arXiv and medarXiv Datasets Relative to Baseline, Detailing Topic Counts (#T), Outlier Frequencies (#O), and Outlier-to-Topic ($\frac{\#O}{\#T}$) Ratios. Bolded values represent the setting selected by the **AHAM** score. The base column indicates the results of the non-adapted sentence-transformer model.

Steps	Domain	#T	#O	$\frac{\#O}{\#T}$	Lev	BERT	Cos	Domain	#T	#O	$\frac{\#O}{\#T}$	Lev	BERT	Cos
Base	/	15	43	2.87	0.32	0.86	0.25	/	15	43	2.87	0.32	0.86	0.25
10		20	32	1.60	0.32	0.86	0.31		11	45	4.09	0.32	0.86	0.25
20		4	29	7.25	0.45	0.88	0.45		5	10	2.00	0.39	0.85	0.45
30	ArXiv	8	35	4.38	0.33	0.87	0.36	MedarXiv	11	12	**1.09**	0.32	0.88	0.41
40		**19**	**5**	**0.26**	**0.31**	**0.85**	**0.30**		4	23	5.75	0.29	0.87	0.42
50		19	27	1.42	0.35	0.87	0.38		5	9	1.80	0.32	0.86	0.47

RQ2. Which is more effective for meta-literature analysis: broad domain-specific adaptation or niche-specific knowledge? The trajectory of the AHAM objective revealed that both lexical and semantic distance scores peaked at step 20 for the arXiv and medarXiv datasets, indicating a two-phase process when adapting: obfuscation of knowledge and iterative refinement and improvement. We noticed that by utilizing broad domain knowledge, the adaptation process took more steps however it yielded a lower AHAM score (0.26) with 88.4% reduction in outliers. On the other side the medarXiv adaptation, found the local optimum after 30k steps, yielding a score of 1.09 and 72% reduction of outliers. The results suggest that if possible to identify what domain is best suited to transform from, then adapting such domain would enable the model

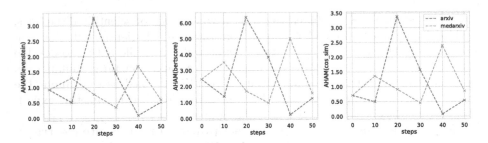

Fig. 2. Assessment of the optimization objective's trajectory and similarity, conducted on intervals of 10k steps within a total budget of 50k steps.

to generalize better. In the case of medarXiv, we noticed that the model would specialize earlier for the topics correlated with a certain domain.

RQ3. What is the relationship between the domain adaptation granularity and performance? Integrating the insights from the AHAM objective with the similarity scores reveals that the specific stages in the modeling process significantly influence the development of topic similarities. Both datasets exhibit a tendency to diverge and then re-converge, which may reflect a common stabilization point in topic similarity as the models are refined. Evaluating the AHAM objective trajectory reveals that the function is non-convex, which makes it difficult to optimize. The results indicate a non-linear and complex progression of topic modeling characterized by both convergence and divergence, suggesting that topic evolution is influenced by the interplay of the dataset characteristics.

RQ4. How do the topics evolve when the model is iteratievly adapted to a specific domain? The results of the AHAM optimization (Fig. 2) suggest that extensive training with both models enhances the subsequent topic modeling's understanding of the domain it aims to represent. It is crucial to select an adaptation dataset that is closely aligned with the specific niche observed. Given the observed lexical and contextual similarities, the AHAM aspects seem to serve as reliable scores for assessing the efficacy of the topic modeling performance of the domain-specific adaptation of the model.

6   Qualitative Evaluation

We observed that document partitioning via domain adaptation significantly improves cluster validity. Manual evaluation by literature-based discovery (LBD) experts of clustering outcomes-baseline, arXiv-specific, and medarXiv-specific models-revealed distinct patterns. The baseline model identified a broad "Discovery through Statistical and Knowledge-Based Methods" cluster, while domain-adapted models yielded more focused clusters like "BioMed Text Mining" and "Knowledge Linkage Discovery," alongside specialized topics such as "Kostoff's cluster" and "Biomedical Connections Discovery." Figure 3 presents the topic

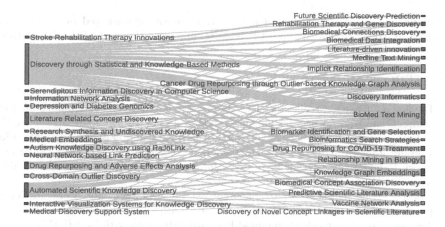

Fig. 3. Evolution of topics between the first and best step (after 40 steps) of domain adaptation following AHAM objective, on the arXiv domain.

evolution between the initialtopics and the topics identified after the domain adaptation selected by AHAM.

7   Conclusion and Further Work

In this study, we introduce AHAM, a methodology and score designed to refine topic modeling through domain adaptation, enhancing the BERTopic framework for meta-literature analysis applied to the task of literature-based discovery. Our main contributions include a novel LBD corpus, the AHAM methodology and heuristic. The AHAM heuristic aims to reduce outliers and improves the discriminability of topics, and advocates incremental domain adaptation. This approach, informed by expert prompts, emphasizes a careful balance in adaptation to minimize outliers, leading to well-defined topics verified by domain experts. The optimization process for achieving the lowest AHAM score is intricate, indicating that extended domain adaptation does not always lead to the most effective topic modeling configuration.

Future directions include extensive testing of domain adaptation's efficacy in topic modeling across various fields, leveraging the AHAM objective. Further, we aim to assess the benefits of in-domain data training, explore model adaptability across diverse languages, particularly those with limited resources, and investigate the influence of the size of the generative language model and different sentence-transformers on topic identification.

Acknowledgements. The authors acknowledge the financial support from the Slovenian Research and Innovation Agency through research core funding project Knowledge technologies (No. P2-0103) and research projects: Research collaboration prediction using literature-based discovery approach (No. J5-2552), Embeddings-based techniques for Media Monitoring Applications (No. L2-50070) and Hate speech in contemporary conceptualizations of nationalism, racism, gender and migration (No. J5-3102). A Young Researcher Grant PR-12394 supported the work of the first author.

References

1. Devlin, J., Chang, M.W., Lee, K., Toutanova, K.: BERT: Pre-training of deep bidirectional transformers for language understanding. In: Burstein, J., Doran, C., Solorio, T. (eds.) Conference of the North American Chapter of the Association for Computational Linguistics: Human Language Technologies, Volume 1 (Long and Short Papers), pp. 4171–4186. Association for Computational Linguistics, Minneapolis, Minnesota (Jun 2019)
2. Gordon, M.D., Dumais, S.: Using latent semantic indexing for literature based discovery. J. Am. Soc. Inf. Sci. **49**(8), 674–685 (1998)
3. Grootendorst, M.: Keybert: minimal keyword extraction with bert (2020)
4. Grootendorst, M.: BERTopic: neural topic modeling with a class-based TF-IDF procedure (2022)
5. Hofstätter, S., Althammer, S., Schröder, M., Sertkan, M., Hanbury, A.: Improving efficient neural ranking models with cross-architecture knowledge distillation (2020)
6. Kastrin, A., Hristovski, D.: Scientometric analysis and knowledge mapping of literature-based discovery (1986–2020). Scientometrics **126**(2), 1415–1451 (2021)
7. Koloski, B., Pollak, S., Škrlj, B., Martinc, M.: Out of thin air: is zero-shot cross-lingual keyword detection better than unsupervised? In: Language Resources and Evaluation Conference, pp. 400–409. European Language Resources Association, Marseille, France Jun 2022
8. Lampinen, A., et al.: Can language models learn from explanations in context? In: Goldberg, Y., Kozareva, Z., Zhang, Y. (eds.) EMNLP 2022, pp. 537–563. Association for Computational Linguistics Dec 2022
9. Lavrač, N., Martinc, M., Pollak, S., Pompe Novak, M., Cestnik, B.: Bisociative literature-based discovery: lessons learned and new word embedding approach. N. Gener. Comput. **38**(4), 773–800 (2020)
10. McInnes, L., Healy, J., Astels, S.: HDBSCAN: hierarchical density based clustering. J. Open Source Softw. **2**(11), 205 (2017)
11. McInnes, L., Healy, J., Saul, N., Großberger, L.: UMAP: uniform manifold approximation and projection. J. Open Source Softw. **3**(29), 861 (2018)
12. Min, B., et al.: Recent advances in natural language processing via large pre-trained language models: a survey. ACM Comput. Surv. **56**(2), 1–40 (2023)
13. Muennighoff, N., Tazi, N., Magne, L., Reimers, N.: MTEB: massive text embedding benchmark. In: Vlachos, A., Augenstein, I. (eds.) European Chapter of the Association for Computational Linguistics, pp. 2014–2037. Association for Computational Linguistics, Dubrovnik, Croatia (May 2023) https://doi.org/10.18653/v1/2023.eacl-main.148
14. Pan, J., Gao, T., Chen, H., Chen, D.: What in-context learning "learns" in-context: disentangling task recognition and task learning. In: ACL 2023, pp. 8298–8319 (2023) https://doi.org/10.18653/v1/2023.findings-acl.527
15. Raffel, C., Shazeer, N., Roberts, A., Lee, K., Narang, S., Matena, M., Zhou, Y., Li, W., Liu, P.J.: Exploring the limits of transfer learning with a unified text-to-text transformer. J. Mach. Learn. Res. **21**(140), 1–67 (2020)

16. Reimers, N., Gurevych, I.: Sentence-bert: sentence embeddings using siamese bert-networks. In: Conference on Empirical Methods in Natural Language Processing. Association for Computational Linguistics (Nov 2019)
17. Reimers, N., Gurevych, I.: Making monolingual sentence embeddings multilingual using knowledge distillation. In: 2020 Conference on Empirical Methods in Natural Language Processing. Association for Computational Linguistics (Nov 2020)
18. Sang, S., Yang, Z., Wang, L., Liu, X., Lin, H., Wang, J.: SemaTyP: a knowledge graph based literature mining method for drug discovery. BMC Bioinf. **19**(1), 193 (2018). https://doi.org/10.1186/s12859-018-2167-5
19. Sebastian, Y., Siew, E.G., Orimaye, S.O.: Emerging approaches in literature-based discovery: techniques and performance review. Know. Eng. Rev. **32**, e12 (2017). https://doi.org/10.1017/S0269888917000042
20. Škrlj, B., Koloski, B., Pollak, S.: Retrieval-efficiency trade-off of unsupervised keyword extraction. In: Pascal, P., Ienco, D. (eds.) Discovery Science, pp. 379–393. Springer Nature Switzerland, Cham (2022). https://doi.org/10.1007/978-3-031-18840-4_27
21. Swanson, D.R.: Fish oil, Raynaud's syndrome, and undiscovered public knowledge. Perspect. Biol. Med. **30**(1), 7–18 (1986)
22. Thakur, N., Reimers, N., Rücklé, A., Srivastava, A., Gurevych, I.: BEIR: a heterogeneous benchmark for zero-shot evaluation of information retrieval models. In: Thirty-fifth Conference on Neural Information Processing Systems Datasets and Benchmarks Track (Round 2) (2021)D
23. Thilakaratne, M., Falkner, K., Atapattu, T.: A systematic review on literature-based discovery workflow. PeerJ Comput. Sci **5**, e235 (2019)
24. Touvron, H., et al.: Llama 2: open foundation and fine-tuned chat models (2023)
25. Vayansky, I., Kumar, S.A.P.: A review of topic modeling methods. Inf. Syst. **94**, 10158101582 (2020)
26. Wang, K., Reimers, N., Gurevych, I.: Tsdae: using transformer-based sequential denoising auto-encoder for unsupervised sentence embedding learning. In: EMNLP 2021, pp. 671–688. Association for Computational Linguistics, Punta Cana, Dominican Republic (Nov 2021)
27. Wang, K., Thakur, N., Reimers, N., Gurevych, I.: GPL: generative pseudo labeling for unsupervised domain adaptation of dense retrieval. In: Conference of the North American Chapter of the Association for Computational Linguistics: Human Language Technologies, pp. 2345–2360. Association for Computational Linguistics, Seattle, USA (Jul 2022)
28. Wang, Q., Downey, D., Ji, H., Hope, T.: Learning to generate novel scientific directions with contextualized literature-based discovery (2023)
29. Wei, J., Wang, X., Schuurmans, D., Bosma, M., Xia, F., Chi, E., Le, Q.V., Zhou, D., et al.: Chain-of-thought prompting elicits reasoning in large language models. Adv. Neural. Inf. Process. Syst. **35**, 24824–24837 (2022)

Author Index

Printed in the United States
by Baker & Taylor Publisher Services